T0296889

Time Series for Data Scientists

Learn by doing with this user-friendly introduction to time series data analysis in R. This book explores the intricacies of managing and cleaning time series data of different sizes, scales and granularity, data preparation for analysis and visualization, and different approaches to classical and machine learning time series modeling and forecasting. Readers will learn to apply these methods in R and interpret the results.

A range of pedagogical features support students throughout the book, including end-of-chapter exercises, problems, quizzes and case studies. The case studies are designed to stretch the learner, introducing larger data sets, enhanced data management skills, and R packages and functions appropriate for real-world data analysis.

On top of providing commented R programs and data sets, the book's companion website offers extra case studies, videos and solutions to the exercises. Lecture slides are also available for instructors.

Accessible to those with a basic background in statistics and probability, this is an ideal hands-on text for undergraduate and graduate students, as well as researchers in data-rich disciplines.

Juana Sanchez is Senior Lecturer in Statistics at the University of California – Los Angeles. She is Editor of the Datasets and Stories section of the ASA's *Journal of Statistics and Data Science Education* and is the author of *Probability for Data Scientists* (2020).

Time Series for Data Scientists

Data Management, Description, Modeling and Forecasting

JUANA SANCHEZ

University of California – Los Angeles

CAMBRIDGE
UNIVERSITY PRESS

Shaftesbury Road, Cambridge CB2 8EA, United Kingdom

One Liberty Plaza, 20th Floor, New York, NY 10006, USA

477 Williamstown Road, Port Melbourne, VIC 3207, Australia

314–321, 3rd Floor, Plot 3, Splendor Forum, Jasola District Centre,
New Delhi – 110025, India

103 Penang Road, #05–06/07, Visioncrest Commercial, Singapore 238467

Cambridge University Press is part of Cambridge University Press & Assessment,
a department of the University of Cambridge.

We share the University's mission to contribute to society through the pursuit of
education, learning and research at the highest international levels of excellence.

www.cambridge.org
Information on this title: www.cambridge.org/highereducation/isbn/9781108837774
DOI: 10.1017/9781108942812

First published 2023

A catalogue record for this publication is available from the British Library.

ISBN 978-1-108-83777-4 Hardback

Additional resources for this publication at http://timeseriestime.org/

To Alberto

Contents

Preface

The volume, velocity and variety of time series data produced and available nowadays, and the enormous growth in data science as a profession have expanded the demand for time series analysis skills to include less experienced learners and researchers. This book trains such learners and researchers to access, assess the quality of, clean, manage, describe, model and forecast time series data with classical and contemporary time series methods, using R [159]. The reader with a basic introductory background in undergraduate statistics and probability will find in this book an accessible and hands-on introduction to the practice of classical and contemporary analysis of time series in the time domain, and a brief introduction to the frequency domain. The book grew out of a series of lecture notes and computer programs developed over many years for a one-quarter course on introductory time series analysis for undergraduate students.

Machine learning methods were not created for time series data, and progress has been slow in adapting them for time series. The reader's skill set required by machine learning methods is much larger than that needed for the application of classical methods, as concepts of clustering, algorithms and computational statistics are essential. Thus, some machine learning concepts are introduced very gently from the beginning of the book, in all chapters, in an intuitive way, not only to introduce them to the reader but also to emphasize the importance of knowing the classical methods in order to understand and use more knowledgeably the contemporary ones. Enough contemporary references are provided for the reader to expand in any direction of interest.

The book is organized into three parts and 10 chapters, some of them with their own appendix, all of them with exercises, end-of-chapter problems, a quiz for self-assessment and, in most chapters, a case study. The case studies increase the skill set learned in the chapters by introducing larger data sets, data management skills, and R packages and functions appropriate for the granularity and scale of contemporary data sets. The R script files and data sets used in each case are mentioned throughout the chapters. Those and solutions to the exercises, video tutorials and other supplementary materials can be found at the following website: http://timeseriestime.org/.

Part I contains the three most important chapters of the book because it familiarizes the reader with time series data, potential data cleaning problems, and the most important attributes or features of time series data used in classical and contemporary time series analysis. Mastering those chapters will help avoid the most common mistakes beginners make downstream. Teachers of introductory statistics that only want to familiarize students with time series data and basic forecasting concepts will find this part meets their needs. Part II is about the classical theory of stochastic univariate stationary time series processes and their properties, the generation of data from those processes and statistical inference and forecasting. Modeling and forecasting with Autoregressive Integrated Moving Average methods is the core chapter in this part. It is this part and the next that require some background in introductory statistics and probability. Part III is about multivariate time series modeling and covers Hidden Markov Models, Vector Autoregression, Time Series Regression models and supervised machine learning models. Regretfully, Bayesian time series, which is a natural habitat for many time series methods, is not covered in depth in this book. Bayesian statistics and time series analysis are rarely offered in undergraduate programs at universities. To keep this book introductory and hands-on without further prerequisites, we had to compromise and leave some topics out.

This book uses real data to teach the practice of time series analysis via examples. Analyzing time series data of reasonable size requires software. The programs provided in this book are written in R. The reader that has already been exposed to R will find them easier to follow. Those unfamiliar with R will first have to download R from www.r-project.org/ and then download the popular interface called RStudio for easier execution of the R script files provided throughout this book. RStudio can be downloaded from https://rstudio.com/products/rstudio/. In the Rstudio website, the reader will find many tutorials. Books at the level of [11], [203] or [159] are highly recommended.

I am indebted to my students at UCLA who have taken the introductory time series course with me for over 20 years. Their questions and their hard work and creative course projects have always helped me improve my lecture notes on which this book is based. I am also thankful for the supportive teaching environment that my colleagues of 24 years at UCLA's Statistics Department have provided and for the financial support that allowed me to hire three undergraduate research assistants, William Foote, Luan (Leroy) Le and Shiqi Liang to help in the accessing and pre-processing of some of the competitions' data sets, testing Prophet, making a few of the graphs and updating some R code. This project would not have been completed without the continued encouragement and patience of my husband, Alberto Candel. Last, but not least, a very special thanks goes to the reviewers for their very helpful suggestions, to Senior Editor Natalie Tomlinson, editorial assistant Anna Scriven and CUP's production team for their editorial assistance.

Part I

Descriptive Features of Time Series Data

Most modern introductory statistics books start with descriptive data analysis of well-curated and clean data: how to summarize the data, what plots to use, how to interpret plots and how to discriminate among the types of summaries and plots that are appropriate for the different types of data. For a person never exposed to the study of time series or autocorrelated data of any kind, that is a great way to start. Part I focuses mostly on the description of well-curated and clean time series data to get the beginner up and running as quickly as possible. But Part I also exposes the beginner to other types of time series data that would be impossible to learn from without the essentials learned in Part I.

Through description and plotting, the reader becomes familiar with attributes of time series data that will be used to model and forecast in other parts of the book: regular patterns (such as trends, seasonality and cycles), random terms of different nature, autocorrelations of the random terms, partial autocorrelations, cross-correlations between several time series and others. Mastering the interpretation of time plots, spaghetti plots and seasonal box plots, with different types of time series, the beginner learns to appreciate those attributes and the many different ways in which they manifest themselves depending on the granularity of the data. Skills in smoothing and decomposing are also learned in this part of the book.

While doing all of this, the reader is gradually introduced to Machine Learning (ML) vocabulary and gets immersed in some unsupervised ML practices, such as clustering of time series. "Features" is a very general word in ML, referring to attributes of the time series. Sometimes the attributes of interest are summary statistics, such as, for example, autocorrelations; sometimes the attributes are new variables derived from the raw time series, such as, for example, a variable containing only the month of the observation in the time series. Part I will help the reader gather the most important features of either kind that we look for in a time series, thus it also prepares the reader for ML. It is important that the beginner realizes that the same concepts receive different names depending on the discipline in which they are studied.

The motivation for Part I is twofold. On the one hand, it is to teach the concepts that have defined classical time series analysis for decades. On the other hand, it aims to bridge the gap between classical analysis and our contemporary realities regarding the analysis of time-stamped data.

We cannot emphasize enough how important it is to spend time learning the material in Part I. This part prepares the beginner for downstream applications of classical and modern Machine Learning in time series analysis methods.

1 Introduction to Time Series Data

1.1 Introduction

Daily recording of the health status of an individual, recording of speech or electromagnetic signals, weekly recording of COVID-19 cases, annual measurement of the teenage population, temperature recorded every day, air quality recorded every minute, amount of gorgonzola cheese produced quarterly, daily solar activity index (sunspot numbers) and daily popularity of soccer are all examples of data collected over time. The data without the time stamp, however, would be useless.

Recording the time at which a measurement was made can greatly expand the value of the data being collected. We have all heard of the flight data recorders used in airplane travel as a way to reconstruct events after a malfunction or crash. A modern aircraft is equipped with sensors to measure and report data many times per second for dozens of parameters throughout the flight. These measurements include altitude, flight path, engine temperature and power, indicated air speed, fuel consumption, and control settings. Each measurement includes the time it was made. In the event of a crash or serious accident, the events and actions leading up to the crash can be reconstructed in exquisite detail from these data. [41]

> *Time series data* are numerical data indexed by a time stamp, and collected over time to observe changes over time. The chronological order of the observations is of extreme importance. We are interested not only in the particular value of the observations but also in the order in which they appear.

Almost every area of scientific inquiry is concerned with data collected over time, that is, with time series. The basic property of time series analysis is that it is concerned with repeated measurements on the same phenomenon at different times. Because of this, the analyst must take into account the correlation between successive observations. This is in marked contrast with the data analyzed in elementary statistics courses where one assumes that the data are made up of independent and identically distributed observations obtained by randomly sampling from some population or populations. The presence of correlation makes the analysis of time series data and the interpretation of the results much more difficult than in the independent case. [131]

> The systematic approach by which one goes about answering the mathematical and statistical questions posed by time series data is commonly referred to as time series data analysis.

Time series data analysis is important to understand the past and to use that understanding to predict the future. Past time series data is sometimes called *historical data*. Predictions of future events and conditions have traditionally been called *forecasts*, and the act of making such predictions has traditionally been called *forecasting*.

> The practice of forecasting uses historical data in order to predict unknown future data.

The word "forecast" and *"prediction"* are used interchangeably, with "forecast" being used more often among business practitioners and meteorologists. The term *prediction* is more widely used than "forecast," and *predictive analytics*, which is necessary in all fields, from bioinformatics to popularity predictions, is as popular a term as "time series analysis."[1]

Nowadays time series data is part of everyone's daily interaction with the Internet. Consider, for example, what *Google Trends* [53] in Figure 1.1 says about the interest in soccer (measured by intensity of Google searches for the term "soccer" in Google) in Spain, Argentina and England between January 1, 2004 and June 24, 2020. The graph displays time series data that help answer the following questions, among others: (i) Which country has been the most interested in soccer during the time analyzed? (ii) What is common to the interest in soccer in all three countries over time? (iii) What do we predict will happen to interest in soccer in those countries after June 24, 2020? (iv) Is Figure 1.1 credible? (v) Can we identify the largest peaks in Figure 1.1 as special events, for example, FIFA World Cup tournaments?

Compare Figure 1.1 with Figure 1.2, a *time window* of the former. Many individuals looked at Google Trends in the early months of 2020 to detect the impact of the COVID-19 epidemic of 2020 on aspects of life after the economic shutdown, focusing on short time intervals, like 12 months. Figure 1.2 suggests that the COVID-19 pandemic slightly decreased soccer interest. But looking at interest during 12 months does not tell the whole story about the interest in soccer over time. Figure 1.1 tells us more, because it shows that the decline is part of a long-term declining trend that started long before the pandemic. Distinguishing between short- and long-term *trends* is addressed effectively by time series data analysis methods.[2]

This chapter introduces time series *metadata*, time series *data cleaning*, basic *data description* and *feature generation*, and preliminaries of time series analysis and forecasting. Metadata refers to the information that tells us what the data measures, the

---

[1] Beginners often are confused by the word "prediction" when they see it used for forecasts and for in-sample model fitting. When we fit a model to known data, we are predicting, estimating, the true expected values of the time series for the time span in which we know the data. The error in our prediction is of interest, and we can calculate it because we know the data. When we forecast, we are predicting future values that are unknown, using what we learned in our prediction of the known data. In time series we say that one prediction is "in-sample" and the other is "out-of-sample."

[2] Google trends are sometimes used as predictors in regression models. Google produces massive numbers of time series and has such high demand for forecasting at scale that Google has its own proprietary in-house automated forecasting tool. Unfortunately, it is not open source like the automated forecasting tools of others such as Facebook, Wikipedia and other giant tech companies.

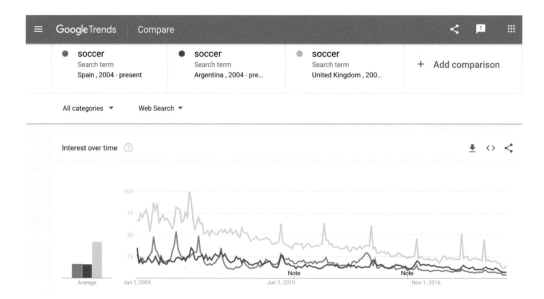

Figure 1.1 Interest in soccer in Spain, Argentina and England between January 1, 2004 and June 24, 2020. Data source: Google Trends. Google and the Google logo are registered trademarks of Google LLC, used with permission.

data quality and prior preprocessing done by others. Trivial as it may seem, ignoring the metadata is often the most common mistake made by beginners. Cleaning the collected, stored and accessed data or rearranging it for the purposes of the time series analysis sought is the second most important task, which, like metadata, is often ignored. Basic data description involves creating *time plots*, *seasonal box plots* and *spaghetti plots*, which help us describe the underlying attributes of the data and determine the important features to record. Feature generation requires thinking what in the description of the data lends itself to a simple numerical summary that together with other summaries would suffice to represent the time series in compressed form. But feature generation refers also to the process of extracting new variables from the raw time series, variables that then can be used in supervised ML methods. Prior to all of the preceding, there is data management, which involves collecting, storing and accessing the data first of all and using *Time Series Database Management Systems (TSDBMS)*. In all of the preceding, software plays a crucial role. The first Base (native) R software functions for time series and some small R programs will also be learned in this chapter. Native or Base R refers to the time series functions that do not require installing any package in R.

Time series data should not be confused with *cross-sectional data*, which are values of a variable or variables observed at one point in time in different locations or individuals. The location domain prevails in cross-section data, while the temporal domain prevails in time series data. For example, starting salary for graduates in each of the different regions of the United States in the spring of 2017, or number of occupied

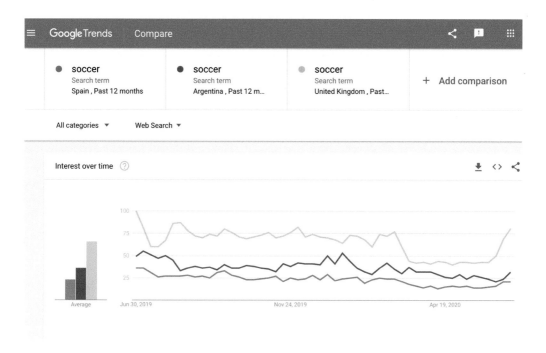

Figure 1.2 Interest in soccer in Spain, Argentina and England between June 30, 2019, and June 24, 2020. Data source: Google Trends. Google and the Google logo are registered trademarks of Google LLC, used with permission [53].

beds in each of all hospitals in the United States last month are cross-sectional data. Cross-sectional data may be correlated in the location scale, and there is a field of Statistics called *Spatial Statistics* that studies these types of data and their spatial correlations [31]. Cross-sectional data helps us detect anomalies in space.

In the rest of this chapter, Section 1.2 introduces a few of the many sources of time series data open for the reader's use. Section 1.3 emphasizes the importance of being aware of the data quality and doing data cleaning before embarking on data analysis. Section 1.4 introduces common attributes to look for in time series data. *Decomposition*, the practice of extracting those aspects when they are present, is introduced in that section. Base R code for time series that will allow the reader to get started quickly with time series data description is described in Section 1.5. Visualizations of time series data are discussed next in Section 1.6. Motivation for decomposing a time series is discussed in Section 1.7. A simple application, showing how the attributes observed (also known as features in ML) can be used in unsupervised ML is presented in Section 1.8. Frequency, how often we record the values of a time series variable, is of crucial importance. Section 1.9 is dedicated to describing types of frequencies and how they could be handled by Base R time series functions. Some comments about R and how its dedication to time series data analysis has grown over the years can be found in Section 1.10, and references to other books on time series that use R also can be found in Section 1.10. Problems, a quiz and a case study concludes the chapter.

1.1.1   Exercises

Exercise 1.1   After reading Section 1.1 and becoming acquainted with the information on Google Trends there, visit Google Trends [53] and do the following:

(a) Enter a search term or topic of interest at the moment of reading this exercise and look at the plot of long-term trends (many years of data) and separately the plot of short-term trends (last year only). For example, in Section 1.1, we looked at interest in soccer. The reader should search a different topic and compare three different countries.

(b) Compare the long-term and short-term interest in the three countries of your choice. Answer the questions asked about the soccer trends but referring to your topic. Notice that to search for trends in the same topic in several countries, it is necessary to use the same topic in all three, as we can see in Figure 1.1, and then follow the instructions in

    support.google.com/trends/answer/4359550

(c) Click on the question mark icon in the website where the plot will be and find out what the numbers you are observing measure (i.e., find the metadata).

(d) Describe the message conveyed by the trends in your topic. That is, what answers do the trends allow us to answer?

(e) Is there some reasonable explanation for what the trends show? Explain. Answering this question may entail familiarity with the topic chosen. If that familiarity is not there, maybe a little research on the topic will help.

Exercise 1.2   *Gapminder trends* is a website providing time series plots of several health, social and environmental variables across the world. There are many types of charts in that website, but in time series you should always visualize time series with line plots that have time on the horizontal axis. To access Gapminder trends go to
    www.gapminder.org/tools/#$chart-type=linechart

By clicking on the name of the variable on the vertical axis you can find several variables that you can choose from. Select the Environment tab and CO2 Emissions per person, for example. Click on the "find" icon and select countries. Select Congo, Bangladesh, Germany and Maldives. This will give a time plot with several trends, one for each country.

(a) Describe the story told by the plot about the differences in trends of CO_2 emissions per person among these countries.

(b) What is a possible explanation for what is observed in the time plots of the trends? Why? Conduct some research that will help answer this question.

1.2   Where to Find Time Series Data?

A question often asked by beginners is, "Where do I find time series data?" Nowadays, we may encounter interactive visuals of time series generated by the *Internet of Things (IoT)* in many websites that house databases. Google Trends in Section 1.1 was an example. Gapminder is another example. We are able to obtain the data behind the interactive visuals by simply downloading data stored in files. But there are many more sources of time series data. Some sources that will be used in this book are:

- *Application Programming Interface (API)*, for extracting data from a TSDBMS.
- Time series data repositories used as benchmarks for researchers designing new time series methods.
- Numerous R packages that allow us to connect to publicly available databases without downloading their time series to our computer.
- Sensor data from smart cities, medical devices and other time-stamped data [196].
- Signals from radio, medical devices, speech, radars.
- Time series data analysis competitions such as the M competitions, some Kaggle competitions, DataExpo, among others.
- Twitter temporal word analysis [194], Wikipedia's daily page views [144, 122, 93], Uber Movement [195].
- Citizen Science websites [25, 123] and government microdata [201].

APIs are a way to access open data from government agencies, companies and research organizations, without having to download the data into our computer. An API provides the rules for software applications to interact. In some cases, the API first reads the organization's API documentation. Many organizations will require entering an API key and using this key in each of the API requests. This key allows the organization to control API access, including enforcing rate limits per user. API rate limits restrict how often a user can request data (e.g., an hourly limit of 1,000 requests per user exists for NASA APIs). API keys should be kept private, so when writing code that includes an API key, it is important to not include the actual key in any code made public (including any code in public GitHub repositories) [148] . For example, the case study of this chapter in Section 1.14 will show how to access Quandl's API [120] hosted at Nasdak Data Link [128]. There are APIs for climate and weather data [134], bioinformatics and many other domains. Practically, every data-producing entity has time-stamped data.

Forecasting competitions have been driving a lot of progress in time series *forecasting at scale* with big data. Examples are the M competitions and the Kaggle competitions [81, 74, 86, 33, 82]. Their data sets have been analyzed by many and are good resources for learning. The R package `tscompdata` [74] contains large data sets used in forecasting competitions.

Time series data repositories that curate data sets are widely used to test some time series methods, such as the UCI Machine Learning Repository and the UEA & UCR Time Series Classification Repository [39, 2]. Not all data sets in those repositories are time series.

There are large and small time series data sets for all tastes in many places, and we will become familiar with several sources throughout this book. We now give some examples of time series obtained from some of those sources.

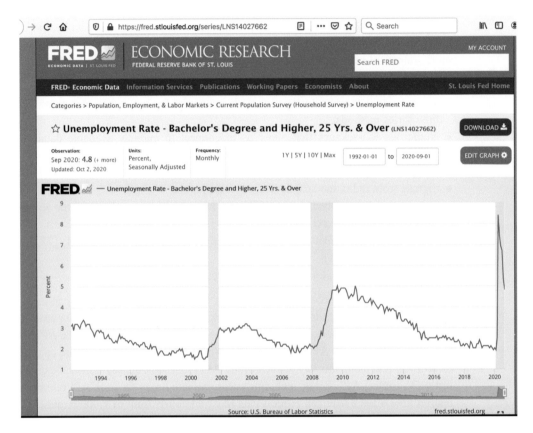

Figure 1.3 US Bureau of Labor Statistics, Unemployment Rate - Bachelor's Degree and Higher, 25 Yrs. & Over [LNS14027662], 1992-01-01 to 2020-09-01, retrieved from FRED, Federal Reserve Bank of St. Louis; http://fred.stlouisfed.org/series/LNS14027662. FRED® graphs and images provided courtesy of the Federal Reserve Bank of St. Louis. ©2020 Federal Reserve Bank of St. Louis. All rights reserved. FRED® and the FRED® logo are the registered trademarks of the Federal Reserve Bank of St. Louis and are used with permission.

Example 1.1   Compare Google Trends time series of Figures 1.1 and 1.2 with data published by Federal Reserve Economic Data (FRED) [49], a TSDBMS which publishes data and interactive online graphs of historical data collected by several government surveys, censuses and other institutions. Their data is curated and is of high quality. Figure 1.3 shows *monthly seasonally adjusted* unemployment rate for workers with bachelor's degree and higher [14] viewed in FRED's graphical interface. We found out by searching in the "Search FRED" window on the upper right-hand corner of the graph for the word *unemployment* and selecting, from the long list of series, the bachelor's degree data shown in Figure 1.3.

Two things stand out in the metadata given with the graph: the units are percent, the frequency is monthly and the series is seasonally adjusted. Additional metadata says

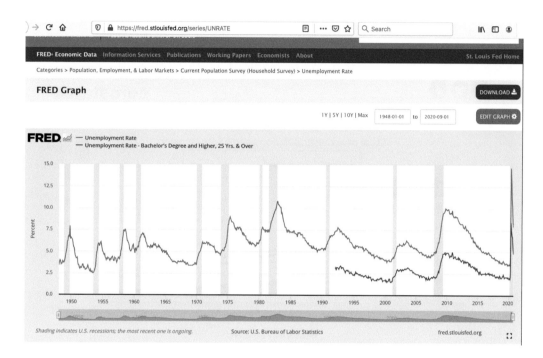

Figure 1.4 US Bureau of Labor Statistics, Unemployment Rate, 1948-01-01 to 2020-09-01, retrieved from FRED [49], Federal Reserve Bank of St. Louis; https://fred.stlouisfed.org/series/UNRATE. Superimposed is the time series from Figure 1.3. UNRATE is also monthly and seasonally adjusted.

that the series was obtained from the Current Population Survey, a very reliable source of economic data.

Comparison of social and economic time series data among different groups of society helps shape government policies and helps us understand the dangers of learning only from the most aggregated data, which hides differences in economic conditions among groups. We can display another time series from FRED's website, the overall unemployment rate, a rather aggregate figure. We do that by clicking on the "Edit Graph" button on the right side of the image in Figure 1.3 and selecting the "ADD LINE" tab. We obtained the code name of overall unemployment prior to that, doing a general search in the "Search FRED" space on the upper right-hand corner of Figure 1.3. We direct the reader's attention now to Figure 1.4.

Something is obvious in Figure 1.4: the time series for the bachelor's degree unemployment rate is shorter than the aggregate unemployment rate for all groups. That means that they started collecting data for this group much later than for the aggregate unemployment rate of the whole population. What message is Figure 1.4 conveying about the benefits of education? How are those with a bachelor's degree faring compared to the overall population, according to Figure 1.4?

When exploring time series in FRED, the reader needs to be aware that some variables are policy variables, others are indexes, others are nonseasonally adjusted data. In other words, paying attention to the metadata, and the nature of the data, before

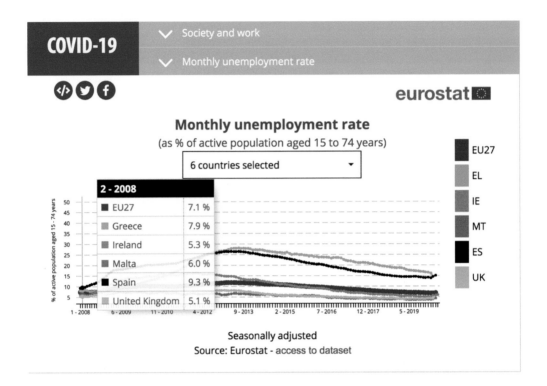

Figure 1.5 European Commission. Eurostat [44], June 24, 2020. Seasonally adjusted unemployment rate for Greece, Ireland, Malta, United Kingdom, Spain and the whole European Union. Source: Eurostat, © European Union, 1995–today. Reprinted with permission.

embarking in any time series analysis or reaching any conclusions is of utmost importance to avoid wrong conclusions.[3]   □

FRED data can be read into R using the Quandl API (see Section 1.14) or can be saved to the hard drive, after downloading from FRED, as `.csv` and then imported into R.

Example 1.2 Eurostat, government offices, and international organizations such as the OECD have interactive websites to visualize time series and they allow downloading of their data files. For example, Figure 1.5 shows the time series for unemployment rate in several European countries. Again, the time series shown are seasonally adjusted. What do you conclude about unemployment trends in those European countries after seeing Figure 1.5?

---

[3]Individuals working with time series as a profession are knowledgeable of the domain where the data come from (financial analysts know finance, political scientists know political science terms, and so on), but beginners not familiar with the context of the data used for learning, for example a biologist using Economics data in an exercise, is more likely to make mistakes with Economics data than with Biology data sets. The metadata can help reduce the risks of making mistakes.

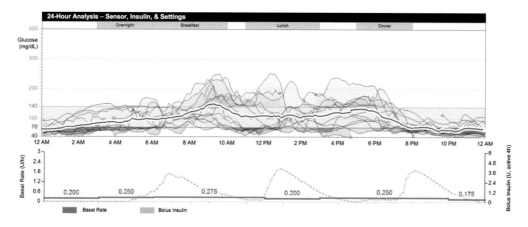

Figure 1.6 Fourteen time series of blood glucose level detected by a sensor, and insulin delivery from an insulin pump for a Type I diabetic person wearing a pump and blood glucose sensor. The gray band reflects what is considered normal glucose levels.

The Eurostat app called "data explorer" allows the reader to download time series for many years and countries, being an additional source of international socioeconomic data. Some of their data can also be obtained via Quand's API.   □

We notice that Figures 1.3, 1.4 and 1.5 have time on the horizontal axis, the way it should be in a plot of the path of the time series over time.

Example 1.3   The National Centers for Environmental Information [133] allows visualization of precipitation, snow, ice, temperature and many other climatic variables for climatic regions or nationally. Not all plots of historical time series at [133] are time plots with time on the horizontal axis as in Figures 1.1, 1.2, 1.3 and 1.4. NOAA's website goes beyond providing data and interactive plots like FRED and Eurostat do. For example, were we interested in the flooding danger for different cities in the shore of Lake Erie, time series of Lake Erie levels around those cities would be important to predict flooding, and because of that, time series for those cities and plots of forecasts are provided by NOAA [135].   □

Example 1.4   Were we to try to determine the effect of carbohydrate intake on glucose level measured by a glucose sensor, and insulin pump delivery for a Type I diabetic person throughout the course of several days, time series like those in Figure 1.6 would be relevant [174]. Time series like those can be compared with time series on carbohydrate intake, time series reflecting exercise done, and subjective measurements of stress level to study how those time series are interrelated [115]. But the sensors are not recording things at the same time, resulting in a lot of breaks in the data set. Medical device companies analyze millions of patients' data like that in Figure 1.6 every day in order to learn to improve their medical devices.   □

Example 1.5   Not all studies of time series are done with the intention of forecasting the future. In many time series that we do not easily observe, the objective is to process what comes out of transmitters of information or signals to extract information concerning the characteristics of what the signal represents. For example, measurements of heart and skeletal muscle in the form of electrocardiographic and electromyographic voltages are signals that are processed to extract information concerning the characteristics of cardiac function and muscular function. Speech, the current that goes through a transistor, radars and other tasks relevant to Engineering and applied science are signals that need to be extracted from noise to obtain information about what is being transmitted. The sounds in music are time series [42]. Many of these signals are continuous, and their study is known as *signal processing*, which is a synonymous name for time series analysis, as Shiavi [178] indicates. The latter name arose within the field of applied mathematics, and the former had been used much before the time series name arose. Due to the purpose of signal analysis and the continuous-in-time and continuous-in-amplitude wave form nature of the time series in Engineering and some applied sciences, the time series analysis in those fields is done mostly in the frequency domain. The reader is referred to Shiavi [178] for an excellent introduction to Engineering Signal Processing.   □

The reader will have noticed that all the examples displayed in images seen so far in this chapter have time on the horizontal axis, sorted in chronological order from oldest to more recent dates.

By exploring other time series from some of the sources cited in this section before continuing to read this chapter, the reader will gain an appreciation of the differences among time series and the different effects that the passage of time has on them.

We will be referring throughout the book to places where the reader can find more data. Of course, the best place to look for time series data is in the reader's own domain of expertise, as nothing helps learning and sound time series analysis more than domain expert knowledge. There is also web scraping [90].

1.3   Time Series Data Cleaning

Not all time series data is well curated and clean. Thus preliminary investigation of the quality of data is necessary to prevent devastating consequences and wrong conclusions in downstream analysis.

In this section, we describe some of the most common inspections needed prior to time series analysis.

The first thing we must check when given a time series data set is that the observations come accompanied by a time index and that they are sorted in chronological order. That seems an obvious requirement. But quite often this is violated.

For those that have always worked with independent observations and have never been exposed to time series data, the requirement of a time index often comes as a

> Whatever the source of the time series data, the time index, or the temporal scale, which indicates the time when the measurement was taken, must always be part of the data and must be sorted in chronological order.

surprise, but that is the first thing they must be aware of. We may sort independent observations in any order and the analysis will not be affected. But changing the order of a time series data set will render the data useless for analysis.

Example 1.6 The following small set of observations was extracted from the AEP energy consumption time series that comes with a Kaggle competition's data set on energy consumption in the Eastern United States [127]. Look at it in detail. Do you notice anything strange?

```
. . . . . . . . .       . . . . . . . .       . . . . . . . .
2004-12-31    22:00:00      14045
2004-12-31    23:00:00      13478
2005-01-01    00:00:00      12892
2004-12-30    01:00:00      14097
2004-12-30    02:00:00      13667
. . . . . . . . .       . . . . . . .       . . . . . . .
. . . . . . . . .       . . . . . . .       . . . . . . .
```

As we can see, the data are not sorted in chronological order. The year 2005 is reported after December 2004, but then December 2004 is reported after January 2005. This means that sorting the data in temporal order must be done before proceeding any further. □

In addition to sorting the data in temporal order, we must inspect the *sampling interval*. Are the observations equally spaced or irregularly spaced in time?[4] The sampling interval could be a year, a month, a week, a day, an hour, a second and so on. The unemployment time series seen in Section 1.1 are monthly, meaning that there is one observation per month, and all months have a measurement. The frequency of observation in a year, then, is 12 (there are twelve measurements per year). Time series data are ideally recorded at equally spaced time intervals, as is typical in most time series and forecasting applications. Base R's built-in time series applications are built for equally spaced time series. But sometimes the sampling interval in a time series is not constant or there are gaps in the time stamp. When the sampling interval is not constant, the analysis gets more complicated than when it is. R's packages such as zoo and xts and ML methods were built for the purpose of analyzing irregularly spaced time series. The gaps in the time stamp are not necessarily due to, for example, sensors breaking or accidental omission in the recording of a value. Gaps in the time

---

[4]Many methods studied in this book will require that the observations are equally spaced in time, and actions will be taken to adjust for the case when they are not. For other purposes and methods, observations irregularly spaced in time may not pose a problem, and those also will be discussed throughout the book.

stamp could be due to the nature of what is being observed. For example, stock prices are recorded during the five trading days of the week. We cannot claim that the data are observed daily since there is no measurement during the weekend days.

Other obvious violations such as having duplicate time stamps (as in, for example, daylight savings time change), also must be resolved. Checking unusual events, such as, for example, holidays, or outliers that may represent measurement errors or actual phenomena in the past is also necessary.

Data cleaning also involves determining whether strange codes for missing data appear with the data, for example, "999" to represent that an observation is missing. If we use R, those need to be converted to NA, otherwise R will think that those numbers are part of the data values.

Checking that there are no embedded summaries of the data within the raw data also is good practice if the data has been obtained from tables published by others. In this case, the summaries must be removed.

Other inspections to do could depend on the software we use. For example, if the software does not accept commas in the numbers (R doesn't), those should be removed before downstream data analysis.

Example 1.7   This example starts at the website listed in the caption of Table 1.1. Imagine that we were looking for passenger data on domestic and international flights and ran into that website and found that data set. Table 1.1 shows only part (a "window") of a much larger collection of monthly time series data that was recorded by the Bureau of Transportation Statistics [136] to observe growth or decline and sources of variation over time in the number of passengers in the US. There are three time series in Table 1.1. Two columns contain two variables indicating the month and year, which will help us construct a time index. The year is in one column, and the month of the year (1 = January, 12 = December) is in another column. Each time series has a single observation per month.

The time format in Table 1.1 is the one used by the US Bureau of Transportation Statistics in the Excel files that the public may download and use. That format and several slight variants of it are used all around the world. But be aware that not all time series data has time reported that way.

There are no missing observations, at first glance. And the data appears in chronological order (meaning that we do not find the year 2004 data listed before the year 2002, or similar things).[5] But we also notice that the numbers have commas between them. And we notice that there is some interruption in display of the time series: every year the table shows a line for the total number of passengers for that year. If interested in the monthly number of air passengers, we would have to clean the data and remove that line, using software, in order to not interrupt the chronological order in which the monthly observations appear. We would also have to remove the commas to make the data acceptable to R.   □

---

[5]When dealing with big data, it will not be so easy to find out whether the time series are in chronological order. Nielsen [132] dived into the 311 hotline data set from the New York City Open Data Portal and found that the data were not sorted in chronological order by looking at only 10 rows of the data.

Table 1.1 A portion of a data set containing the number of passengers on domestic and international flights in the US. Source: Bureau of Transportation Statistics T-100 Market data (www.transtats.bts.gov/Data_Elements.aspx?Data=1).

Year	Month	DOMESTIC	INTERNATIONAL	TOTAL
2002	10	48,054,917	95,78,435	57,633,352
2002	11	44,850,246	9,016,535	53,866,781
2002	12	49,684,353	10,038,794	59,723,147
2002	**TOTAL**	**551,899,643**	**118,704,850**	**670,604,493**
2003	1	43,032,450	9,726,436	52,758,886
2003	2	41,166,780	8,283,372	49,450,152
2003	3	49,992,700	9,538,653	59,531,353
2003	4	47,033,260	8,309,305	55,342,565
2003	5	49,152,352	8,801,873	57,954,225
2003	6	52,209,516	10,347,900	62,557,416
2003	7	55,810,773	11,705,206	67,515,979
2003	8	53,920,973	11,799,672	65,720,645
2003	9	44,213,408	9,454,647	53,668,055
2003	10	49,944,935	9,608,358	59,553,293
2003	11	47,059,495	9,481,886	56,541,381
2003	12	49,757,124	10,512,547	60,269,671
2003	**TOTAL**	**583,293,766**	**117,569,855**	**700,863,621**
2004	1	43,815,481	10,252,443	54,067,924
2004	2	45,306,644	9,310,317	54,616,961
2004	3	54,147,227	10,976,440	65,123,667
2004	4	53,253,194	10,802,022	64,055,216
2004	5	53,030,873	10,971,254	64,002,127
2004	6	56,959,142	12,159,514	69,118,656

Example 1.8 The UCI Machine Learning Repository contains many time series data sets. There are different types of files reflecting the distinct ways of tracking information across time. In particular, we are interested in the *air quality* data set [111]. This data set can be downloaded to the reader's computer as a .csv (comma-separated values) file, but at the time of writing, there was some irregularity with that file. Thus we suggest downloading the Excel file and importing it into R instead. A copy of that file can be seen under the name *AirQualityUCI.xlsx* on the book's website. The program *ch1airquality.R* contains the code used in this example.

According to the documentation, the data set contains 9,358 instances of hourly averaged responses from an array of five metal oxide chemical sensors embedded in an Air Quality Chemical Multisensor Device. The data were recorded from March 2004 to February 2005 (one year). There are 12 variables, excluding the time index ones. According to the documentation (the metadata), missing values are tagged with the "−200" value. But R's missing value code is NA. This means that we will have to worry about changing that tag to NA if we want to use R. But first we count how many of those "−200" values are in each variable. The time series with the most missing values are NMHC(GT) with 8,443 out of 9,358 observations, followed by CO(GT) with 1,683, NOx(GT) with 1,639 and NO2 with 1,642. All other time series have the same number of missing values, the lowest number, which is 366.

After figuring out how many missing values there are, we must convert them to NA, the missing data symbol used by R. After converting the "−200" to NA there should be the same number of missing values as "−200" values.   □

We will see several instances of data cleaning throughout the chapters in this book. What we learn in the data cleaning process is useful information to be used in subsequent analysis.

1.4   Components of a Time Series

The notion that a time series consists of several *unobserved components* with different periodicities was introduced into Economics a long time ago. Since then, the main concern in time series analysis has been to disentangle those components before proceeding with any statistical modeling or ML approach.[6] Methods for the extraction of these components are studied in Chapter 2, and they allow us to create new variables (features), each of which represents a component. We will just introduce the components here. Contemporary methodology used in automated forecasting at scale used by high-tech companies' functional models is inspired by the old tradition of decomposing time series.

The classical components of a time series that involves humans and nature in general depend on the granularity of the data. If the time series is annual or collected at *low frequency*, it could have a *long-term trend* (reflecting a long-term pattern usually caused by growth or decline of external factors such as, for example, population). Annual data sometimes also have *irregular cyclical waves* longer than a year around the long-term trend,[7] and most time series have *random fluctuations* with hidden autocorrelations that must be found within noisy behavior. When the time series contains monthly observations, it could have the components already mentioned and perhaps, in addition to those, *regular seasonal* variation (annually recurring phenomena, patterns that complete within a year and repeat every year; for example, seasonal increase in employment in the holiday seasons, called *annual seasonality*). If the time series contains daily measurements, we could observe weekly recurring phenomena such as, for example, smaller traffic during the weekend days (*weekly seasonality*) in addition to an annual seasonality and to the components already mentioned. For hourly data, we could perhaps add *daily seasonality*, as when, for example, household electricity use is larger in the evening. These components may occur alone or all together or in any other combination, depending on the frequency of measurement of the data.

---

[6] In signal processing, such as that for audio signals, the main concern is to decompose the signal time series into components with different frequencies, as in different sounds in a piece of music.

[7] Such cyclical variations are found in market economies, where sometimes those cycles are believed to be related to which party wins elections [95].

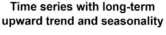

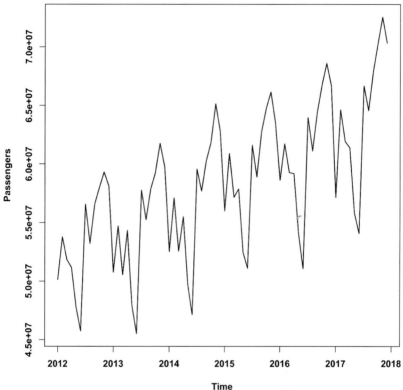

Figure 1.7 A time plot of a window of a monthly domestic passengers time series from January 2012 to December 2017. Time series data should always be plotted with a time plot in order to see all relevant patterns.

Components that are easily observable at times, without further analysis, are the trends and the seasonalities. The *random components* left after removing those usually contain hidden signals that are not so easy to interpret, and understanding and modeling these signals is often the goal of time series analysis. Some call them the "hidden signal in the noise."

Example 1.9   Everyone expects that if population is growing, there will be an upward long-term trend in the number of domestic passengers, as illustrated in Figure 1.7. The image corresponds to data set *ch1passengers.csv* and can be reproduced with program *ch1passengersplots.R*. Fluctuations in the state of the economy may translate into up-and-down short-term trends or cycles taking place within five years, more or less. We would need more years of data than those plotted in Figure 1.7 to determine whether there are cycles. But the annual seasonality can be detected with a few years of monthly data. It shows itself as the recurring increase or decrease above or below the long-term trend. Figure 1.7 displays annual seasonality.

We cannot see any other pattern, signal or not, clearly in Figure 1.7, but there could be one hidden below the regular long-term trend and regular seasonality. The trend and strong seasonality, when present in a time series data set, obscure a hidden signal that could be of interest. □

A technique or model used to analyze time series must be able to account for long-term trend and seasonality if they are present, either to remove them before analyzing the random component (operations called *detrending* and *seasonal adjustment* respectively) or to incorporate them into the model to account for them, as most classical and modern functional models used to forecast at scale do. Data-producing agencies such as FRED sometimes seasonally adjust the data. The seasonally adjusted unemployment time series of the BLS of Figures 1.3 and 1.4 are a product of complicated iterative seasonal adjustment filters intended to give users data that has removed the annual seasonal bumps in unemployment. Not all time series exhibit all the components mentioned; therefore, no single model or technique is a best model or technique for all time series. For example, not all trends are like those in Figure 1.7.

Example 1.10   Many financial and economic time series, for example, have volatility and trends that do not resemble the trends in other time series. The time series of the stock of Apple shown in Figure 1.8 is an example of a time series with a stochastic trend. □

> One of the most important problems to be solved in data science is that of trying to match the appropriate predictive model to the pattern of the available time series data. That depends on where the data comes from. Methods appropriate for daily Facebook events data may not be appropriate for daily incomplete environmental data. That is why a thorough investigation of the components of a time series is crucial for modeling.

Other terms for the different components of a time series are no-memory series, short-memory series and long-memory series. When looked at in the frequency domain, we talk about high-frequency and low-frequency waves.

- No-memory series or white noise series have observations that are independent. Their frequency is very high.
- Long-memory series are series that appear as almost deterministic functions of time, although some trends are stochastic, like random walks. Long-memory series are equivalent to long-term trend and low frequency. In these series, the dependence on the past does not die away quickly.
- Short-memory series lie between white noise and long-memory series, occur often in the physical and engineering sciences and comprise the bulk of time series that can be most usefully analyzed by classical time series methods. Observations closer in time are more similar than those far apart in the data.

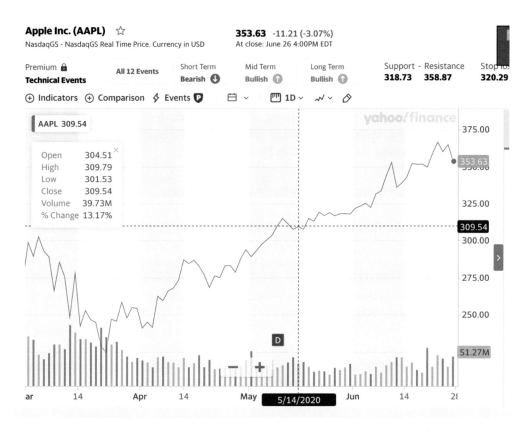

Figure 1.8 Apple Stock time series. Source: Yahoo Finance (https://finance.yahoo.com/chart/AAPL?p=AAPL). Trends in financial time series will require special time series models.

The components of a time series are also classified as stationary or nonstationary. The former reflect the short-memory series and the latter the long-memory one, but they both have a very precise mathematical meaning that we will study in Chapters 3 and 5.

1.4.1   Exercises

Exercise 1.3   Consider Figure 1.4. Instructions to create it are provided in Section 1.2, Example 1.1. Search in FRED for seasonally adjusted unemployment rate for other groups and add the time series to an image like that in Figure 1.4. For example, one time series could be the unemployment for people 20 years and over in Missouri, another for the same population group in Florida. Compare the unemployment trends of the different groups and compare them all to the overall unemployment rate. Reading Section 1.4 before doing this exercise is highly recommended. Use components to describe the series.

Exercise 1.4   After reading Section 1.2, which provides instructions to navigate FRED's website, search in FRED for nonseasonally adjusted unemployment rate for college graduates with master's degrees 25 years and over and for nonseasonally adjusted unemployment rate for college graduates with doctoral degrees 25 years and over. FRED has a separate series for "men" and "women." Plot for both the master's and Ph.D. categories, a total of four time series. Is there something in the time series that leads you to realize what nonseasonally adjusted means? Point out what that is and describe the differences, if any, among the four time series. Comment on the observable components of the time series. Do you expect any signal hidden under the observable components? Why? Explain.

1.5   Hands-on Base (Native) R Code for Time Series

A time series data set could arrive to our hands pre-formatted by the software as a time series object of some class or just raw from the source. If the time series data has been previously stored and managed by a specific Time Series Database Management System (TSDBMS) [40, 57, 139, 141] or a software such as R, the reader needs to be familiar with that software and use the commands allowed by that software to clean, manage and analyze the time series.

R has several class formats for time series. There is the native `ts()` format, and the more recent `zoo()`, `tsibble()` and other formats that have been born out of necessity to handle data sets of different levels of granularity and different qualities. Beginners, particularly those not too familiar with R, benefit from learning time series analysis using native `ts()` and the types of time series analysis that objects of this class allow, and that is why it is used a lot to introduce the topics presented in this book. The `ts()` class works very well with well-curated and clean data sets. But other classes will be introduced as needed and as the complexity of the data call for them.

Example 1.11   R software has already made the time series `JohnsonJohnson` (a time series that is part of base R data sets) a time series object (`ts()`).[8] An object like this is already formatted inside R in a way that makes it very easy to extract the information using native R 's functions. Of course, if the data has been stored in another software (e.g., Stata, SAS, proprietary business software), those packages will have their own native time series class object but we can import them into R and make them a `ts()` object. The reader is encouraged to open RStudio and execute the following statements one by one, applied to the quarterly time series `JohnsonJohnson`. Because the data is a `ts()` object, the functions will work:

- View the data first of all. It is small, thus typing its name will show you the whole data set and how nicely formatted it is by the `ts()` class in R. Notice the time index. The data actually presents three features: the quarter `Qtr1 Qtr2 Qtr3 Qtr4`, the year `1960, ...., 1980`, and the value of the series each quarter of each year. We could create three variables (features in ML language) conveying the same information. Each variable would have 84 lines (21 years times 4 quarters per

---

[8] Base R 's data sets can be accessed by typing `library(help = "datasets")`. There are some that are time series and can be used by the reader to practice the commands introduced in this section.

year). But creating those three features would not be necessary unless we are not planning to do our analysis within R.

JohnsonJohnson

- `class(JohnsonJohnson)`
  will return `ts()`.
- `frequency(JohnsonJohnson)`
  will return 4, meaning that there are four observations for each year of data. This means that the time series is *quarterly*. Quarterly data is time series data that has been recorded four times per year, once per quarter of the year (quarter 1 being January to March, quarter 2 April–June, quarter 3 July–September, quarter 4 October–December). The `frequency()` function assumes that whatever you write in it is the number of observations per year.
- `start(JohnsonJohnson)`
  will return `1960 1`, the first year and the first quarter present in the time series, that is, the first observed value in the time series corresponds to the first quarter of 1960 (the quarter corresponding to January, February and March).
- `end(JohnsonJohnson)`
  will return `1980 4`, that is, the last observed value, the most recent, corresponds to the last quarter of 1980 (the quarter containing the months October, November, December).
- `length(JohnsonJohnson)`
  will return 84. If we do the accounting, that means that no quarter was skipped, no observation is missing.
- `ts.intersect(JohnsonJohnson, sunspots)`
  will return an error because `sunspots` is a monthly time series and JohnsonJohnson is a quarterly time series. You cannot intersect time series with different frequencies. The function `ts.intersect()` is to find out what dates are common in different time series of the same frequency. It is a convenient function when you have multiple time series and want to keep only years of data that are common to all.
  Contrast the preceding with the outcome of

`ts.intersect(UKgas, JohnsonJohnson)`

where `UKgas` is another Base R quarterly time series. The results give a multivariate time series of length 84 containing two variables with the measurements, and another two categorical variables with the features "year" and "quarter" representing the time index.

- `window(JohnsonJohnson, start=c(1962,2),`
              `end=c(1971,3))`
  will return a time series containing only data between the second quarter of 1962 and the third quarter of 1971 (1962:2–1971:3), inclusive. This is a subset (a window) of the JohnsonJohnson data that we started with. It is common to look at

intervals of the data, particularly when there are many observations and a time plot will not show much in that case, or when splitting the data into a training set and a test set. Time series data is always a subset by time intervals.

- `sum(is.na(JohnsonJohnson))`
 will give an indication that there are missing values that R recognizes (`NA`) in the time series if it gives a value larger than 0. Notice that this will not indicate whether the main assumption we are making when using a `ts()` class, that the data is recorded at equally spaced time intervals and there are no gaps, is satisfied. A time series could be devoid of `NA` and yet present gaps in the time index (for example, jump from February 2020 to April 2020, for monthly data).

 If there are `NA`, we will need to think further and determine whether to aggregate the data, or study only a period where there are no missing values, or some other action, such as use other R data classes for these cases. However, if there are no missing values, we must still check for gaps and equal interval sampling, which require a little more sophisticated data management.
- To view the quarter corresponding to each observation, we type:

`cycle(JohnsonJohnson)`

This produces a categorical variable or label for each observation. This variable could later be used as a feature in supervised ML.
- To extract a variable that indicates the time (both the year and quarter) of the observations we type:

`time(JohnsonJohnson)`

This produces the year represented as the year number and the fraction of the year (.00 for first quarter, .25 for second quarter, .50 for third quarter, and 0.75 for the fourth quarter).
- The function `aggregate()` allows us to aggregate the observations in a time series. For example, we might want to convert quarterly data to annual data. ☐

If the time series is raw data coming from a source that has recorded the values in some format (Excel, or some other form), without making the time series any particular software-specific class of time series, and without a time index in it, but with metadata indicating what that index should be, then the reader can read the data into R and make it one of the time series formats that R can work with. The following example illustrates this point.

Example 1.12   Suppose someone gives us the following numbers of amber alerts in the country: 10, 11, 5, 10, 15, 6, 8, 11, 13, 14, 1, 10, and this data does not need cleaning. Suppose also that the data donor indicates that the first observation corresponds to the first quarter of 1910 (1910:1), and the last observation is for the 4th quarter of 1912 (1912:4). Then we could enter the data into R and tag it with the name `entered.by.hand.data` to create an R `ts()` object. The reader should copy, paste or type the following code in R and run it line by line to see the outcome of its execution.

```
entered.by.hand.data=c(10, 11, 5, 10, 15, 6, 8,
                11, 13, 14, 1, 10)
 # make a ts object
amber=ts(entered.by.hand.data, start=c(1910,1),
                end=c(1912,4), frequency=4)
class(amber)
amber # to view the data in ts format
```

This will create a time series object called `amber` that is now time-indexed in a way that R functions that act on `ts()` class understand, and to which we can apply any of the tasks done to the `JohnsonJohnson` time series. Notice that, because the time series is recorded four times a year, the frequency is entered as 4. □

Example 1.13 We revisit now the monthly number of passengers in Table 1.1, mentioned in Example 1.7 and partly visualized in Section 1.4, Figure 1.7. For this example, we look at the amount of domestic passengers in the US from October 2002 to June 2019. After cleaning the table with the data in Excel, removing the headings and the annual record, the resulting data set is *ch1passengers.csv*. We use the R code given in file *ch1passcode.R* to read it and to clean it. There are no missing observations. But there are commas, strange characters. For this reason, R, by default, will read the passenger data in Table 1.1 as it is showing in the table, interpreting it as if it was character data, not numerical data, and we would not be able to run time series analysis using R. The `ts()` class only accepts numeric variables and will give an error. The commas need to be removed in order to be able to use R. Notice that other software might welcome data with the commas, but not R.

After we conduct the preliminary data cleaning work, we create an R `ts()` object. As a `ts()` object, we can then apply to it most of R's functions that work with the `ts()` class. □

1.5.1   Exercises

Exercise 1.5 Import into R the Unemployment rate - Bachelor's degree and higher, 25 years and over, which is series LNS14027662 in FRED. Make it an R time series object and then do a time plot of it. Double check that the `ts()` object was created properly by using the R functions introduced in this section.

Exercise 1.6 We are often interested in comparing time series or in determining where time series intersect.

(a) After saving as a `.csv` file, import into R the UNRATE time series and LNS14027662.
(b) After making both time series `ts()` objects, find the intersection of the two series.
(c) Create a new time series that contains a window of the UNRATE of your choice.

1.6   Time Series Data Visualization

Once we have inspected and cleaned the time series data, making sure that the time index is sorted chronologically, that there are no duplicate dates or gaps in the time index, that no strange symbols are embedded in the data and that errors do no exist, the next step involves careful visual scrutiny of the recorded data or windows of it (if the data set is very large) with plots [179]. This section introduces some of the most important exploratory plots.

   We start with *time plots*, which are plots of the data against time.

1.6.1   Time Plots

Joining the points of a plot of the time series against time by lines helps us see the *trend in mean* and the the recurring, seasonal sawtooth patterns if there are any. A series with a regular sawtooth pattern is described as a *seasonal series*. Seasonal series are common in social and economic data, atmospheric data and some biological data. Figure 1.7, seen earlier in the chapter, is a time plot that indicates that the number

---

A time plot is a **line** plot of time series data against **time** and is the correct way of displaying a time series. The observations are connected through lines in order to appreciate the patterns over time. The time plot must reflect the data in the order in which it was measured, February after January, March after February, 9:00 AM after 8:00 AM, Tuesday after Monday, and so on. If data for some time interval is not available, then special adjustments to the time plot functions must be done in order to be able to plot the time series without showing that data.

---

of passengers is increasing over time (has an upward trend) and there is an annual seasonal recurring pattern. The line plot not only lets us see the trend in mean but it also lets us see the recurring sawtooth pattern that reflects seasonality.

1.6.2   Plots Not to Do

We mention now scatterplots just to warn the reader not to use them when visualizing the time series against time. Beginners, used to the analysis of independent observations in data sets, jump to the conclusion that a scatterplot is as good a plot for a time series as for independent observations. But that is incorrect if the goal is to visualize the historical data sequence. Scatterplots may be appropriate for other tasks.

In time series data analysis, a scatterplot is **not** an appropriate way to visualize the data over time. The scatterplot does not let us see seasonal sawtooth patterns.

   Figure 1.9 is a scatterplot of domestic passengers over time. Some patterns can be seen in Figure 1.9. There are several bands of points, and we see some sort of up-and-down regular pattern. But we do not plot time series data like in Figure 1.9 because this type of plot hides the seasonal and other patterns in the time series.

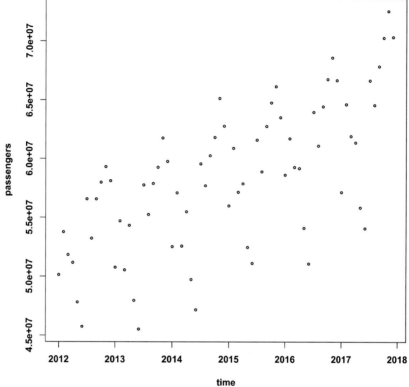

Figure 1.9 People do not plot time series as scatterplots because the sequential nature of the data is not appreciated this way. This scatterplot is showing the same data as Figure 1.7, a window of a monthly domestic passengers time series from January 2012 to December 2017. But unlike in Figure 1.7 we cannot see the changes of the data over time.

Typically, time series data is plotted against time with the points joined up by lines, as in Figure 1.7, because with this type of plot the sawtooth pattern jumps out at us as the points get connected up. The major things we see in Figure 1.7 are an overall trend upwards and the seasonal sawtooth pattern, the pattern that repeats every year.

1.6.3   Visualizing the Trend

We could trace by hand or visualize the long-term trend in mean, to make it more obvious to the untrained eye, as in Figure 1.10, where we can now see a curve superimposed on the correct time plot. We will explain later how that trend was calculated. That trend curve is there to help see the overall long-term trend in mean without the clutter of the seasonal and other random details of the time series.

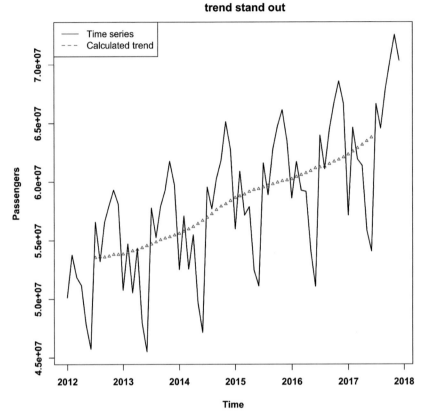

Figure 1.10 Once a time plot is obtained, we can emphasize the most prominent aspect of it, the seasonal pattern and the upward trend. We overlaid an estimate of the trend on the time plot. Chapter 2 will show the reader how to obtain estimates of trend.

We should not expect trends to be linear or increasing all the time. There are all kinds of trends: increasing, decreasing, random, of wave form, and many other descriptions.

1.6.4   Visualizing the Variability

Figure 1.10 does not indicate that the variance of the time series increases or decreases or changes over time in a way proportional to the trend. We see the seasonal swings, but they are not becoming vertically wider and wider as we move to the right or the left or any other part of the plot. That means that the growth is not due to the seasonal component or that we are looking at too narrow a window of the data to appreciate changes in variance.

Contrast Figure 1.10 with Figure 1.11, which shows the monthly number of occupied hotel rooms. As we can see, the amplitude of the seasonal swing increases

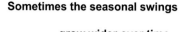

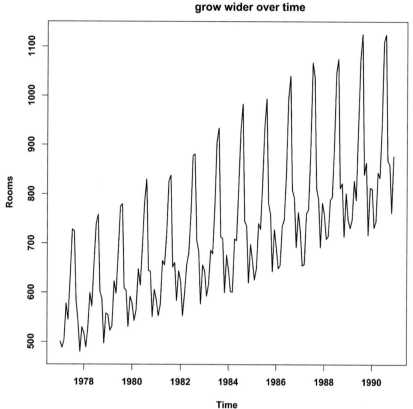

Figure 1.11 The number of occupied hotel rooms has increasing variability, proportional to the trend, indicated by an increasing amplitude of the seasonal swing. Data from [8].

proportionally to the trend, indicating that the variability of the time series is increasing proportional to the trend.

Long time series allow us to see more patterns than short ones. Perhaps if we had looked at the whole length of the passengers series, we could have noticed a change in variance, if there was one.

1.6.5   Seasonal Time Plots

We now study the seasonal pattern observed in Figure 1.10 in more detail. Seasonal time plots allow us to do that. These plots slice the time series plot into as many time plots as years and plot those slices in the same image. Since we have six years of data, there are six line plots to look at in one image, as in Figure 1.12, where we plotted the data against month (1 being January, 2 February and so on). There are 12 observations in a year if the time series is recorded monthly.

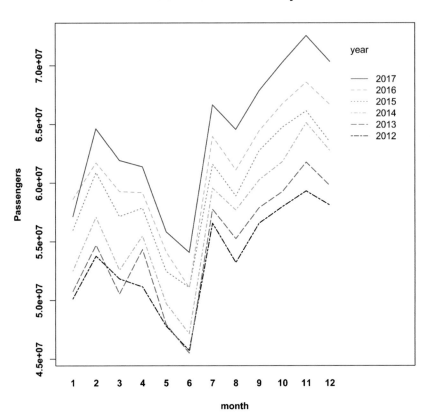

Figure 1.12 Plotting the data against month, when the data are reported monthly, we can learn more about the seasonal pattern by slicing the time series into as many time plots as years we have. This is informally called a *spaghetti plot*.

We observe that every year the passenger numbers are bigger in November (Thanksgiving break in the United States, when flying is very common) and lowest in June. There is more traveling in the second half of the year. With the exception of June, there is an upward trend each year. There are also two clear-cut time periods. The first part has higher values in March. In addition to this seasonal pattern, we can see that the curve gets bigger each year.

Seasonal time plots are sometimes called spaghetti plots.

1.6.6   Seasonal Box Plots

A less intuitive plot that will be more familiar to those with a statistics training is a *seasonal box plot* like that in Figure 1.13. A seasonal box plot for monthly time series makes a box plot of all the January values, another box plot of all the February values and so on. For example, the box plot for January will display the value of domestic

We see that June has the lowest
median number of domestic passengers

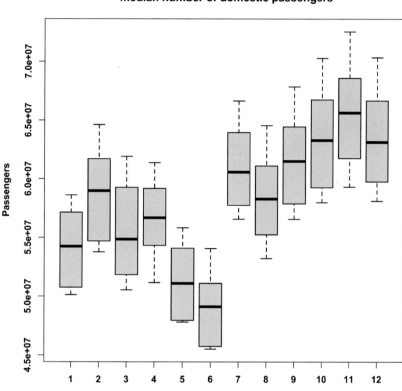

Figure 1.13 A seasonal box plot makes a box plot of the January values across all years (first box on the left), the February values across all years (second box) and so on.

passengers in all Januaries of the years 2012–2017. The domestic passengers value of 43,032,450 of January 2003 and the value 43,815,481 of January 2004 that we can see in Table 1.1 are represented in the box for January, among other January values. The box plot for month 2 (February) will have the February passenger values for the years 2012–2017.[9] Figure 1.13 conveys the same message as Figure 1.12 but with specific summary statistics such as the median, first quartile, third quartile, and maximum and minimum values per month, a more disaggregated information than the value of those statistics for the whole of 2012–1017. In contrast with Figure 1.12, the box plots do not show the year. Seasonal box plots are not time plots.

Seasonal patterns are more common in some time series than others. They are not always present. They are common in economic, energy, meteorological, biological, traffic and other data. Seasonal patterns are distinguished from cycles, which usually represent fluctuations of more than three years duration.

---

[9]Box plots are very widely used in Statistics. For a correct description of what each part of a box plot means, see [37, 108, 190].

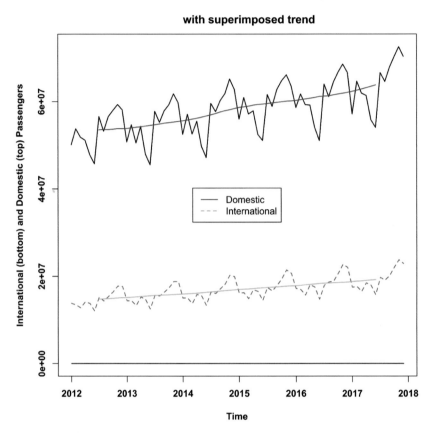

Figure 1.14 Comparing several time series in one plot is often useful. But if the scale is too different, they should not be plotted in the same image as we do here.

1.6.7   Multiple Time Series Plots

Until now, we have been looking at a single time series at a time. That is a first step in time series data analysis. Often, the relation between a number of jointly measured time series is of interest. We address those in Chapters 8 and 9. The technical term for jointly measured time series is *vector time series*. Consider, for example, the relation between domestic and international air passengers of Table 1.1. Figure 1.14 shows the international and domestic passengers series from the BTS data, with both line plots in a single graph, and a trend estimate superimposed. Figure 1.14 illustrates both the advantages and disadvantages of plotting multiple time series in the same graph.

The big advantage is that everything is on the same scale. So you can say at a glance which are the big travelers. The fact that domestic passengers are higher jumps right out at us, as does the fact that both types of passengers are growing steadily, but domestic seem to be growing a little faster. And we can see that both series have a seasonal

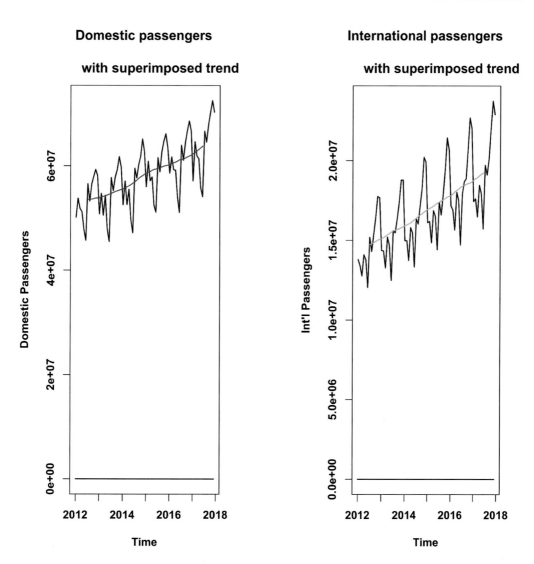

Figure 1.15 Two separate time plots make it easier to compare patterns in two related time series. The plots have the raw data and a smooth trend estimate.

pattern. The big disadvantage is that information from international passengers with a much smaller number is squashed down at the bottom.

A complementary way of displaying the information gives each type of passengers its own graphs, as seen in Figure 1.15. Now we can see the trend for each series much more easily and can compare the trends. But to tell anything about relative sizes we have to rely on the scales. The growths are similar, but the seasonal patterns are not. International passengers have a less extreme seasonal pattern (in the hundred thousands). We have to go to the scales to see that the seasonal swings are strongest for the domestic passengers, which tend to have a more extreme seasonal variation on the order of a million passengers. Variability increases faster in the international passengers series as well.

Program *ch1passengersplots.R* has code to obtain Figures 1.14 and 1.15.

Seasonal Time Plots are Not Multiple Time Series Plots

We studied seasonal time plots in Section 1.6.5. They are very useful to study the seasonal patterns of one time series in more detail than we can see in the time plot of the whole time series. It may appear to the reader that seasonal time plots are multiple time series plots. They are and they aren't. They are because seasonal time plots and multiple time series plots are both time plots. They aren't because in seasonal time plots we are observing a single time series (only one variable) sliced by cycle (Figure 1.12). In multiple time series plots we are studying more than one time series (more than one variable), as in Figure 1.14.

1.6.8   More Types of Plots in the Temporal Scale

Several authors have created other interesting two-dimensional and three-dimensional visualizations. See, for example, chapter 3 of [132], where, for example, plots appropriate for multiple time series that have different amounts of observations (Gantt charts), per month curves of year-to-year time series, visualizations in different perspectives and others are discussed. The reader will find scatterplots of different colors in those multidimensional presentations, which is why they should not be looked at until a proper time plot, a line plot of the data against the temporal scale of the kind used in Figure 1.7, is seen. The warning against scatterplots given in Section 1.6 should not be forgotten by the reader.

It is also the case that R's graphical capabilities allow the creation of multicolored and more advanced plots. Those familiar with the `tidyverse` package [203, 204] can practice the plots presented in this chapter and others presented, for example, in [76], using the `tidyverse` environment. Plots in most packages just extend the capabilities of Base R. The same plots could be obtained with programming in Base R. The more granularity in the data, the more types of plots are available to us, and it might be more convenient to use packages. We will be exposed to some of those other plots throughout the chapters in this book.

1.6.9   Exercises

Exercise 1.7   The US Bureau of Transportation Statistics analyzes data like that seen in Table 1.1. An article they wrote titled "2007 Traffic Data for U.S. Airlines and Foreign Airlines U.S. Flights" [197] comments on how they use the data. Access the article and observe the plot in it. What did the BTS choose to display? Why are we not seeing the sawtooth pattern that we saw in Figure 1.7? Which of the procedures mentioned in this chapter was used to make the data look like that? Explain.

Exercise 1.8   Watch the video at [205] and compare the patterns in the time series analyzed in that video to the patterns in Figures 1.11 and 1.7. The author in the video identifies some points of time that stand out as different from the general patterns. Which are those times and what is the reason for their standing out? Can you see any points that stand out in the domestic passengers time series that we analyze in Section 1.6? If you can, can you find an explanation for them?

Exercise 1.9   Program *ch1passengersplots.R* was used to obtain the results in Section 1.6. It allows us to analyze a time series to construct a time plot, a spaghetti or seasonal time plot and a seasonal box plot. The program refers to the passengers data described in Section 1.6. Run that program and observe what it is producing. After that, use the same program, but modifying names, removing lines that may not be needed, and changing, adding or removing arguments of functions as needed to construct a time plot, a spaghetti plot and a seasonal box plot of the number of occupied rooms time series seen in Figure 1.11. The data is *rooms.txt*.

The rooms data set is a monthly time series starting in January 1977. The following questions are about this time series.

(a) What year and month does the original time series end? What is the frequency of the time series? What is the length of the time series? Are there missing values in the series?

(b) Obtain a window of the data ranging from January 1980 to December 1988 (inclusive).

(c) Write your interpretation of what you see in the time plot, the spaghetti plot and the seasonal box plot of the window of the data.

Exercise 1.10   The Johnson & Johnson data set, `JohnsonJohnson`, is one of the time series that come with Base R. We talk about it in Example 1.11. The time series contains quarterly earnings per share for the US company of that name, recorded from the first quarter of 1960 to the last quarter of 1980. Thus, it is quarterly data, which means that there are four observations recorded each year. There are 84 quarters or 21 years of data. This data set is of class `ts()`, R's native time series data class.

Use the R code in program *ch1passengersplots.R* that produced the plots found in Section 1.6, adapted to this data set, to do a time plot, a spaghetti (seasonal time) plot and a seasonal box plot of the `JohnsonJohnson` data set. Describe what the plots are saying about the time series.

Exercise 1.11   Watch the video at [206]. What is the difference between the story on visitors to New Zealand given in the video and the story on airline passengers in the US given in Section 1.6?

Exercise 1.12   The annual high and low values of the Amazon river discharge between 1962 and 1978 can be seen in the data set *amazon.txt*.

Make the data set an R multiple time series object and do a time plot of the two series like that in Figure 1.14, without the trend line. What are the time series features of this data set?

Exercise 1.13   The data set *precipitation.txt* contains monthly precipitation in the state of Wyoming between October 1921 and September 2005. A total for each year and the year are also in the data set. Is this a multivariate time series data set or a univariate one? If you had to do a time plot, would it be a univariate time plot like the type of plot in Figure 1.7 or like the type of plot in Figure 1.14? Program *ch1precipitation.R* will be helpful in working on this problem.

1.7   Why Worry about Trends and Seasonality?

Why do people worry about identifying and estimating patterns, like the ones we've been seeing? Mainly because they want to take them and project them into the future. They want to use them for forecasting what's going to happen next so that they can plan for this anticipated future, how to adapt to it or take advantage of it. This might be having the right levels of resources in place, perhaps budgeting to employ additional staff to cover peak periods [205]. However, we are also interested in seeing the overall patterns because we may want to temporarily remove them from the time series in order to be able to see hidden patterns that are not so obvious and that might represent short-term correlations that are as relevant as the overall and recurrent patterns to forecast the future. A complete model for the time series may need in its specification not only the trend and seasonal patterns but also the pattern hidden in the random component of the time series.

In future chapters we will show how to break a seasonal series into component parts, with a particular emphasis on the seasonal components and the trend, a process called *decomposition*. We'll also talk about using stochastic models for what remains after removing the trend and the seasonal pattern. And we will talk about putting all the pieces together to account for all the patterns before forecasting the future.

> Trends in mean, seasonal patterns, increasing variance, sudden outbursts of variance change and cycles are the first things to look for in a time plot of a time series that has passed the data quality inspection. Once those trends are identified and modeled, what remains (the random term) is inspected to see what hidden signal remains in order to model and forecast its future values. After describing the overall features of the time series, then we can move to seeking detailed disaggregated information, such as patterns within each year, or week, or day.

1.8   Feature Generation for ML Applications

When there are millions of observations and/or many time series at once, not necessarily large, the reader can imagine how difficult it would be to visualize and to do the exploration that we have described in this chapter, particularly because each time series could have different discontinuities, missing values at different times, different measurement protocols, length and overall quality and components. For example, Uber users use an app in their phone to make requests for transportation. There are millions of Uber users. Learning about the behavior of users, even a random sample of 20,000 users, over the course of a year would be a huge task with data that are very clean and equally spaced in time. Most of those large Uber time series are far from that, which makes the task harder (for example, discontinuity, because unless the person is in a major city, not everyday is Uber called for transportation). A solution must be found. One possible solution in ML methods, which were not designed with time series data in mind, has been to encapsulate each time series into a set of numerical *features* or summary statistics and categorical labels. Fortunately, all the things that we have learned in this chapter are features of time series, and our learning about them is called in ML "feature generation." We will illustrate feature generation with a small data set in Example 1.14 and very simple features. When analyzing large data sets, feature generation is automated.

With features, we can *cluster* the time series into subgroups containing member time series that have very similar features within the group but very different across groups. Cluster analysis falls under the umbrella of *unsupervised ML* methods that were invented to be used with data that are not time series. By working with summary statistics, which, as we said, are also known in ML jargon as "features," cluster analysis can be applied to time series [132]. The clustering could help Uber, for example, learn from the temporal behavior of millions of users, and figure out what distinguishes different groups of users. Readers unfamiliar with cluster analysis will find introductions in data mining books such as, for example, [59] or [102].

Example 1.14   The data set *HistoricalAprilJulyRunoff.csv* is a multivariate annual time series data set containing the average April–July runoff, in cubic feet per second, of 25 California rivers between 1930 and 2005. We will use a subset of five years of this set of time series to show graphically their nature and to keep this example simple. Figure 1.16 shows the five years of runoff for the 25 rivers. The R program *ch1riverscode.R* was used to do the plot and to obtain the summary statistics or features. The reader may ignore for now the code used to reshape the data found in that program, if it is too advanced. Notice also that the missing values were made NA, the symbol for missing values in R.

River runoff is affected by precipitation and snow pack, and those also fluctuate a lot (not seen here). The Bend, Feather and Shasta rivers have the highest runoff throughout the years plotted. Most rivers in California have medium (500 to 1000) or low runoff (less than 500). Five years is not enough to observe a trend; we would need to plot all the years given to do that. We cannot tell from this plot whether California is drying out or not.

We use, however, all years (1930–2005) of data to obtain the features. For the purpose of keeping this example simple, we use a few features that the reader may be familiar with from introductory statistics studies, but in future chapters we will be able to extract many more attributes. The more features that make sense for the data we use, the more useful the analysis will be. For the rivers, we just selected the mean, standard deviation, minimum and maximum of each time series and then the number

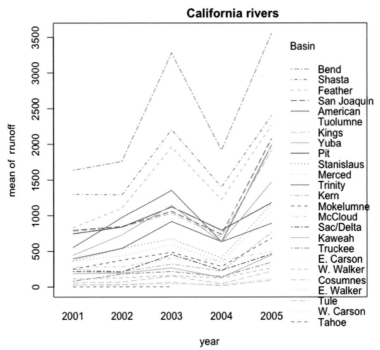

A few years of river runoff

Figure 1.16 Multiple time series plot of five years (a window) of river runoff for 25 rivers in California. This is just a window of the much longer time series.

of observations that are above two standard deviations from the mean and the number of observations one standard deviation below the mean. The runoff does not have a trend overall, so these summary statistics are appropriate for the purposes of illustrating the method. Table 1.2 contains the features for each river. Notice that we have reduced a data set that had a time index, and 25 time series in it, each with 76 observations, some of them missing, to a data set that is not a time series data set, with 25 observations and 6 variables, and no missing values.

Without assuming that the reader has had training in cluster analysis, we apply the k-means method (with one stroke of code in R), which will measure the distances of the row vectors containing the river features, to the vector of cluster means. Rivers will be allocated to the cluster that has mean closest to the row of the river. The Euclidean distance is used to compute the distance of each row to the cluster means. Figure 1.17 shows the cluster allocation, with one color and one symbol for each cluster.

In cluster 1 (dot symbol), we find rivers Trinity, Pit, Yuba, American, Stanislaus, Tuolumne, Merced, San Joaquin, Kings, and Kern, which, according to Figure 1.17, are rivers with intermediate runoff (in the middle, so to speak) and in the middle in the number of times the runoff was beyond two standard deviations from the mean.

Table 1.2 Features of California rivers' runoff. With features we can use unsupervised machine learning methods such as cluster analysis to determine rivers that have features in common and label the rivers accordingly. This data set of features is not a time series data set.

Basin	mean	sd.	min	max	$+2sd$	$-1sd$
Trinity	641.12	294.28.	116.62.	1593.35	4	10
Sac/Delta	292.66	141.48	63.42	711.20	4	8
McCloud	397.87	132.07	184.67	748.03	3	9
Pit	1037.95	332.54	480.10	2097.72	3	8
Shasta	1795.64	660.68	764.00	3525.31	3	9
Bend	2453.62	989.65	943.00	5075.46	2	7
Feather	1822.91	962.31	391.85	4676.00	4	11
Yuba	1036.31	501.31	199.88	2424.09	4	12
American	1275.89	658.01	228.96	2912.26	3	13
Cosumnes	127.56	93.13	7.96	362.84	6	10
Mokelumne	463.32	219.65	101.59	1038.00	3	13
Stanislaus	710.88	351.49	115.51	1636.18	3	12
Tuolumne	1210.10	559.48	301.02	2645.28	3	13
Merced	623.45	332.30	123.29	1587.46	3	11
San Joaquin	1233.34	641.00	261.91	2898.00	4	10
Kings	1203.78	633.85	274.49	3112.61	2	13
Kaweah	283.91	170.07	61.72	799.70	3	10
Tule	63.26	56.19	2.36	259.14	3	3
Kern	452.69	328.28	84.39	1657.07	2	6
Truckee	261.81	147.88	52.42	712.73	3	12
Tahoe	1.39	0.82	0.17	3.57	4	15
W. Carson	54.12	26.40	12.06	135.21	3	12
E. Carson	184.80	88.67	42.57	406.72	5	12
W. Walker	149.87	63.97	34.79	303.33	3	13
E. Walker	62.13	45.11	6.66	209.04	3	7

In cluster 2 (triangle symbol), we find Sac/Delta, McCloud, Cosumnes, Mokelumne, Kaweah, Tule, Truckee, Tahoe, W. Carson, E. Carson, W. Walker and E. Walker, which, according to Figure 1.17 are rivers with small average runoff, and large variability in how often they go beyond two standard deviations from the mean.

Cluster 3 contains Shasta, Bend and Feather, rivers with large average runoff.

This information on cluster allocation helps confirm for the whole length of the time series (which we have not seen) what is obvious in the small window of data shown in Figure 1.16. Without plotting the whole length of the time series, it would have been impossible to deduce the clustering observed with cluster analysis.

Imagine having thousands of very long time series, which we could not plot even at smaller scale. In those cases, as in this example, clustering based on features would be very useful. □

Cluster analysis of features of time series would also be an appropriate way to identify groups of Uber customers with similar characteristics, or groups of locations based on their pollution measurements over time and many other cases. New labels

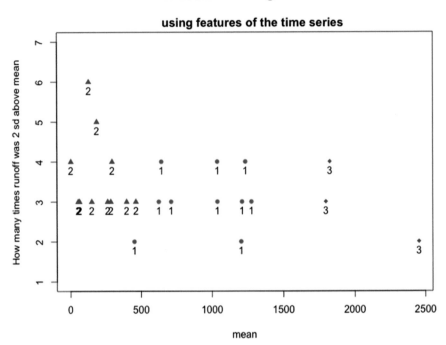

Figure 1.17 Allocation of California rivers' complete time series (a time window of which was depicted in Figure 1.16) to different clusters based on "features" presented in Table 1.2. The labels in the plot represent the cluster numbers. See names of rivers in each cluster within the text.

for groups our minds never thought existed could be identified. Patterns out of the ordinary could be discovered.

Working with features instead of the raw time series helps avoid the problems that missing observations, different lengths of series, different observation times for each time series and other issues bring to time series analysis. These problems have become more prevalent with the indiscriminate proliferation of time-stamped data brought by the IoT. Examples of resources for the reader in ML learning time series data analysis are Nielsen [132] and Maharaj, D'Urso and Caiado [112]. But the analysis that can be done with just summaries or features is very different from the analysis of the whole time series. If we have well-curated and clean data without any of those problems, we apply classical time series methods that use the uncompressed time series.

1.8.1   Candidate Features for Time Series

Nielsen [132] suggests several features that could be easily extracted from time series. We will enumerate some of them that could be extracted from time series seen in this chapter, and the reader can go back to the content of this chapter and ask, "What would those features' values be in the plots seen?"

- Frequency
- Slope at different time intervals (for trend or cycles encapsulation)
- Mean values for each morning, midday, and evening (if hourly data)
- Mean and variance if the time series is stationary
- Mean and variance at different intervals if the time series is not stationary
- Number of local maxima and minima
- Smoothness of the time series
- Earliest occurrence in the time series of the minimum or maximum values
- Length of the longest consecutive subsequence that is above or below the mean
- Percentage of the data points that are within a fixed window of values close to the median of the data

There are many more features that could be extracted to compress time series data. Software usually have programs to do the job for us. For example, the `tsfeatures` package in R, developed by Rob Hyndman and colleagues [79], and the `feasts` package [137]. As the reader learns more time series methodology in future chapters, these packages will become more applicable. It does not make much sense to calculate features whose concept has not yet been learned.

1.8.2　Exercises

Exercise 1.14　Uber Movement [195] shares anonymized data, aggregated from over ten billion trips carried by Uber drivers, to help urban planning around the world. Uber movement speed data is an example of such data. The site provides data and tools for cities to more deeply understand and address urban transportation challenges. Travel times and speed data can be obtained from Uber for many travel segments in several cities of the world. The data are time series, one for each travel segment. But some travel segments are widely traveled continuously, while others are traveled infrequently. Thus, we end up with some very short time series for some parts of the city and some very long ones for other parts. The times of observation (the rides) do not happen at consecutive times either. So classical time series analysis would be hard to implement. We could use instead "features" of the time series, like we did with the River data in Section 1.8. The time series data set used in this exercise contains a fragment of a data set downloaded from Uber. Access the data set *ch1uberSF2000-1.csv* and, using the program *ch1uber.R*, do the following:

(a) Select three or four segments that interest you, some shorter than others. Delete all other segments.
(b) Think about what type of "features" would be meaningful to use to summarize the segments, so that you can obtain a "features" data set.
(c) Create the features data set and, making sure that you only have numerical data, use the `kmeans` function in R to cluster the time series. Identify which travel segment is in which cluster.

(d) Notice that we see different things depending on which features we plot to visualize the clusters. Not all pairs of features will display clearly the clustering. Thus, before plotting, pay attention to which feature is most different across the clusters. Select those for plotting. Provide a plot like that in Figure 1.17 with the clusters in the Uber data. See hints in program *ch1uber.R*.

1.9   Time Indexing

No matter what software is used, time indexing is the most important aspect of a time series. The sequential nature of time series requires that the user be able to extract, link, or split by time index. Base R's `ts()` class has some limitations when it comes to handling highly granular data, such as hourly or weekly data, which is so prevalent nowadays due to the IoT and sensors. Packages have been developed in order to address those limitations. Several concepts regarding time indexing help make a beginner more cautious when approaching time series data of the `ts()` or `mts()` classes. We summarize here information that will be useful [75, 71].

The *frequency* of a time series is the number of observations per cycle (normally a year, but sometimes a week, a day or an hour). This is the opposite of the definition of frequency in physics, or in Fourier analysis, where *period* is the length of the cycle, and *frequency* is the inverse of period. When using the `ts()` function in R, the following choices should be used.

Data	Frequency
Annual	1
Quarterly	4
Monthly	12
Weekly	52

That would be if the need is to see patterns that repeat annually in a particular quarter, month or week.

Actually, there are not 52 weeks in a year, but $365.25/7 = 52.18$ on average, allowing for a leap year every fourth year. But most functions that use ts objects require integer frequency.

If the frequency of observations is greater than once per week, then there is usually more than one way of handling the frequency. For example, data with daily observations might have a weekly seasonality (frequency = 7) or an annual seasonality (frequency = 365.25). Similarly, data that are observed every minute might have an hourly seasonality (frequency = 60), a daily seasonality (frequency = $24 \times 60 = 1,440$), a weekly seasonality (frequency = $24 \times 60 \times 7 = 10,080$) and an annual seasonality (frequency = $24 \times 60 \times 365.25 = 525,960$). If using ts objects, then the user needs to decide which of these is the most important.

An alternative is to use a `msts` object (defined in the `forecast` package)[78, 77], which handles multiple seasonality time series. Then one can specify all the frequencies that might be relevant. It is also flexible enough to handle noninteger frequencies.

Frequencies of Data	Minute	Hour	Day	Week	Year
Daily				7	365.25
Hourly			24	168	8,766
Half-hourly			48	336	17,532
Minutes		60	1,440	10,080	525,960
Seconds	60	3,600	86,400	604,800	31,557,600

It is not necessary to include all of these frequencies, just the ones that are likely to be present in the data. For example, any natural phenomena (e.g., sunshine hours) is unlikely to have a weekly period, and if our data are measured in one-minute intervals over a three-month period, there is no point in including an annual frequency [75, 71].

Example 1.15 `# time series starting on the  2nd Quarter of`
`1959 ts(1:10, frequency = 4, start = c(1959, 2))`

`# daily data where one day stands out, we think`
`print ( ts(1:10, frequency = 7, start = c(12, 2)),`
`calendar = TRUE)`

`# hourly data where one hour of day stands out, we think`
`print ( ts(1:48, frequency = 24, start = c(110, 6)),`
`calendar = TRUE)`   □

> Indicating a frequency as an argument in the R function `ts()` makes sense when there are no gaps in the data. If, for example, only trading days get trade volume recorded, we may not assume that freq = 7 (assuming that a cycle is a week).

1.9.1 Time Series Databases

The IoT has made time series grow in importance. The need for effective ways to collect, store and access large-scale time series data for analysis is growing exponentially. The theory behind time series databases and practical methods for increasing their efficiency is growing as a separate field related to time series data analysis. There are resources to acquire skills in this important area of time series, for example [40]. However, anyone needing to become proficient in handling or even designing databases for time series data must be aware that without knowing the concepts introduced in this book and similar books, that is, without knowing the theory and application of time series data analysis, engaging into the design or use of a database management system will be risky.

1.9.2   Exercises

Exercise 1.15   Suppose your goal is to analyze the airline passengers but for the period January 2002 to December 2005. Compare the story behind the data in this period with the story told in this chapter for 2012–2017. Refer to the code in code files *ch1passengersprogram.R* and *ch1passengersplots.R* to see the location of the data set and convert it to an R `ts()` object before doing the work requested.

1.10   About R

The programs accompanying this book are written in R [159], sometimes in more than one version of this software. R is a high-level programming language with several functions that perform statistical analysis of time series. There is Base R, which is R core and contains many functions that have settled in the language and are there to stay, and there are "packages." The latter are maintained by users of R, and extend the capability of Base R. Some packages repeat what Base R offers but may do so with different wrapper function names, enhancing some aspects such as, for example, graphics, or displaying results in alternative ways. A person new to R or to time series analysis might find this combination of core Base language and packages unsettling. For this reason, as much as possible, the code in most of the programs is written in Base R. But to satisfy those more advanced in the R language, we also offer versions using packages that are in the mainstream practice of time series analysis with R.

The reality of time series data analysis with R is that, although all the basic introductory ideas of classic time series analysis of well-curated and clean data can be introduced with Base R, a lot of packages need to be accessed in order to cover some modern demands of time series analysis brought by the volume, velocity and variety of data available today. In writing this book, we have tried to use Base R to introduce the concepts of time series analysis with clean and regularly spaced data in order to accelerate the learning of the main concepts. But we have supplemented that with code that accesses other packages in order to bring the most contemporary uses of time series analysis to the reader's attention.

R functions and all code in the book are written in `courier` font. But we have avoided writing too much code within the main text. Most of the code can be found in the accompanying R scripts that are posted on the book's website and referred to within the book. The reader not familiar with R software but with a laptop or desktop and access to the Internet is encouraged to download R and the interface RStudio [163], go over the tutorials found at stats.idre.ucla.edu/r/ if unfamiliar with R, and use the R script window in RStudio to execute the commands given throughout this book. The RStudio website also has tutorials for those who do not know R and additional information on RStudio access.

To get an overview of the range of packages and resources that R offers for time series, it is worthwhile visiting Open Data CRAN Task View [69]. The reader will notice there that Base R plays a very important role in all aspects of time series analysis.

It is also worth noting how there are different classes for time series data. Authors create those classes because they could be more appropriate for the data sets they study. Another thing that is not noticeable but that is very common in R is that when an author writes a book, the author tends to create a new package. That author's package sometimes does the same things that Base R might do, for example to show a plot, but the author wraps Base R with functions that modify the output of Base R.

It is for this reason that we have done the majority of the analysis in the chapters of this book with Base R, but we have also invoked packages that would do the job at hand better, depending on the nature of the data. The reader will get an overview of the many ways that R can be used for time series analysis. But this book is not about R, but about time series. Other software such as Python, or SAS or STATA, or RATS could be used to do the analyses presented in this book.

1.11 Other Books on Time Series Using R

There are many books that introduce time series using the software R. The author of this book has used these books throughout the many years teaching time series and has, of course, been influenced by their writers. Many of these books are too advanced for undergraduate students but present many good examples and exercises.

The very first book using R for time series analysis known to the author of this book was *Practical Time Series Analysis* by Janacek [85], which is appropriate for undergraduate students. At the time, Base R did not even have the functions that it now has. Janaceck created functions to be used for the book. Those functions are highly recommended for those who like to see what is behind the functions used by R. Some of Janacek's functions appear as exercises and are used with the author's permission.

Cryer and Chan [32] published the first edition of their book in 2008, creating their own package, TSA, for the book [192]. It is in its second edition at the time of writing this book.

Cowpertwait and Metcalfe [30] published a very accessible and practical introductory time series book appropriate for undergraduates in 2009, using mostly Base R.

Shumway and Stoffer [179] published the third edition of their book and incorporated R code for the first time.

In 2014, Derryberry [35] wrote a book titled *Basic Data Analysis for Time Series with R*. This is a basic introduction to R, doing regression with R and basic time series data analysis.

Hyndman and Athanasopoulos' third edition of the book titled *Forecasting: Principles and Practice* is one of the most recent additions to the list [76]. The book is very introductory and has the advantage that it uses the most modern class of time series object, the tsibble, and the fable [138] set of packages developed by the author and colleagues, all relying on tidyverse [204]. Readers familiar with the latter may explore versions of the examples presented throughout this book in their tsibble version. The earlier version of their book used the package `forecast` [78], which used Base R.

There are many other excellent books on time series such as Chatfield, and many others that do not use R that will be mentioned throughout this book.

1.12   Problems

Problem 1.1   Answer the following questions after studying this chapter.

(a) What is time series data?
(b) Why are people interested in time series data?
(c) What is monthly data?
(d) Why do people plot time series data with points joined up by lines instead of using scatterplots?
(e) What, besides trends, is another form of pattern that is very common in time series data of humans and nature?
(f) What steps might be needed before our analysis software can work with the data we import?

Problem 1.2   R comes with two time series that are already R's `ts()` class: `Air-Passengers` and `JohnsonJohnson`. Use the R program *ch1problem1.R* to do the work required to answer the following questions.

(a) What is each of those time series reporting?
(b) Find the start date, the end date and frequency of each of the time series. It would be helpful to review Example 1.11 and Section 1.5 before answering.
(c) Do separate time plots of the two time series after reviewing Section 1.6. Describe what the time plots tell us about the time series and compare the two plots. What are the main features of each of the time series?
(d) Describe what you see when you use the R statement `cycle(AP)` to obtain the seasonal index of the time series. Describe this index and explain what it means. Describe also what you see when you use the R statement `time(AP)` to obtain the time index, and explain what it means.
(e) Obtain the seasonal box plots and spaghetti plots of the two time series and describe what those plots are conveying about the time series.

Problem 1.3   Use program *ch1problem1.R* and add to it *ch1problem2.R* or simply start with the latter in order to do this problem. One of the things we often do with time series, particularly if there are missing values, is to aggregate the time series.

(a) Aggregate the AP and JJ time series of Problem 1.2 to make them total annual first. Then aggregate to get average annual values. You will do the latter in two ways.
(b) Obtain the time plots of the aggregated time series. Add to the code given in program *ch1problem2.R* meaningful arguments `ylab` and `main`.

(c) Describe what was learned about the time series and explain how and why the new time plots look different from the plots obtained in Problem 1.2.

Problem 1.4 Consult the R code given in program *ch1passengersprogram.R* and use the data file *ch1passengers.csv*. Table 1.1 contains three time series. We described in Section 1.6 domestic air passengers, one of the three series. Repeat the analysis we did in that section, but for the international airline passengers time series. Give the story behind the same window of the time series used for domestic passengers.

Problem 1.5 Sometimes we are interested in studying only a subset of the data set. The correct practice is to select consecutive values of the time series, to preserve the order and to follow the rule "one observation for each time observed."

(a) Download the time series labeled CAUR from FRED. This is the seasonally adjusted unemployment rate in California. Download all the observations available in FRED. Read that data into R and select the window January 1997 to March 2003.
(b) Obtain a time plot of the window. Compare it with the time plot of the whole time series. Does the data in the window help to conclude the same that can be concluded when looking at the time plot of the whole data set? Explain.

Problem 1.6 The article "Modelling calls and effects" [210] uses multiple regression analysis to model and forecast calls to a center.

(a) Read that article and summarize its content.
(b) Which features of the airline passengers of Problem 1.4, if any, can be found in the data analyzed by Zelin?
(c) How did the article by Zelin model the seasonal part of the time series?
(d) How did it model the trend?
(e) Are the modeling of trend and seasonality in Zelin's article appropriate ways to model those components? Make sure to review Chapter 1 before answering.

Problem 1.7 The UCI Machine Learning Repository contains many time series data sets [39]. In particular, we are interested in the air quality (Air+Quality) data set, which contains 9,358 instances of hourly averaged responses (March 2004 to February 2005) from an array of five metal oxide chemical sensors embedded in an Air Quality Chemical Multisensor Device. The data set was introduced in Example 1.8, and the R program used there is program *ch1airquality.R*. The measuring device was located on the field in a significantly polluted area, at road level, within an Italian city. Ground Truth hourly averaged concentrations for CO, Non Methane Hydrocarbons, Benzene, Total Nitrogen Oxides (NOx) and Nitrogen Dioxide (NO2) were provided by a colocated reference certified analyzer. Additional documentation is provided on the website indicated. Use program *ch1airquality.R* and the data set AirQualityUCI.xlsx to answer the following questions.

(a) Read the data into R and make it a time series object of `frequency` 24, `start=c(1,1)`. If we use `ts()` applied to a data set with several numerical random variables, then it is a multiple time series object, to the eyes of R, an `mts`.

(b) Which variable has the most missing values? Which has the least? Write down an R program that will find out and convert the missing values to missing values that R will understand, if any.

(c) Use the `aggregate()` function of R to obtain the daily average instead of hourly observations.

(d) Do a time plot of a variable of your choice using the hourly observations and another plot of the same variable using the daily average observations. Is there something you miss by using daily average instead of hourly observations? Describe what you see in the plots using the language and concepts learned in this chapter.

(e) Is there some kind of seasonality in the data? Is there a trend? If there is seasonality, select a window of the data and study it using spaghetti plots (seasonal time plots) and seasonal box plots.

Problem 1.8   The M competitions, organized by the International Institute of Forecasters [4, 5], bring together hundreds of beginners and seasoned researchers and practitioners to advance the methodology for massive forecasting of hundreds of time series. The package `tscompdata` contains some of those competitions' large data sets. Visiting the github space for the package [74], the reader will find some examples and some code to access some of the variables in those data sets. For this exercise, study that github site and replicate the example provided by the author of the package.

Problem 1.9   Suppose we are Google's data scientist and are tasked with classifying countries according to their interest in soccer over a long period of time. Based on attributes of time series that we have paid attention to in this chapter, what features would you select to compress the time series in a way that the features by themselves can help distinguish among the time series and the raw data can be ignored?

1.13   Quiz

Question 1.1   A time series is

(a) data that contains only dates of occurrence of events, in chronological order.

(b) numerical data collected in several locations at one particular point in time.

(c) numerical data with an index representing the time of collection and a variable containing the values collected over time to observe changes over time.

(d) the name of a baseball series in Singapore.

Question 1.2   Why are people interested in time series data?

(a) To understand the past in order to predict the future

(b) To understand the correlations between values of a variable in different locations at a point in time

(c) To take a random sample of the time series and summarize it

(d) To have some measure of cross-sectional values of a variable

Question 1.3   What is monthly data?

(a) Data recorded once every year
(b) Data recorded twelve times a year
(c) Data recorded every day of the month
(d) A time series containing the months of the year

Question 1.4   Why do people plot time series data with points joined up by lines instead of using scatterplots?

(a) To obscure the message in the time series
(b) To remove the trend
(c) To remove the seasonal component
(d) To make it possible to view seasonality and cycles in the data

Question 1.5   What are typical patterns in a time series? Select all that apply.

(a) Trends in mean
(b) Seasonal trends
(c) Cycles three or four more years long
(d) Irregular up-and-down movements that do not repeat regularly

Question 1.6   What work needs to be done sometimes before our analysis software can meaningfully allow us to do time series operation with the data we import? Select all that apply.

(a) Acquire the data
(b) Inspect how missing data is coded
(c) Remove strange symbols from the data such as comma and dollar signs
(d) Determine the starting and ending points of the time series

Question 1.7   A window of a time series is

(a) a subset of consecutive values of the series.
(b) randomly selected values of the time series.
(c) the first 10 and last 10 values of the time series.
(d) the minimum, first quartile, median, third quartile and maximum of the time series.

Question 1.8   A TSDBMS is

(a) a Totally Specialized Device Based on Maximum Strength.
(b) a method of estimating a model for a time series.
(c) a method of measuring the accuracy of the forecast obtained from a time series.
(d) software specialized in storing, managing and analyzing time series data sets.

Question 1.9   Without which of the following are we not able to merge, select subsets, plot properly or extract accurate information about the past of a time series?

(a) The time index of the time series
(b) A value for every single time of the time series

(c) The length of the time series

(d) The frequency of the time series

Question 1.10 When we are interested in comparing time series that have different length and that may have discontinuities, which of the following could be potential solutions to the problem that this creates for using classical time series methods? Select all that apply.

(a) We could set the `freq` argument in the `ts()` function to account for the proper frequency, if the missing dates occur with the same frequency (for example, every saturday and sunday is missing in a daily data set).

(b) We could summarize features of the time series and create a data set that is not a time series data set, and then use machine learning methods.

(c) We could always, in all cases, just ignore the missing value, by telling R to make them NA and ignore when analyzing the data.

(d) We could just look at the data set file and guess what the missing values are.

1.14   Case Study: Using APIs to Access Time Series

In this book, we often use R's built-in functions (what we call Base R) because they are the core of most of the user-contributed R packages, they are the simplest to use, and they use language that is easy to understand. But when it comes to finding time series data sets in databases outside of R that are current and live and to accessing them directly within R, Base R does not have a way of connecting to time series databases directly without a lot of programming. For example, the unemployment time series UNRATE and LNS14027662 from FRED that we saw in Section 1.2 could not have been accessed directly from within R since Base R does not have direct access to FRED. Instead, we would have had to go to FRED and download their data to our computers. Thankfully, R users have created packages programmed to connect to a variety of databases external to R with just a few lines of code without ever leaving R. We will see several of these packages and databases throughout this book. This case study is about a package to access thousands of time series via a particular API.

1.14.1   Former `Quandl's`, Now `Nasdaq Data Link` API to Access Economic Data Sets

An employee in a financial company is in charge of keeping an eye on national and international markets trends, using not only financial data but also other economic data. The number of economic time series that this employee must look at is enormous. The employee uses a web API to do that [99], provided by former Quandl.com and accessible via the `Quandl` package in R [120].[10] The new host of the Quandl app is

---

[10]On September 8, 2021, Quandl announced that the Quandl technology platform was being transformed into a new global solution: Nasdaq Data Link. However, because none of the former access to Quandl's data sets has changed, we may use the same code that was being used with Quandl.

https://data.nasdaq.com. The web API connects to former Quandl's database, which in turn connects live to up-to-date data sets from many different databases, lots of time series data, including FRED data, data from the European Central Bank, the Central Bank of Brazil Statistical Database, the United Kingdom Office of National Statistics and the National Bureau of Statistics in China, to name just a few of the ones that are freely available to anyone. Data from all those databases and many others can be accessed directly through the R package Quandl [120]. The data cannot be accessed without a KEY provided for free when you register. Once the KEY is entered, there is free access to the web API with less scope than licensed access but still extensive enough to get a lot of time series. API keys should be kept private, so if you are writing code that includes an API key, be very careful not to include the actual key in any code made public (including any code in public GitHub repositories) [148]. Program *Quandl.R* contains the code used in this section.

When training interns for summer jobs, the employer makes interns get a free account from Quandl.com in order to see how at ease they feel with time series and live data. The employer usually starts with a simple exercise. Among the time series the interns must access are the two unemployment data sets from the FRED database seen in Section 1.2 but more updated for a longer time period.

Recall that in Section 1.2 we produced a time plot of the monthly seasonally adjusted unemployment rate in the United States compared with the unemployment rate of those with a bachelor's degree. We did that within FRED's website, using FRED's interactive graphical capability. We did not use R for that. The employer makes the interns do something similar to what we have in Figure 1.4 but using the data accessed with the API, using R code and adding to it the unemployment rate in the UK. That is much more than we can do in the interactive graphs built inside FRED. And it is different from downloading the data and then importing it into R. Here are the instructions given to an intern candidate:

- Go to the data.nasdaq.com website and register for a free account to access the thousands of free data sets to get started. This is a required step, because no data can be extracted from the databases from any platform without a KEY to access the web API.[11] A look at the information about the package in R using ?Quandl() confirmed in the documentation that this step is needed. If you do not do this step, you will get the following error message when running code from the package.

```
Error: { "quandl_error": { "code": "QELx01", "message":
"You have exceeded the anonymous user limit of 50 calls
per day. To make more calls today, please register for a
free Quandl account and then include your API
  key with your requests." } }
```

- After explaining how to set the arguments in the function Quandl, the following code was given to the candidate to run. Replace "key goes here" with the KEY

---

[11]A KEY is not needed in all APIs used to access databases located outside R from within R. Some open data resources do not require it. But Quandl does.

obtained when registering in Quandl.com, but do not publish the key. See Section
1.14.2 for an explanation of the arguments included in the `Quandl ()` function.

```
install.packages("Quandl")
?Quandl() # get information
library(Quandl)
# enter the key you got where it
# says "key goes here"
Quandl.api_key("key goes here")
ts2= Quandl(code="FRED/UNRATE",
                type="ts",
                collapse="monthly",
                order="asc",
                end_date="2020-07-31",
                meta=TRUE)
head(ts2)
tail(ts2)
class(ts2)
str(ts2)
plot.ts(ts2,
        ylab="Unemployment rate",
        main="FRED data obtained with\n   Quandl's web
API")
start(ts2)
end(ts2)


ts3= Quandl(code="FRED/LNS14027662",
                type="ts",
                collapse="monthly",
                order="asc",
                end_date="2020-07-31",
                meta=TRUE)
head(ts3)
tail(ts3)
class(ts3)
str(ts3)
lines(ts3,lty=6, col="red")
start(ts3)
end(ts3)
```

- The intern was then asked to explain what the time plot is indicating and to
  remember what the employer had said each argument of the `Quandl` function is.

- Encouraged by the success in accessing the most current unemployment data, the employer then decided to ask the intern to access monthly unemployment rate, seasonally adjusted, from the United Kingdom, using the database of the UK National Statistical Office. The intern went to Quandl.com and searched until it was learned that the database code for that institution is UKONS and the monthly seasonally adjusted data set name within the database is MGSX_M. So the intern created this additional chunk of code in R.

```
ts4= Quandl(code="UKONS/MGSX_M",
                type="ts",
                collapse="monthly",
                order="asc",
                end_date="2020-07-31",
                meta=TRUE)
head(ts4)
tail(ts4)
class(ts4)
str(ts4)
lines(ts4,lty=4, col="blue")
start(ts4)
end(ts4)
```

All this gave the image in Figure 1.18, produced by the candidate with the code in program file *Quandl.R*

The employer knows that figures similar to those obtained inside the FRED website could have been obtained, with time slider and all, and could have challenged the intern to give them, but that routine is not part of Base R, and it is more important to know that the intern can create built-in programs in R. More packages that are of interest to the company can be learned later. Exercise 1.19 at the end of this case study can be used to obtain one such graph.

The intern learned quickly, with the help of the code provided, and soon started helping the employer access thousands of times series from multiple national and international databases using only the Quandl() API, and, of course, using the commercial license, not the free account that the intern had.

1.14.2   The Quandl() Function

Quandl() is the R function in the package of the same name [120] that accesses the web API that puts the user in contact with all the databases that Quandl() accesses. It looks like each source of data has a different code, FRED in the case of the FRED database, and UKONS in the case of the UK Office of National Statistics. The European Central Bank has another, and so does each institution's database. The argument type refers to how the data will be read, in this case as an R ts() object, the built-in

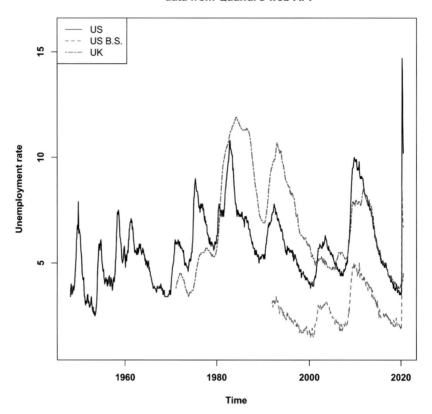

Figure 1.18 Quandl() in R is a web API that allows us to access thousands of time series in hundreds of time series databases around the world from within R. The free version has access to less than the full paid version. In this image we see the same time series on unemployment in the United States seen earlier, and unemployment in the UK.

and predominantly used time series class for time series objects in R and packages. But see the documentation to understand that other classes can be used. The order is asc (ascending) and refers to the time index order. (Obviously, we do not want to sort the values of a data set. That would make the temporal order disappear.) The argument meta is perhaps the most important, because without the metadata, there is no idea as to whether the data is seasonally adjusted or not, whether it is monthly or not, or other details. Notice how the employer and the intern made use of the str() function to get the attributes of the time series data sets.

Notice also what happened to the UK unemployment time series. Even though the employer requested up to July 2020, the UK time series ends in May 2020.

Eager to see if anything was lost by accessing the data via the API instead of going directly to FRED, the intern went to FRED's website directly and searched for the data sets and read from it. Everything, including the metadata, was accurate. The web API is transcribing all the information in FRED. That made the intern trust the information in the UK NSO, but going there to double-check is advisable.

1.14.3   Advantages and Disadvantages

What are the advantages of using Quandl as compared to going directly to the source and viewing their interactive graphs? After all, FRED has very useful ways of looking at the data and comparing time series without doing any programming. But we could not compare unemployment in the United States with unemployment in the UK and other countries if we limited ourselves to FRED's website. Moreover, once FRED's time series are accessed via the API we can do series analysis any time we want with them, something that we could not do inside FRED's website. And we save space in our computer by not having to save the data in our hard drive. Something similar to that happens in UKONS statistics, where one gets to the main page of the institution and will see the plot of the unemployment series there, not as sophisticated as that of FRED, but interactive nonetheless. But data analysis and comparison with other countries cannot be done there either.

Advantages of using a web API like Quandl() in lieu of going directly to the source database are as follows:

- We can compare multiple time series from a disparate number of databases, countries and institutions without having to go individually to each of them, extracting their data, downloading it to our computer, and putting it together in a data set, something that would take a lot of time.
- The data can be read without much programming, and in convenient R formats with which we can do additional analysis with them within R, something that we could not do in a database source like FRED without downloading the data to our hard drive.
- If the required job is to inspect many time series of the nature of the data provided by the API on a daily basis, then API is the way to do it. It would take ages to visit all the individual databases manually.

1.14.4   There Are APIs for Many Subject Matter Data

Peng and colleagues [148] give some additional examples, in particular the riem package, developed by Maelle Salmon, and an ROpenSci package, as an excellent and straightforward demonstration of how we can use R to pull open data through a web API. The riem package allows us to pull weather data from airports around the world directly from the Iowa Environmental Mesonet. Users may request a time series for each location in the world, or a time series for just their local zone over time. Peng and colleagues [148] provide detailed code to access data with this API, and the reader should consult that reference for further details.

> Lessons learned about web API:
>
> - We may not skip some training in time series. How are we to know the arguments to use to read the data from the databases? How would we interpret all the information provided by the metadata? How would we know whether seasonally adjusted data is giving us all the information needed?
> - We need to be familiar with what we are searching. The name of the data set searched and the institution from where we want it must be known. Domain knowledge is as crucial as time series analysis literacy to do a good job.

Peng and colleagues [148] offer other examples of existing R packages to interact with open data APIs, including the following:

- twitteR: Twitter
- rnoaa: National Oceanic and Atmospheric Administration
- RGoogleAnalytics: Google Analytics
- censusr, acs: United States Census
- WDI, wbstats: World Bank
- GuardianR, rdian: The Guardian Media Group
- blsAPI: Bureau of Labor Statistics
- rtimes: New York Times
- dataRetrieval, waterData: United States Geological Survey

Of course, not all APIs offer data in time series format and ready for the user to analyze. The reader should do an exploratory analysis of what time series are available via API and R.

For additional information about using web APIs in R, the reader may consult Peng and colleagues [148], Open Data CRAN task view, and ROpenSci packages, all of which have GitHub repositories. ROpenSci is an organization with the mission to create open software tools for science.

1.14.5   Exercises

Exercise 1.16   After registering a free account with data.nasdaq.com, obtaining an access KEY, and reading this chapter, do the following:

(a) Run the programs given to the intern and the one created by the intern, and find out the start date of each of the time series looked at. Do they all start at the same time?

(b) Collapse the data to annual, and plot again to compare. What is the difference between the plots?

Exercise 1.17   The time series TOTALNSA from FRED can be accessed via the web API Quandl. Access it and modify the code given in this case study to do that. Do a time plot of that time series. Are there similar series in any of the other databases that can be accessed with the `Quandl` API and plotted in the same image? Get them and compare the plots.

Exercise 1.18 Select another database that you can access for free (other than FRED), and use the Quandl API to access it and a time series of your choice. Why did you choose this time series? Use the code given to the intern to provide a plot. Describe the attributes of the time series as an R object.

Exercise 1.19 Access FRED/UNRATE via Quandl API to do a dygraph of it . The code is given here. We use `dyRangeSelector` like the one that can be seen in time plots in FRED.

```
install.packages("dygraphs")
library(dygraphs)
dygraph(ts2,main="U.S. Unemployment rate",
        ylab="Percentage") %>% dyRangeSelector
```

After creating the dygraph, go to FRED's website, find the same time series and get the interactive graph for this time series there. Compare the dygraph obtained with R with that produced inside FRED's website.

Exercise 1.20 The Wikimedia websites (such as Wikipedia) are visited by hundreds of millions of people a year, and so the open data sets of page views contain useful information on the subjects that interest people around the globe. The Wikimedia Foundation has an API for this data that is officially supported and allows a distinction between different types of users and different types of traffic. The `pageviews` R package serves as a client for that API [144].

The reader may use the following small piece of R code to access any page that interests the reader for the time period of interest by simply changing the `article`, `start` and `end` arguments of the function. The reader may also change any of the other arguments. See the documentation at [144] and additional examples in web pages that talk about the package `pageviews` by simply googling. With these data the reader can see the popularity of the subject in Wikipedia, because the data will provide the number of views.

To start with, copy paste the following program into RStudio and see what you get. Then do a time plot of Wikipedia page count data for the "Ada Lovelace" web page in the time period considered. Describe the time plot.

```
install.packages("pageviews")
library(pageviews)

df_lovelace = article_pageviews(project="en.wikipedia",
                                article="Ada Lovelace",
                                start=as.Date('2018-10-01'),
                                cnd=as.Datc('2020-10-01'),
                                user_type=c("user"),
                platform=c("all"),
                granularity=c("daily"))

colnames(df_lovelace)
dim(df_lovelace)
```

2 Smoothing and Decomposing a Time Series

2.1 Introduction

We have seen in Chapter 1 how plotting time series data with points joined up by lines helps us see long-term trends, seasonality and other repeated patterns that help us imagine how the future will be if those patterns persist. We also started learning Base R code to get raw data into R's ts() format, and some R functions for that data class. In this new chapter, we look at a set of techniques for *smoothing* time series and decomposing them. Smoothing per se is "fundamentally an exploratory operation, a means of gaining insight into data without precisely formulated models or hypotheses" [38]. It helps us highlight the most important components or features in a time series. Most of the current chapter is dedicated to one of the oldest approaches to smoothing and decomposing, *classical decomposition* of a time series into its components, based on moving average smoothing. After that, we cover briefly *regression smoothers*, which recast classical decomposition in the form of a regression model that lends itself to forecasting. The discussion continues with *exponential smoothing*, which is based on geometrically weighted moving average smoothers and where the main interest is forecasting using knowledge about the components in classical decomposition when they are present. After exponential smoothing, we acquaint the reader with *Prophet*, a modern modeling approach used by Facebook to do forecasting at scale, and also based on decomposition of the time series. *LOWESS* is then used in the Case Study, providing us with yet one more decomposition method.

The goal of this chapter is to make the reader aware of the way we think about time series as a set of different components, the role of smoothing in finding those components, and how useful all that can be for forecasting. The ideas behind these smoothing and decomposition approaches are the basis of automated forecasting methods used by high-tech companies nowadays to analyze and forecast massive time series data sets produced by their operation and the IoT. Those ideas are also behind statistical models that use past autocorrelations to model and forecast. This chapter, therefore, is very important and is a requirement for anyone aiming at becoming a time series data analysis practitioner.

Section 2.2 in this chapter introduces classical decomposition and defines the components and the assumptions involved in decomposition. Section 2.3 presents *classical additive decomposition* based on moving average smoothing, which is appropriate for time series that display constant seasonal variation. Section 2.4 describes *classical multiplicative decomposition*, which is appropriate for time series that display increasing or decreasing seasonal variation over time. Smoothing can also be treated as a regression problem. For this reason, Section 2.5 introduces the regression decomposition of a time series and ways to combine aspects of smoothing with regression. Regression smoothing will be discussed more extensively in Chapter 9. Once the reader understands classical decomposition, it will be easier to understand exponential smoothing, a popular method presented in Section 2.6 that helps us forecast one step ahead. After understanding the more sophisticated approach to smoothing the trend in exponential smoothing, the reader will find the introduction to *Prophet* easier to understand in Section 2.7. The case study of this chapter introduces yet one more approach to decomposition, Locally Weighted Scatterplot Smoothing (LOWESS) and decomposition based on that smoothing method.

2.2 Classical Decomposition

Classical decomposition decomposes a time series into long-term trend, *seasonal effect* and random component. Decomposition is particularly useful when the time series has those features, trend and seasonality. The decomposition can be additive or multiplicative and can use different smoothing techniques. The key methodology is smoothing of some kind. This chapter shows the calculations laid out for simple cases based on moving average smoothing. By understanding the basics, the reader will be able to understand what more modern, albeit more complex, smoothing techniques do.

The motivation for decomposition lies in the trends and seasonal swings that are common in social, economic, meteorological and other classes of time series data. Decomposition helps data producers separate what is obvious to the naked eye and happens regularly (such as trends and seasonals) from aspects of the time series that appear less regular, the random component. The decomposition makes it easy for entities producing time series data to do detrending and seasonal adjustment. Chapter 1, Figures 1.3, 1.4 and 1.5, have already exposed us to seasonally adjusted time series.

Example 2.1   Consider, for example, Figure 2.1, obtained by applying R's decom-pose() function to a ts() object. The figure shows the multiplicative decomposition of a time series that records monthly number of occupied hotel rooms. The reader may reproduce this graph using program *ch2figsampledecomp.R* and data set *rooms.txt*. Judging by the time plot on the top of Figure 2.1, we would say that occupied rooms will continue to increase, and they will do so more in certain seasons of the year with ever increasing swings. The regularity of the trend and the seasonal swings help us say that. But is there something else that we do not see because it is hidden by the strong regular pattern created by the long-term trend and the seasonal swing? In other words, is there some other interesting information hidden under the regular pattern? Decomposing the time series helps us answer that question, because decomposition helps us isolate the random term, also shown in Figure 2.1.

Decomposition of multiplicative time series

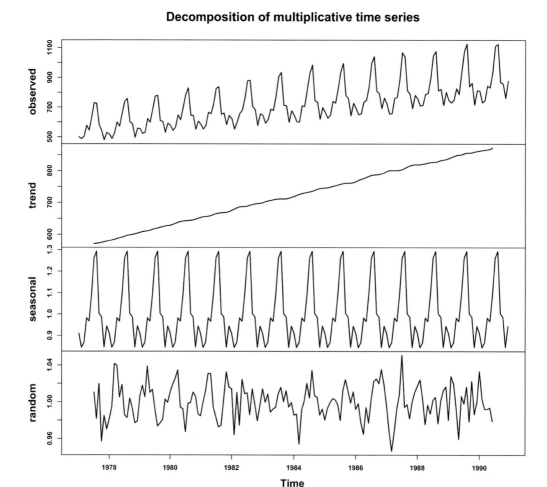

Figure 2.1 For a time series with time plot that reveals an upward trend due to population growth and increase in standard of living, and increasing seasonal variability due to having more individuals, we either transform the data and do additive decomposition or do multiplicative decomposition. The observed data is the same as that in Figure 1.11.

We isolate the random term extracting the trend and the seasonal effect shown in Figure 2.1. That combined operation is called detrending and seasonal adjustment.

It is possible to just seasonally adjust the time series by extracting only the seasonal. Figure 2.2 shows the rooms time series after the regular seasonal effect has been removed. The reader may reproduce that image with program *ch2roomsseasadj.R* and data set *rooms.txt*. We still see the trend and the random term, but the regular seasonal swings have disappeared. A lot of published social and economic time series, like those seen in Figures 1.3, 1.4 and 1.5 in Chapter 1, are published after seasonally adjusting them. Users of time series data should always look for the metadata to see whether seasonal adjustment and/or detrending have been performed on a time series.

Seasonally adjusted rooms

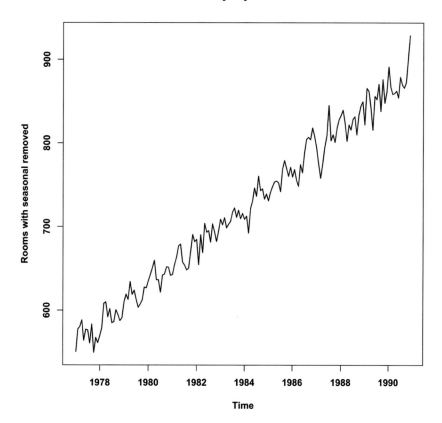

Figure 2.2 The monthly rooms time series after removing the regular multiplicative seasonal effect. In other words, this is the seasonally adjusted rooms time series.

Irregular seasonals, appearing at random, not as a repeating regular pattern, are not uncommon after seasonal adjustment.

The trend and the seasonal effect may also be plotted on the same plot as the data to give an idea of whether the decomposition done was appropriate for the given time series. We see in Figure 2.3 a little underfitting early in the series. The reader may reproduce this image with program *ch2superimposedrooms.R* and data set *rooms. txt*.   □

We did multiplicative decomposition instead of additive decomposition of the occupied rooms time series because we noticed the increasing seasonal variability (seasonal swings getting larger and larger over time).

Many time series do not have trends and/or seasonals, but smoothing techniques are still very useful to extract the obvious from the ragged signals and to isolate what needs further studying with statistical methods.

Rooms with trend

and seasonal effect superimposed

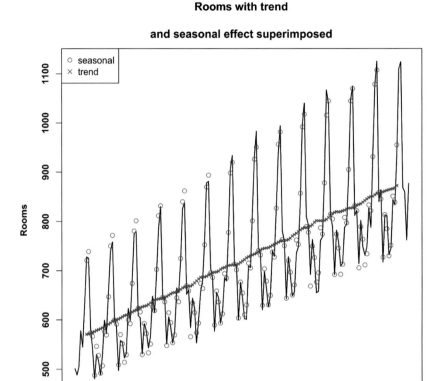

Figure 2.3 Superimposing the trend and the seasonal component on the rooms time series allows us to see how little is left unexplained after the removal of those.

2.2.1   Time Series Components

Classical decomposition models have been found to be useful when the parameters describing a time series do not change over time. Decomposition models have no theoretical basis – they are strictly an intuitive approach that has proved itself useful over time. Decomposition models work well for time series like that in Figure 1.7, the time plot of monthly domestic air passengers in the United States described in Chapter 1, which has a trend and a seasonal component and a pattern that repeats over time.

> The *trend* is the long-term pattern observed in a time series that tells us the direction of change (up, down, polynomial, constant and many other forms).

The trend that we superimposed on the time plot of domestic passengers in Figure 1.10 is a special kind of smoother. It is a moving average trend.

Seasonal swings, the regular sawtooth patterns observed in some economic, mete-oreological and other time series, are the increases or decreases in the value of the time series that are observed regularly every cycle. They must be distinguished from up-and-down movements of the series that do not have a repetitive pattern, which we call the random term. If the time series is monthly, seasonal swings are repetitions of a pattern each year, for example, holiday seasons. If the time series is daily, seasonal swings could refer to patterns that repeat every week, for example, every Saturday. If the time series is weekly, seasonal swings refer to patterns that repeat each month.

We must distinguish the observed seasonal swings from the *seasonal effect* in the time series. The latter is a constant average quantity calculated from the observed seasonal swings. It refers to how much, on average, the number of passengers increases or decreases each month relative to the trend. Swings in the time series may not look identical each year in the observed data, but the seasonal effect is a fixed average that is calculated by some method and is the same each year for a given season. If the cycle considered is a year, and the time series is monthly, then the seasonal effect refers to the constant number that must be added (or multiplied) to each month's value every year. The value for January will be different from that for other months. We learn to calculate the seasonal effect in this chapter.

The *seasonal effect* is a fixed number for each season, an average that is calculated to obtain an estimate of the seasonal swing.

Patterns in an observed time series that take more than a cycle to complete but are not the overall long-term trend in mean are known as *cycles*. Cyclical factors would increase or decrease the expected values above or below levels that would be expected if only long-term trend and seasonal effects were considered. Cycles in time series derived from human activity are not regular, and at times, if the time series is not very long, may give the impression of being long-term trends.

2.2.2 Assumptions about Components

We can think of the seasonal effects as adding to the trend or multiplying the trend. If the seasonal swings are, more or less, of the same magnitude throughout the whole time series, as is the case in Figure 1.7, then we think of the seasonal effects as adding to the long-term trend. If the seasonal swings are larger in some parts of the time series than in others and proportional to the trend, then the seasonal effects are not additive but multiplicative. Figure 1.11 in Chapter 1 displays increasing seasonal variability proportional to the trend. It is not uncommon that the swings are bigger when the trend is higher. When it concerns humans, the more there are, the larger the variability in their actions.

More formally, additive seasonal effects mean that we assume an additive decomposition model for the time series:

$$Y_t = T_t + S_t + \epsilon_t, \tag{2.1}$$

where Y_t is the time series at time t, T_t is the trend value at time t, and S_t is the seasonal effect at time t. The random component of the series is represented by ϵ_t at time t. This component could be white noise or contain additional signal to be found by further analysis.

Multiplicative seasonal effects mean that we assume a multiplicative decomposition model for the time series:

$$Y_t = T_t * S_t * \epsilon_t. \tag{2.2}$$

Once we calculate the trend component, and the additive or multiplicative seasonal effects, we have an estimate of the models proposed earlier, based on classical decomposition smoothers, namely

$$\hat{Y}_t = \hat{T}_t + \widehat{S}_t$$

if additive decomposition is assumed, or

$$\hat{Y}_t = \hat{T}_t * \widehat{S}_t$$

The following are R commands to do classical decomposition of R's `ts()` objects with moving average smoothers and to extract the components:

```
# assuming x is a ts() object
add.decomp.x=decompose(x, type="additive")
multi.decomp.x=decompose(x,type="mult")
# extract additive components
trend=add.decomp.x$trend
seasonal=add.decomp.x$seasonal
random = add.decom.x$random
# multiplicative components may be extracted similarly
```

Be aware that compared with the original time series, the components will have some missing values at the beginning and the end resulting from the smoothing. Further programming may be needed to handle those, depending on what we do next. Notice that if `freq=1` (annual time series) in the `ts()` object that we input to `decompose()`, then the latter will not work. Can the reader guess why?

---

if multiplicative decomposition is assumed. $\hat{Y}_t$ is the estimated (or fitted) value of Y_t, $\hat{T}_t$ is the estimate of the trend, and $\widehat{S}_t$ is the estimate of the average seasonal swing, the seasonal effect. The seasonal effect varies per month or quarter within a year, but it is constant every year. The seasonal effect is not the observed swing. For example, suppose that in a quarterly time series we find that the seasonal effect (the average of the seasonal swings in quarter 1) is 10 for quarter 1, -50 for quarter 2, -15 for quarter 3, and 8 for quarter 4. Then, if the model is additive, we will add 10 to any value of the trend that happens in every first quarter's trend value, subtract 50 from every second quarter's trend value, and so on. Whereas the seasonal swing for the first quarter, for example, may be very different each year, the seasonal effect is a constant value, an average to be applied each year.

2.3   Classical Additive Decomposition

Additive decomposition is useful if the time series has a trend and a constant seasonal effect. In that case, it will be of interest to find the signal remaining in the random term. We will now see the calculations laid out for a simple case. In this section, we will learn to find a moving average long-term trend and an additive seasonal effect. We will do this with quarterly data, with a short enough time series to appreciate the calculations being done by hand. For those inclined to programming, a basic R program that transparently does the same computations done by hand here will be recommended and asked to be used in exercises. Finally, the automated use of R's `decompose()` function will allow us to get the same results that we get by hand or using the transparent code. The reader should be aware that different R packages may use different wrappers for the function used here.

2.3.1   Calculating the Trend of Quarterly Data with a Four-Point Centered Moving Average

If a data set is quarterly, it is good practice to apply a centered four-point moving average in order to see the smoothed trend in the data. We will explain in this section the calculation of moving average and centered moving average.

     Table 2.2 shows the results of applying a centered four-point moving average to a small observed time series, y_t, of acceleration values used by Janacek [85] to illustrate how to approximate the long-term trend using a centered moving average. The column headed *ma4* contains the first step of the calculation, the four-point moving average, and column $\widehat{T}$ contains the centering of that average.

     Focusing on column *ma4* now, we use a four-point moving average because the time series data is quarterly (four seasons per year). If the data were monthly, we would compute a 12-point moving average. The next paragraphs describe how we found the first two numbers in column *ma4*. First,

$$2.000175 = \frac{y_1 + y_2 + y_3 + y_4}{4} = \frac{3.3602 - 3.1769 + 0.3484 + 7.469}{4}.$$

Table 2.1 An example using a four-point and a two-point centered moving average for approximating the trend of the quarterly acceleration time series. Source: [85], p. 4.

Time	y_t	$ma4$	$\hat{T}$
1960q2	3.3602		
1960q3	−3.1769		
		2.000175	
1960q4	0.3484		2.1422
		2.2842	
1961q1	7.469		2.6236
		2.9629	
1961q2	4.4963		3.0096
		3.0563	
1961q3	−0.4621		2.9912
		2.9261	
1961q4	0.7218		3.0187
		3.1114	
1962q1	6.9484		3.5347
		3.9579	
1962q2	5.2374		4.4553
		4.9527	
1962q3	2.9242		5.494
		6.0354	
1962q4	4.7006		6.0262
		6.0169	
1963q1	11.2793		5.844
		6.6719	
1963q2	5.1637		6.5995
		7.527	
1963q3	1.5441		7.3245
		7.1219	
1963q4	12.121		7.488
		7.854	
1964q1	9.6588		8.1567
		8.4593	
1964q2	8.0922		8.3714
		8.2835	
1964q3	3.9653		8.7272
		9.171	
1964q4	11.4177		
1965q1	13.2088		

The second moving average is obtained by dropping the first acceleration value (y_1) from the average and by including the next acceleration value y_5 in the average. Thus, we obtain the second number in column *ma4*:

$$2.2842 = \frac{y_2 + y_3 + y_4 + y_5}{4} = \frac{-3.1769 + 0.3484 + 7.469 + 4.4963}{4}.$$

The third moving average is obtained by dropping y_2 from the average and including y_6 in the average. Thus, we obtain the third number in column *ma4*:

$$2.9629 = \frac{y_3 + y_4 + y_5 + y_6}{4} = \frac{0.3484 + 7.469 + 4.4963 - 0.4621}{4}.$$

Successive moving averages are computed similarly until we include y_{20}. Note that we use the term "moving average" because, as we calculate these averages, we move along by dropping the most remote observations in the previous average and by including the "next" observation in the new average.

The first moving average in column *ma4* leaves an ambiguous correspondence between y_t and the smoothed value. It corresponds to a time that is midway between quarters 2 and 3. The second moving average corresponds to a time that is midway between quarters 3 and 4, and so forth. That is why the column headed *ma4* has its values at no quarter in particular. In order to obtain averages corresponding to time periods in the original acceleration time series, we *center* by applying a two-point moving average to the *ma4* values. The resulting values are in column $\hat{T}$ and they will be the estimate of the trend, T. The *centered moving averages* are two-point moving averages of the previously computed four-point moving averages.

The first centered moving average, the first term in column $\hat{T}$, is the average of the first two terms in column *ma4*. That is,

$$2.1422 = (2.000175 + 2.2842)/2.$$

The second centered moving average is

$$2.6236 = (2.2842 + 2.9629)/2,$$

and so forth.

Altogether, the operation we have done to obtain column *ma4* amounts to the following:

$$ma4_{first} = (1/4)\begin{pmatrix} 1 & 1 & 1 & 1 \end{pmatrix}\begin{pmatrix} 3.3602 \\ -3.1769 \\ 0.3484 \\ 7.469 \end{pmatrix} = 2.1422,$$

$$ma4_{second} = (1/4)\begin{pmatrix} 1 & 1 & 1 & 1 \end{pmatrix}\begin{pmatrix} -3.1769 \\ 0.3484 \\ 7.469 \\ 4.4963 \end{pmatrix} = 2.2842,$$

$$ma4_{third} = (1/4)\begin{pmatrix} 1 & 1 & 1 & 1 \end{pmatrix}\begin{pmatrix} 0.3484 \\ 7.469 \\ 4.4963 \\ -0.4621 \end{pmatrix} = 2.9629,$$

· · · · · · · · ·

The operation we did assigns equal weight to the terms in the original series. We then applied a two-point moving average to the result of that operation:

$$\hat{T}_{first} = (1/2)(1 \quad 1)\begin{pmatrix} 2.1422 \\ 2.2842 \end{pmatrix} = 2.1422,$$

$$\hat{T}_{second} = (1/2)(1 \quad 1)\begin{pmatrix} 2.2842 \\ 2.9629 \end{pmatrix} = 2.6236,$$

$$\cdots\cdots\cdots$$

It is not difficult to see that the two smoothing operations are equivalent to the unequally weighted five-point moving average,

$$(1/8)(1\ 2\ 2\ 2\ 1),$$

because, for example, for the first $\hat{T}$,

$$(1/2)[(1/4)(y_1+y_2+y_3+y_4)+(1/4)(y_2+y_3+y_4+y_5)] = (1/8)(y_1+2y_2+2y_3+2y_4+y_5).$$

Thus, the first centered four-point moving average in column $\hat{T}$ is

$$(1/8)(1\ 2\ 2\ 2\ 1)\begin{pmatrix} 3.3602 \\ -3.1769 \\ 0.3484 \\ 7.469 \\ 4.4963 \end{pmatrix} = 2.1422,$$

the second four-point centered moving average in column $\hat{T}$ is

$$(1/8)(1\ 2\ 2\ 2\ 1)\begin{pmatrix} -3.1769 \\ 0.3484 \\ 7.469 \\ 4.4963 \\ -0.4621 \end{pmatrix} = 2.6236,$$

$$\cdots \text{ and so on.}$$

Four-point moving average means that we average four numbers. In general, an m-point moving average averages m numbers. If m is even, then we need to center. If m is odd, then we do not need to center. For example, the reporting of COVID-19 cases during the COVID pandemic was done with weekly averages, a seven-point moving average, and does not need the centering operation.

The reader will find, in R program *ch2janacek-smoothers.R*, an R function provided as a supplement to [85] and posted in the book's website with permission of the author, that shows the calculation of the moving average just seen. Readers interested in programming their own smoothing procedures may build from that program

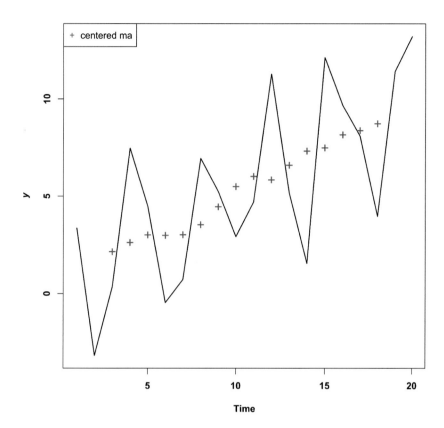

Figure 2.4 A four-point moving average superimposed on the y_t of Table 2.1 obtained with program *ch2janacek-smoothers.R*, used with permission of the author. The symbols represent the moving average trend, column $\hat{T}$ in Table 2.1.

by changing the data set given at the top of the program. Figure 2.4 was obtained by running that program. The results obtained with that program are exactly the same as those obtained by hand. If we use, separately, the decompose() function in R, we would again get the same numbers.

2.3.2   Exercises

Exercise 2.1   Suppose we approximate the trend of a time series by taking a three-point moving average smoother. We usually would not need to center a three-point moving average; but for the purposes of this exercise suppose we do, and then take a two-point moving average of the three-point moving average results. These two smoothing operations are equivalent to which of the following single smoothing operations?

(a) $(1/6)[2, 1, 1, 2]$
(b) $(1/8)[1, 2, 2, 1]$
(c) $(1/6)[1, 2, 2, 1]$

(d) $(1/2)[1, 2, 2, 2, 1]$

Give two practical reasons why we would smooth a seasonal time series with a moving average smoother.

Exercise 2.2   Consider the domestic air passengers time series *ch1passengers.csv* that we studied in Section 1.6 for January 2012 to December 2017. That is monthly data. Obtain the 12-month moving average trend using the same approach as in the *ch2janacek-smoothers.R* program and plot the moving average trend superimposed on the time plot. Compare numerically the smoothed values obtained with that program and those obtained with the decompose() function in R. Review Section 2.2 and Section 2.2.2 before doing the latter.

Exercise 2.3   Demonstrate by hand, using a small data set like the following one, how to construct a 12-point moving average. For quarterly data y_t with an even number of observations, the operation we did could be summarized by

$$(1/8)\,(1, 2, 2, 2, 1)\,.$$

After calculating the values, write the equivalent expression for the operation done with monthly data.

   The data for the exercise starts on January 2005 and ends in February 2006 ($t = 1$ to $t = 14$).

```
5.41,  7.01,  10.94,  13.03,  19.56,  30.59,  22.41,
19.31,  20.56,  19.18,  20.59,  21.62,23.91,25.71
```

Exercise 2.4   In Example 1.11 of Chapter 1, we introduced the *JohnsonJohnson* data set that comes with R. We also introduced basic R functions to find out what type of ts() object it is. The following assumes that you have run the R programs that were used and recommended in that example.

(a) Do a time plot of the time series using the plot.ts() function. Make sure to label the vertical axis according to what the time series measures.
(b) Give the time series the name jj and find the start and end time and the frequency.
(c) Consider the first 12 observations. Hand-calculate a four-point moving average smoothing of the series and then average consecutive pairs in order to get this smoothed series back in step with the original series. Write down a small table with the values of the series (first 10 numbers), and the final smoothed series (that is, the series after the two previous steps requested earlier). Notice that the result is the equivalent of Table 2.1, $\hat{T}$ column.
(d) Compare the values you obtained with those given by the decompose() function in R.

2.3.3   Calculating Additive Seasonal Effects

We continue in this section with the time series of acceleration values used in Section 2.3.1. The seasonal effects are the fixed up or down adjustments to be made to the long-term trend. Recall, the seasonal effect is not the seasonal swing. The former is an average, the latter is the observed swing. To find seasonal effects, we want to find the average of the seasonal swings happening each quarter. Since the model

$$Y_t = T_t + S_t + \epsilon_t$$

implies that

$$S_t + \epsilon_t = Y_t - T_t,$$

it follows that the estimate of $S_t + \epsilon_t$ is $y_t - \hat{T}$. Noting that the values of $y_t - \hat{T}$ are calculated in Table 2.2 (column labeled d), we can find $\widehat{S}$ by grouping the values of d by quarters and calculating an average for each quarter.

Looking at Table 2.2 we notice that in the first quarter of each year, the difference is positive. In the fourth quarter, the difference is mostly negative. In the third quarter, the difference is always negative.

Now take the average per quarter. Averaging quarter 1 values, we obtain

$$\bar{s}_{q1} = \frac{4.8454 + 3.4137 + 5.4349 + 1.5021}{4} = 3.799025,$$

averaging quarter 2 values, we obtain

$$\bar{s}_{q2} = \frac{1.4867 + 0.7821 + -1.4358 - 0.2792}{4} = 0.13845,$$

averaging quarter 3 values, we obtain

$$\bar{s}_{q3} = \frac{-3.4533 - 2.5698 - 5.7804 - 4.7619}{4} = -4.14135,$$

and averaging quarter 4 values, we obtain

$$\bar{s}_{q4} = \frac{-1.7938 - 2.2969 - 1.3256 + 4.633}{4} = -0.195825.$$

The seasonal additive effects are supposed to sum up to 0. If they do not, because of rounding errors, we adjust them by subtracting a quarter of their sum from each value. The sum of the four $\bar{s}$ is -0.3997 and a quarter of their sum is -0.099925.

Let's denote the seasonal effects by $\widehat{S}_{quarter}$, where quarter refers to the quarter (q1, q2, q3, or q4). We calculate them as follows:

$$\widehat{S}_{q1} = 3.799025 + 0.099925 = 3.89895,$$
$$\widehat{S}_{q2} = 0.13845 + 0.099925 = 0.238375,$$
$$\widehat{S}_{q3} = -4.14135 + 0.099925 = -4.041425,$$

and

$$\widehat{S}_{q4} = -0.195825 + 0.099925 = -0.0959.$$

Now we have estimates of seasonal effects that add up to 0.

So, if we think the seasonal effects are fairly constant over time, it makes sense to use this and think that, generally, the January to March figure will be, on average, 3.89895 above what you would expect from the trend $\hat{T}$. The April to June figure is

Table 2.2 Subtracting the estimated moving average trend ($\hat{T}$) from the data y_t as an intermediate step to find the seasonal effects. Source: [85] p. 4. The estimate of trend and the estimate of seasonal effects, the latter in column $\hat{S}$, are shown in this table.

Time	y_t	$ma4$	$\hat{T}$	$d = y_t - \hat{T}$	$\hat{S}$	$\hat{y}_t = \hat{T} + \hat{S}$
1960q2	3.3602					
1960q3	−3.1769					
		2.000175				
1960q4	0.3484		2.1422	−1.7938	−0.0959	2.0463
		2.2842				
1961q1	7.469		2.6236	4.8454	3.89895	6.52255
		2.9629				
1961q2	4.4963		3.0096	1.4867	0.238375	0.238375
		3.0563				
1961q3	−0.4621		2.9912	−3.4533	−4.041425	−1.050225
		2.9261				
1961q4	0.7218		3.0187	−2.2969	−0.0959	2.9228
		3.1114				
1962q1	6.9484		3.5347	3.4137	3.89895	7.43365
		3.9579				
1962q2	5.2374		4.4553	0.7821	0.238375	4.693675
		4.9527				
1962q3	2.9242		5.494	−2.5698	−4.041425	1.452575
		6.0354				
1962q4	4.7006		6.0262	−1.3256	−0.0959	5.9303
		6.0169				
1963q1	11.2793		5.844	5.4349	3.89895	9.74295
		6.6719				
1963q2	5.1637		6.5995	−1.4358	0.238375	6.837875
		7.527				
1963q3	1.5441		7.3245	−5.7804	−4.041425	3.283075
		7.1219				
1963q4	12.121		7.488	4.633	−0.0959	7.3921
		7.854				
1964q1	9.6588		8.1567	1.5021	3.89895	12.05565
		8.4593				
1964q2	8.0922		8.3714	−0.2792	0.238375	8.609775
		8.2835				
1964q3	3.9653		8.7272	−4.7619	−4.041425	4.685775
		9.171				
1964q4	11.4177					
1965q1	13.2088					

about 0.238375 above, July to September is about 4.041425 below, and the October to December figure is about 0.0959 below the trend, on average.

With this thought, we can complete Table 2.2 by repeating the seasonal averages over and over again each year. The results of this action can be seen in column $\hat{S}$.

We are hoping to approximate the series well, in terms of simple, stable components that we can project into the future to form forecasts. If constant seasonal effects give us good approximations to the seasonal swing, that is something that is easy to project forwards.

2.3.4 Estimated Additive Decomposition Model

Our additive decomposition model has now been estimated using moving average smoothers for trend and additive seasonal effects.

Assuming that the seasonal effects are additive, we said earlier that the additive decomposition model was

$$Y_t = T_t + S_t + \epsilon_t. \tag{2.3}$$

The estimated model, using the smoothers and seasonal effects found so far, is

$$\hat{Y}_t = \hat{T} + \widehat{S}.$$

We present the estimated value of the model in column $\hat{y}_t$ of Table 2.2 and display it in Figure 2.5.

As we see in Figure 2.5, the trend plus the seasonal effect (the values in column $\hat{y}_t$ in Table 2.2) are recreating the original series quite closely, but not perfectly. The difference between the data and the fitted additive model is the estimated random term. This random term could have some additional signal in it that is hidden in noise, or it could be just noise. Program *ch2figadddecompfit-Radddecomp.R* will allow the reader to recreate Figure 2.5.

Once the reader understands how additive decomposition is done, the process can be automated by asking R to do it, using the function `decompose()` of a `ts()` object, which will give the smoothed moving average trend, the additive seasonal effects and the random term by default, plus the original data in one plot. For example, the additive decomposition of the time series analyzed in this section is given in Figure 2.6, which also can be recreated using program *ch2figadddecompfit-Radddecomp.R*.

For this series, the random term is small compared to the movement of the trend (which ranges from 2 to 8) and the seasonal effects (which go between about plus and minus 4).

2.3.5 Pros and Cons of Moving Average Additive Decomposition

It is convenient to know the limitations of additive decomposition. Here are several things to take into account.

- Constant average seasonal swings are only useful if the seasonal swings look approximately similar in magnitude every year, which means that the variance of the series is relatively constant. In Section 2.4, we will show an example where the

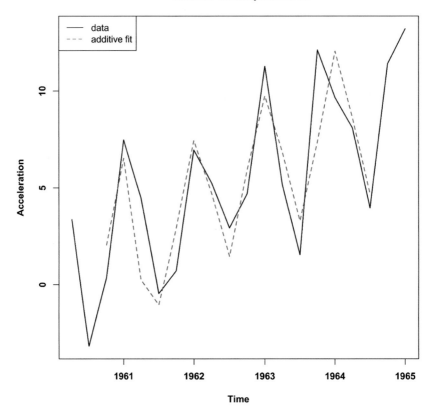

Figure 2.5 The centered four-point moving average plus the additive seasonal effect in Table 2.2 is superimposed on the data.

seasonal effects are obviously not constant. This will require using another type of decomposition, "multiplicative decomposition."[1]

- Additive decomposition models where the trend has been found using a moving average cannot be used for forecasting. The reason being that we would not have data to obtain an estimate of the future trend, since the trend is formed by averaging.

But additive decomposition is very useful to: (i) see the most prominent long-term trend in the data; (ii) calculate the seasonal effects if there is constant seasonal effect in the data; (iii) estimate the random term as the difference between y_t and $\hat{y}_t$. In other words, additive decomposition of a time series gives us a very good idea of the nature of the time series and is a very good starting point for any time series analysis if there is a trend and a constant seasonal effect.

---

[1] In the example just seen, if we look very critically at what trend plus additive seasonal effects is giving, there's a slight suggestion that the swings are too wide some years and not wide enough other years. And in some places, the swings follow the data instead of superimposing closely the original data.

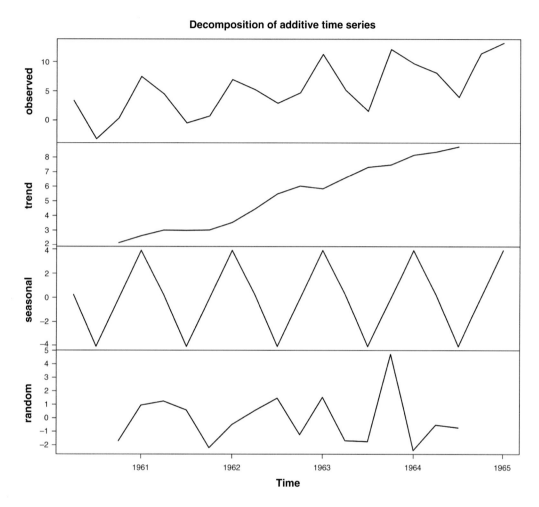

Figure 2.6 Additive decomposition of the acceleration time series of Table 2.2 using R's function decompose() shows the components we obtained separately and the random term $y_t - \hat{y}_t$.

2.3.6 Seasonal Adjustment and Detrending

It is not uncommon to see published data labeled as "seasonally adjusted." We saw that in the metadata corresponding to Figures 1.3, 1.4 and 1.5 in Chapter 1. This, in broad terms, if we applied it to our example, would mean that they are publishing only the $\hat{T}$ plus the random term (they have smoothed out the seasonal swings) by subtracting the seasonal effect.

Notice that if we had used other smoothing methods, for example, regression, LOESS or other, the same decomposition idea applies, but the methods used are more complex.[2]

---

[2]The decompositions and seasonal adjustment methods used by government agencies throughout the world are usually more complex than the one described in this chapter. There are programs to do seasonal adjustment with those methods. See [76] for ways to do that.

> *Seasonal adjustment* means, when doing additive decomposition, that the $\widehat{S}$ has been subtracted from the data and only the trend, cycles (if any) and random terms are left in the data.

> *Detrending* is a term often heard in some literature and data publication outlets. It means, when doing additive decomposition, that the $\widehat{T}$ has been subtracted from the data and only the seasonal swings and the random term are left in the data.

Detrending and seasonally adjusting data that contains trends and seasonality is commonly done to economic time series. Even if decomposition is not used for those purposes, detrending and seasonal adjustment are a necessary practice in order to be able to explore the random component to see if it presents a signal not due to the passage of time and seasonal factors. If after detrending and seasonally adjusting the only thing that remains in the data is independent observations, then no other signal is left, and we are done. But if the random terms are not independent, then we must continue investigating.

Example 2.2   If we try to explain electricity consumption with monthly data, certainly there is a trend (population is growing), and there is seasonal effect (summer and winter), but there must be some other factors determining the consumption of electricity (perhaps people switching to solar power over time, changes in the fees). What is left after we detrend and seasonally adjust the time series, the random term, if properly explored, will tell us whether effects due perhaps to those other factors are present.   □

As the Bureau of Labor Statistics says:

Over the course of a year, the size of the labor force, the levels of employment and unemployment, and other measures of labor market activity undergo fluctuations due to seasonal events including changes in weather, harvests, major holidays, and school schedules. Because these seasonal events follow a more or less regular pattern each year, their influence on statistical trends can be eliminated by seasonally adjusting the statistics from month to month. These seasonal adjustments make it easier to observe the cyclical, underlying trend, and other nonseasonal movements in the series [15].

In fact, governments and other institutions use rather convoluted methods to seasonally adjust and detrend time series. For example, the Census Bureau produces a lot of time series data and uses the X-13ARIMA-SEATS, a seasonal adjustment method produced, distributed, and maintained by the Census Bureau [198].

2.3.7   Other Smoothers

We have seen so far moving average smoothers, which take the average of specific values of the data. But nothing prevents us from smoothing some other way. Smoothing is an area of considerable development.

A variant of the moving average process consists of using the running medians instead of the means. We take a sequence of observations and take the median as the middle value.

$$median\{y_1, y_2, ..., y_s\} = y^*_{(s+1)/2},$$
$$median\{y_2, y_3, ..., y_{s+1}\} = y^*_{(s+1)/2+1},$$

..,

$$median\{y_{n-s+1}, y_{n-s+2}, ..., y_n\} = y^*_{n-(s+1)/2+1}.$$

Running medians of this form tend to give less smooth traces than their running mean counterparts. But they are less likely to be distorted by outliers in the series. This robust behavior can be very useful since in many cases a simple smoother will reveal a few unusual observations [85].

Another smoother is LOWESS (Locally Weighted Scatterplot Smoothing) [26]. The case study in Section 2.10 at the end of the chapter makes use of a LOWESS smoother. There are also nearest-neighbor and kernel smoothers.

2.3.8 Exercises

Exercise 2.5 The nottem time series is one of Base R's data sets. Inspect it and then do an additive decomposition using decompose(). What is the value of the seasonal effect on Quarter 4 of 1929? Do you think that additive decomposition is appropriate for this time series? Why or why not? Section 2.2.2 has R code that helps extract the components.

Exercise 2.6 Do the additive decomposition of the data in Table 2.1 using R's decompose() after making the data a ts() object. Extract the components of the decomposition, namely, the trend, the seasonal and the random component. Double-check that the addition of those three components gives the values of the raw data. Provide at least two examples of such an instance. Section 2.2.2 has R code that helps extract the components.

Exercise 2.7 Use the code in program *ch2mediansmootherexercise.R* based on Janacek's programs [85] to do the following:

(a) Fit a median smoother to the data used to construct Table 2.1.
(b) Plot the time series and superimpose on it the median smoother.
(c) Illustrate by hand how the first four values of your median smoother were obtained.
(d) Comment on what you see in the time plot of the time series and the median smoother.

Exercise 2.8 Consider the observed time series x_t given in Table 2.3. The additive decomposition model is assumed. Complete the column of the table that should contain the seasonally adjusted time series.

Exercise 2.9 To answer the following questions, use program *chapter2ma4trend.R*, or R's decompose().

Table 2.3 Complete this table as indicated in Exercises 2.8 and 2.9.

Time	x_t	ma4	$x_t - \hat{T}$	$\widehat{S}$	Seasonally adjusted
1985 Q2	322	NA	NA		
1985 Q3	144	NA	NA		
1985 Q4	472	450.500	21.500	−25.5185	
1986 Q1	821	474.125	346.875	512.58125	
1986 Q2	408	506.250	−98.250	−24.79375	
1986 Q3	247	538.500	−291.500	−462.26875	
1986 Q4	626	554.750	71.250	−25.5185	
1987 Q1	925	559.500	365.500	512.58125	
1987 Q2	434	567.875	−133.875	−24.79375	
1987 Q3	259	618.750	−359.750	−462.26875	
1987 Q4	681	712.125	−31.125	−25.5185	
1988 Q1	1277	783.500	493.500	512.58125	
1988 Q2	829	837.875	−8.875	−24.79375	
1988 Q3	435	915.500	−480.500	−462.26875	
1988 Q4	940	1009.875	−69.875	−25.5185	
1989 Q1	1639	1078.625	560.375	512.58125	
1989 Q2	1222	1112.625	109.375	−24.79375	
1989 Q3	592	1172.125	−580.125	−462.26875	
1989 Q4	1055	1224.250	−169.250	−25.5185	
1990 Q1	2000	1253.250	746.750	512.58125	
1990 Q2	1278	1320.250	−42.250	−24.79375	
1990 Q3	768	1417.375	−649.375		
1990 Q4	1415	NA	NA		
1991 Q1	2417	NA	NA		

(a) Detrend and seasonally adjust time series x_t of Table 2.3. Plot the result superimposed on the time plot of x_t. Recall that a time plot is a line plot.
(b) Comment on the features observed.

2.4   Classical Multiplicative Decomposition

Consider a time series that exhibits increasing seasonal swings over time. We can decompose it using multiplicative decomposition, which assumes that the time series Y_t is such that

$$Y_t = T_t * S_t * \epsilon_t. \tag{2.4}$$

Notice that this decomposition model employs a multiplicative seasonal factor. Multiplying the trend by the appropriate seasonal factors models the seasonal effect of the time series as being proportional to the trend. That is, the seasonal factor, which is a constant average for each season, is multiplied by the trend (rather than added to

Table 2.4 Soda sales, January 1990 to
December 1992. Data from [8], 360.

Time	y_t	Time	y_t
1990:1	189	1991:7	831
1990:2	229	1991:8	960
1990:3	249	1991:9	1152
1990:4	289	1991:10	759
1990:5	260	1991:11	607
1990:6	431	1991:12	371
1990:7	660	1992:1	298
1990:8	777	1992:2	378
1990:9	915	1992:3	373
1990:10	613	1992:4	443
1990:11	485	1992:5	374
1990:12	277	1992:6	660
1991:1	244	1992:7	1004
1991:2	296	1992:8	1153
1991:3	319	1992:9	1388
1991:4	370	1992:10	904
1991:5	313	1992:11	715
1991:6	556	1992:12	441

the trend). This may seem a contradiction, but remember that the seasonal effect is the average of all seasonal swings for a given season.

To illustrate how to obtain a multiplicative decomposition of a time series, consider a business selling soda that has recorded monthly sales for the previous three years (in hundreds of cases). The data can be seen in Table 2.4 and its plot in Figure 2.7. Program *ch2sodasales.R* contains the code that created this plot and other plots in this section. In addition to having a trend, the soda sales time series possesses regular seasonal swings, with sales being greater in the summer and early fall months and lowest in the winter months. It is reasonable to assume that the multiplicative decomposition model (Equation (2.4)) holds. We now summarize the calculations needed to find the estimates of the trend and the seasonal components of this model.

Table 2.5 contains part of the time series data in a column headed by y_t. The remaining columns of the table contain the multiplicative decomposition calculations that we will explain shortly. Notice that the first six and the last six observations of y_t are not showing in Table 2.5 for lack of space. They are used in the computations. So, for complete reference the reader should look at Table 2.4 when doing the 12-point moving average described in what follows.

We start by calculating a 12-point moving average to get an estimate of the trend. This can be seen in column *ma*12. The first moving average is the average of the first 12 sales values in Table 2.4.

$$447.833 = \frac{189 + 229 + 249 + 289 + 260 + 431 + 660 + 777 + 915 + 613 + 485 + 277}{12}.$$

**Soda sales 1990:1-1992:12
have seasonal and trend**

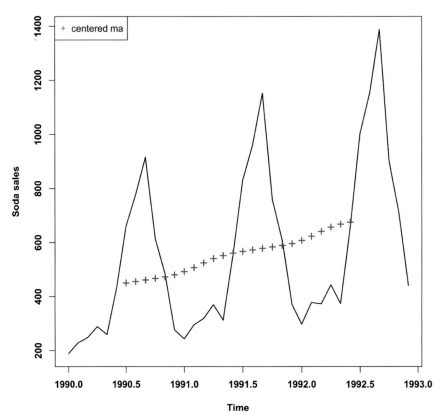

Figure 2.7 Soda sales data used to illustrate the multiplicative decomposition calculations. The centered 12-point moving average is superimposed.

We calculate a 12-point moving average because the sales time series is monthly (12 observations per year, one each month), so we want to smooth out the seasonal effect. When the data was quarterly, for example, the acceleration data in Table 2.1, we used a four-point moving average.

The second moving average is obtained by dropping the first observation from the average and including y_{13} in the calculation:

$$452.417 = \frac{229 + 249 + 289 + 260 + 431 + 660 + 777 + 915 + 613 + 485 + 277 + 244}{12}.$$

The third moving average is obtained by dropping y_2 from the average and by including y_{14} in the average. We obtain

$$458 = \frac{249 + 289 + 260 + 431 + 660 + 777 + 915 + 613 + 485 + 277 + 244 + 296}{12}.$$

Successive moving averages are computed similarly.

Because the number of terms averaged is even, 12, the value of the first moving average corresponds to a time that is midway between periods 6 and 7, the second corresponds to a time midway between periods 7 and 8, and so forth. In order to obtain averages corresponding to time index in the original sales time series, we calculate centered moving averages. The centered moving averages are two-point moving averages of the previously computed 12-point moving averages. They can be seen in column $\widehat{T}$ of Table 2.5. Each value in that column is the average of the value of $12ma$ before and the $12ma$ after. Thus, the first centered moving average is

$$\frac{447.833 + 452.417}{2} = 450.1,$$

the second centered moving average is

$$\frac{452.417 + 458}{2} = 455.2,$$

and so on. Successive centered moving averages are calculated in a similar fashion. If the first moving averages had been calculated using an odd number of time series values, the centering procedure would not be necessary.

Model 2.4 implies that

$$S_t * \epsilon_t = \frac{Y_t}{T_t},$$

which implies that

$$\widehat{S}_t = \frac{Y_t}{\widehat{T}_t}.$$

To estimate the seasonal effect, we first group the values of $S_t * \epsilon_t$ by month and calculate an average for each month.

The average of the two January values is

$\widehat{S}_1 = \frac{0.495 + 0.490}{2} = 0.4925,$

the average of the two February values is

$\widehat{S}_2 = \frac{0.583 + 0.607}{2} = 0.595,$

and so on. The other months' average values are, respectively, 0.595, 0.6795, 0.564, 0.9845, 1.4655, 1.6915, 1.9885, 1.306, 1.028, 0.5995.

The averages are then normalized so that they add to $L = 12$, the number of months in a year. This normalization is accomplished by multiplying each value of $\widehat{S}_t$ by the quantity $\frac{12}{\sum_{t=1}^{12} \widehat{S}_t} = \frac{12}{11.9895} = 1.0008758$. This normalization process results in the estimate $\widehat{S}_t$ shown in Table 2.5. For example, the first $\widehat{S}_t$ in the table, which corresponds to the month of July, is found by multiplying $1.4655 * 1.0008758 = 1.466783 \approx 1.467$.

Using the estimate of the trend in column $\widehat{T}$ and the estimate of the seasonal effect in column $\widehat{S}$, we have now an estimate of Y:

$$\widehat{Y}_t = \widehat{S}_t * \widehat{T}.$$

Table 2.5 Multiplicative decomposition of soda sales. First lines.

Time	y_t	$ma12$	$\hat{T}$	$\frac{y_t}{\hat{T}}$	$\hat{S}$	$\hat{y}_t = \hat{S}_t * \hat{T}_t$
		447.833				
1990:7	660		450.1	1.466	1.467	660.2967
		452.417				
1990:8	777		455.2	1.707	1.693	770.6536
		458				
1990:9	915		460.9	1.985	1.990	917.191
		463.833				
1990:10	613		467.2	1.312	1.307	610.6304
		470.583				
1990:11	485		472.8	1.026	1.029	486.5112
		475				
1990:12	277		480.2	0.577	0.6	288.12
		485.417				
1991:1	244		492.5	0.495	0.493	242.8025
		499.667				
1991:2	296		507.3	0.583	0.596	302.3508
		514.917				
1991:3	319		524.8	0.608	0.595	312.256
		534.667				
1991:4	370		540.7	0.684	0.680	367.676
		546.833				
1991:5	313		551.9	0.567	0.564	311.2716
		557				
1991:6	556		560.9	0.991	0.986	553.0474
		564,833				
1991:7	831		567.1	1.465	1.467	831.9357
		569.333				
1991:8	960		572.7	1.676	1.693	969.5811
		576.167				
1991:9	1152		578.4	1.992	1.990	1151.016
		580.667				
1991:10	759		583.7	1.300	1.307	762.8959
		586.75				
1991:11	607		589.3	1.030	1.029	606.3897
		591.833				
1991:12	371		596.2	0.622	0.6	357.72
		600.5				
1992:1	298		607.7	0.490	0.493	299.5961
		614.917				

We can see the values of the $\hat{Y}_t$ in Figure 2.8. To seasonally adjust (or equivalently, remove the seasonality), the operation to do is to divide y_t by the $\hat{S}_t$.

Once the reader understands how multiplicative decomposition is done, the process can be automated by asking R to do it, using the function `decompose()` of a time series object, but this time using the argument `type="mult"`. This will give the

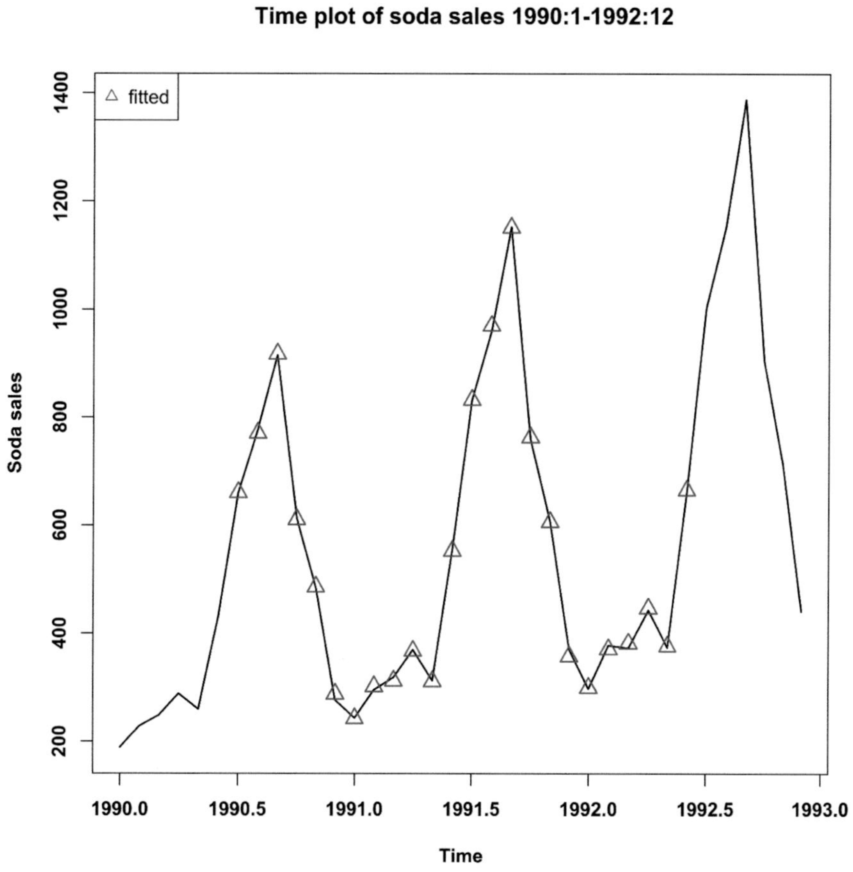

Time plot of soda sales 1990:1-1992:12

Figure 2.8 Multiplicative decomposition model fitted to soda time series.

smoothed moving average trend, the multiplicative seasonal effects and the random term, plus the original data in one plot. For example, the multiplicative decomposition of the time series analyzed in this section is given in Figure 2.9.

2.4.1 Exercises

Exercise 2.10 In the soda time series discussed in Section 2.4, multiply the trend, seasonal and random components. Compare the numbers obtained with those in the raw time series. What can we conclude?

Exercise 2.11 In the soda time series discussed in Section 2.4, what is the value of the seasonal effect for June 1991? Do all June months have the same seasonal effect calculated for them?

Exercise 2.12 The data set in R called `AirPassengers` is monthly data. Double-check that the first five terms of the multiplicative decomposition that you obtain using

Decomposition of multiplicative time series

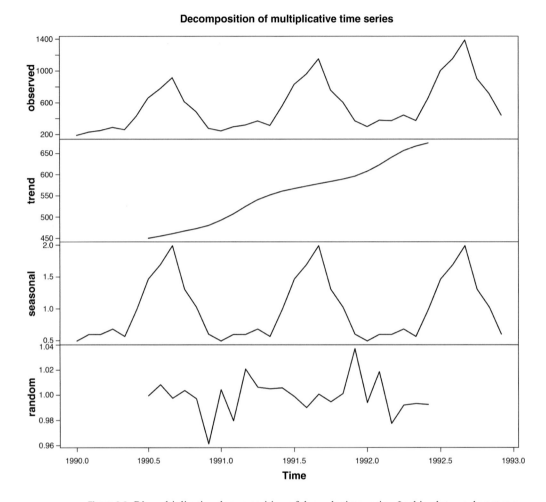

Figure 2.9 R's multiplicative decomposition of the soda time series. In this plot, we do not see the $\hat{y}$, only the parts. If we multiplied the random part times the seasonal part times the trend part, we would get the fitted multiplicative decomposition model that we obtained in Table 2.5, seen in Figure 2.8.

R's `decompose()`, using the argument `type="mult"`, is what we would get if we did it by hand as we did in this chapter. Are they the same? Show the calculations.

Exercise 2.13 The available monthly historical data of a time series containing the number of rooms occupied, starting on January 1977, can be found in data file *rooms.txt*. We saw the multiplicative decomposition done with *ch2figsampledecomp.R* produced in Figure 2.1. This problem assumes that we have run the programs recommended and used in Chapter 1.

(a) Take the log of the rooms time series and do an additive decomposition of the logged time series. Compare with the multiplicative decomposition of the unlogged data. Based on the analysis, does taking log and doing additive decomposition give the same result as not logging and doing multiplicative decomposition? Explain.[3]

(b) Do a seasonal box plot of the logged time series and another of the unlogged one. Are the conclusions reached from the two seasonal box plots the same? Explain.

2.5   Regression Smoothers

Not being able to forecast with classical decomposition models because we use moving average smoothers to estimate the trend can be frustrating. For that reason, some practitioners and textbook authors use decomposition to estimate the seasonal effect, and then deseasonalize the data using those effects in order to estimate the trend with linear regression (thus using regression smoothers).[4] Then, if multiplicative decomposition was used, the seasonal factors estimated with multiplicative decomposition are multiplied to the trend [8]. It would be possible to forecast with a model like this if we are willing to assume that the seasonal factors will not change in the future.

An alternative use of traditional linear regression is to model the seasons with dummy variables, which are binary variables, as indicators of the season and to model the trend with polynomials in time.

In this approach, let's assume that there is only trend and random component present in the time series. Then the trend can be fitted by a polynomial trend, for example, a simple line as follows:

$$T = constant + \beta_1 t + \epsilon_t,$$

where t is the time index.

> We extract the time index of a `ts()` object using the command `time()` applied to the `ts()` object.

If the time series presented regular seasonal trend, and it is quarterly, for example, the seasonal effects are modeled with a dummy variable component modeled as

$$S = constant + \beta_2 q_2 + \beta_3 q_3 + \beta_4 q_4,$$

where only three dummy variables are included, because the effect of quarter 1 will be embedded in the constant term, and $q_i = 1$, $i = 2, 3, 4$ if the observation is for quarter i and 0 elsewhere. The β coefficients will indicate by how much the seasonal effect of quarter i differs from the effect of quarter 1 [68].

---

[3]Usually, if the seasonal effect is proportional to the trend, taking the log of the data will make the seasonal effect similar each year. That is why we would expect that additive decomposition would be appropriate for the logged data.

[4]Those not remembering regression may find a brief review of simple linear regression and references in Chapter 9's Appendix, Section 9.9.

Putting it all together, the model assumed is

$$y_t = \beta_0 + \beta_1 t + \beta_2 q_2 + \beta_3 q_3 + \beta_4 q_4 + \epsilon_t,$$

where β_i, $i = 0, \cdots, 4$ are unknown constant parameters to be estimated. We may use ordinary least squares to obtain an estimate of this model, namely

$$\hat{y}_t = \hat{\beta}_0 + \hat{\beta}_1 t + \hat{\beta}_2 q_2 + \hat{\beta}_3 q_3 + \hat{\beta}_4 q_4.$$

Example 2.3   In this example, we will look at the same data that we analyzed on acceleration in Section 2.3.1. We will use regression smoothers, which consist of a a polynomial in time to represent the trend, and dummy variables for the seasonal effects. In program *ch2regsmoothers.R*, we consider two alternative coding ways to estimate a model like this for a hypothetical data set. We can see the result of applying the code in Figure 2.10.

An advantage of regression smoothers like the one discussed is that we may use the model for forecasting, assuming that the same seasonal and trend patterns will follow into the future. Figure 2.10 shows the forecast for four quarters ahead. Recall that forecasts are predictions out-of-sample. When using regression to forecast, the future values of the independent variables must be given. Hence, a dataframe with values of the trend variables and the dummies must be known.   □

An advantage of regression smoothing is that other variables can be added to the model, in addition to the trend and seasonality. To a large extent, a lot of the forecasting at scale nowadays is done with regression smoothers that feature the seasonality, trend, holidays and other markers and then other variables that are believed to affect the time series of interest.

We will have much more to say about regression smoothing in Chapter 9.

2.5.1   Exercises

Exercise 2.14   Compare the seasonal effects for acceleration obtained with the moving average that appears in Table 2.2 and the coefficients of the dummy variables obtained with regression using the first part of the code in program *ch2regsmoothers.R*. Is there any difference? If there is, to what do we owe that difference? Explain.

Exercise 2.15   Use the seasonal effects obtained in Table 2.2 for the acceleration data and add them to an estimate of the trend obtained with simple linear regression,

$$T = constant + \beta_1 t + \epsilon_t,$$

where t is the time index. Forecast four steps ahead using the estimated trend and the additive estimated seasonal effects. Compare the forecast that you obtain this way with the one obtained with the regression specification that contained the dummy variables to represent seasonality. Use program *ch2regsmoothers.R* as a starting point.

Regression smoothers and forecast

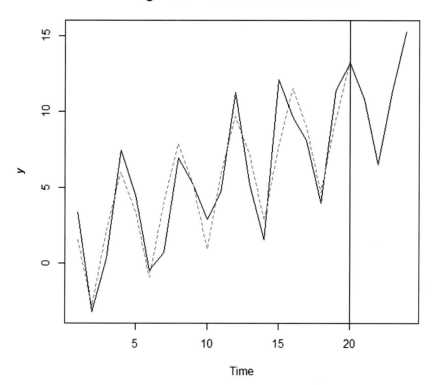

Figure 2.10 Forecast obtained for the acceleration data of Section 2.3.1. The forecast is obtained using regression smoothers, and corresponds to the values past time $= 20$. The discontinuous line is the fitted model, the $\hat{y}_t$.

Exercise 2.16 This exercise is a version of another exercise in Bowerman and O'Connell [8]. Suppose that sales of outboard motors by the Power Drive Corporation are seasonal. Sales are lower in the first quarter, highest in the second quarter, moderately high in the third quarter, and moderately low in the fourth quarter. Furthermore, assume that sales exhibit a linear trend given by

$$T = 500 + 50t,$$

where $t = 0$ is considered to be the fourth quarter of 2019.

(a) If trend alone is considered, what are the values of outboard motor sales expected in the four quarters of 2020?

(b) Sales are seasonal. We can model the seasonal behavior of sales by defining seasonal factors. Suppose that the seasonal factors for quarters 1, 2, 3, and 4 are $\hat{S}_1 = 0.4, \hat{S}_2 = 1.6, \hat{S}_3 = 1.2$, and $\hat{S}_4 = 0.8$, respectively. Assuming that these seasonal factors are multiplicative, and that we consider both trend and seasonal effects, what are the expected values of sales in the four quarters of 2020?

(c) If multiplicative seasonal factors remain constant over time, they allow us to model increasing seasonal variation. Calculate the expected outboard motor sales in 2021, that is, when $t = 5, 6, 7, 8$. What is observed after doing that?

2.6   Exponential Smoothing

Exponential smoothing is a more sophisticated smoothing method than moving averages. It also has more ambitious goals than classical decomposition, as it smooths while predicting the next value of the time series according to an optimality criterion. It is this aspect of the method that makes it suitable for forecasting, which is something that classical decomposition models with moving averages cannot do. The nature of the time series determines whether *simple exponential smoothing, trend-corrected exponential smoothing* or *trend and seasonal exponential smoothing* is needed.

Exponential smoothing is a forecasting method that applies unequal weights to the time series observations. This unequal weight is accomplished by using a smoothing constant that determines how much weight is attached to each observation. The most recent observation is given the most weight. Older observations are given successively smaller weights [8].

In contrast to classical decomposition but as is the case in regression smoothers, *exponential smoothing* methods [80] allow us to forecast. Forecasts derived from exponential smoothing methods are weighted combinations of past observations, with weights that decrease exponentially as the observations get older.

Exponential smoothing methods have been around since the 1950s [80]. They follow an optimality criterion familiar to anyone with an introductory background in Statistics, namely they minimize the mean square error and arise as optimal solutions for methods that would not appear to the reader to look at all like exponential smoothing. Exponential smoothing is also intuitive and very popular in business forecasting. Until recently, exponential smoothing, like the classical decomposition methods based on moving averages, lacked a statistical framework that produced both prediction intervals and forecasts. However, the state space approach [80] has provided this framework to exponential smoothing. In this section, we study the non-state space approach to exponential smoothing.

2.6.1   Simple Exponential Smoothing

Simple exponential smoothing is a method of time series smoothing that assumes that the mean μ_t of a time series at time t is a weighted average of the value of x_{t-1} of the time series at time $t - 1$ and the value of μ_{t-1}.

We consider that the time series model generating the observed x_t is

$$X_t = \mu_t + W_t, \tag{2.5}$$

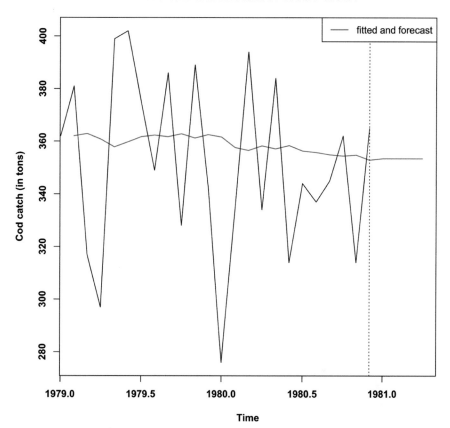

Figure 2.11 Simple optimal exponential smoother of cod catch data and forecast four months ahead using the Holt–Winters (HW) simple exponential smoother.

where W_t are independent and identically distributed random variables with mean 0 and constant variance, that is, white noise, and μ_t is the mean of the series at time t, which changes slowly over time.

Suppose we start with an initial estimate $\hat{\mu}_2(1) = 362$. This is the estimate of μ_2 made at time $t = 1$. Thus,

$$\hat{x}_2(1) = 362.$$

If the value of the cod time series at $t = 2$ is $x_2 = 381$, there is an error of 19.

The updating or *smoothing equation* for $\hat{\mu}_3(2)$ (the estimated value of μ_3 made at time $t = 2$) is

$$\hat{\mu}_3(2) = \alpha x_2 + (1 - \alpha)\hat{\mu}_2(1) = \hat{\mu}_2(1) + \alpha(x_2 - \hat{\mu}_2(1)), \tag{2.6}$$

where α is a smoothing constant between 0 and 1 and $\hat{\mu}_3(2)$ is the forecast of μ_3 made at time $t = 2$. The general smoothing equation is

$$\hat{\mu}_{t+1}(t) = \alpha x_t + (1 - \alpha)\hat{\mu}_t(t - 1) = \hat{\mu}_t(t - 1) + \alpha(x_t - \hat{\mu}_t(t - 1)). \tag{2.7}$$

Simple exponential smoothing is appropriate when there is no trend and there is no seasonal component.

Substituting Equation 2.7 into Equation 2.5, we get

$$\hat{x}_{t+1}(t) = \hat{\mu}_{t+1}(t). \tag{2.8}$$

The general estimated model is

$$\hat{x}_{t+1}(t) = \alpha x_t + (1 - \alpha)\hat{\mu}_t(t - 1) = \hat{\mu}_t(t - 1) + \alpha(x_t - \hat{\mu}_t(t - 1)).$$

This is known as the *smoothing or updating equation*. Notice that by repeated back substitution, we can also express the mean process as

$$\mu_t = \alpha x_t + \alpha(1 - \alpha)x_{t-1} + \alpha(1 - \alpha)^2 x_{t-2} + \ldots,$$

that is, the mean today is a linear combination of current and past observations, with more weight given to the more recent observations.

The output α is the value of α that minimizes the sum of square errors of the one-step-ahead prediction in-sample (in the training sample), $\hat{\alpha}$. The Sum of Square Errors (SSE) is then calculated as follows:

$$SSE = \sum_{t=1}^{n}(x_t - \hat{x}_t)^2,$$

and the root mean square error of the fitted model (or estimate of the standard deviation of the time series) is

$$RMSE = \sqrt{\frac{\sum_{i=1}^{n}(x_t - \hat{x}_t)^2}{n}}.$$

The α controls the rate at which the predicted value adjusts. Smaller values of α adjust the forecasts more slowly (a lot of smoothing). Larger values, close to 1, result in hardly any smoothing (fitting noise). This parameter must be estimated using the time series data. The value of n is the training data set size.

Base R has code to do exponential smoothing, and we use that code to do simple exponential smoothing of the cod catch data shown in Figure 2.11. We do not split the data into a training set and a test set, as our goal is only to demonstrate how the smoothing works.[5]

---

[5] The training set also receives the name of *learning set* in machine learning circles.

The R command to do Holt–Winters simple exponential smoothing of a `ts()` object (i.e., without any pretransformation) is

`HoltWinters(variablename, beta=FALSE , gamma=FALSE)`

where `variablename` is the name of the `ts()` object. This is very general code. When fitting only simple exponential smoothing (no trend, no seasonal), the parameters `beta` and `gamma` are set to `False`). Implicitly, that means that alpha is the only parameter changing. The default is `alpha=T, beta=T`, and `gamma =T`. By not mentioning alpha, we are telling R to use the default, `alpha=T`.

Example 2.4   The Bay City Seafood Company owns a fleet of fishing trawlers and operates a fish processing plant. In order to forecast its minimum and maximum possible revenues from cod sales and to plan operations of its fish processing plant, the company desires to make both point forecasts and prediction interval forecasts of its monthly cod catch (measured in tons). The company has recorded the monthly cod catch for the previous two years, 1979 and 1980. The data set is *cod.txt* [8].

The first things we ask before using exponential smoothing methods are: what do the data look like? Is there a trend? Are there seasonals? Is the mean changing slowly up and down?

When the cod time series is plotted in Figure 2.11 using program *ch2simpleexpsmooth.R*, we can see that it fluctuates randomly around a constant average but we can also think of it as an average that seems to be changing with time, although it is hard to discern the direction of these changes. The plot suggests that there is no trend or seasonal pattern. Since there is no trend or seasonal pattern, simple exponential smoothing could work.

We fitted a simple exponential smoothing model to the cod data with code given in program *ch2simpleexpsmooth.R* and obtained the following output, included here only because it is hard for the beginner to make sense of this output at first.

The `alpha` is the estimate of the parameter α, giving the following updating smoother:

$$\hat{x}_{t+1}(t) = 0.04627398 x_t + (1 - 0.04627398)\hat{\mu}_t(t - 1).$$

The in-sample (the training sample's) sum of squared residuals, obtained by prompting the software to provide it, at the optimal α, is $SSE = 28357.97$ and the root mean square error $RMSE = \sqrt{\frac{28357.97}{24}} = 34.37415$, which R does not provide directly.

The coefficient

```
a 353.4506
```

is $\hat{x}_{1981:1}(1980: 12)$, that is, the forecasted value for January 1981 made in December 1980. Since there is no data past 1980:12, that forecast is a forecast proper, an out-of-sample prediction.

```
#### OUTPUT
Holt-Winters exponential
smoothing without trend and
without seasonal component.

Call:
HoltWinters(x = cod.ts, beta = 0, gamma = 0)

Smoothing parameters:
alpha:  0.04627398
beta :  0
gamma:  0

Coefficients:
[,1]
a 353.4506
#First out-of-sample forecast
#
```

We prompt R for the fitted values, and we obtain the values in Table 2.6 to double-check.

Notice how the value 353.4506 does not appear as the last fitted value in Table 2.6. Rather, we can see that it is a forecast of January 1981, $\hat{x}_{t+1}$ by observing the following:

The last fitted value is $\hat{x}_{24} = 352.8902$, while $x_{24} = 365$. Thus, we can predict the first out-of-sample forecast as

$$\hat{x}_{25} = 352.8902 + 0.04627(365 - 352.8902) = 353.4506,$$

which is the value that we see given in the R output shown in the frame under a (Coefficients part).   □

The reader is encouraged to check how the updating equation works for other values in the table.

By minimizing the SSE, we obtained the optimal smoother, which will capture the signals in the data, not the noise.

Simple exponential smoothing can be viewed as an adaptive-forecasting algorithm or, equivalently, as a geometrically weighted moving-average filter. Simple exponential smoothing is most appropriate when used with time series data that exhibit no linear or higher-order trends but do exhibit low-velocity, nonperiodic variation in the mean.

For a long time, it was believed that simple exponential smoothing was an ad hoc method, without statistical meaning (i.e., no standard errors could be attached to coefficients). But not long ago it was found that it is related to modern time series methods. For example, simple exponential smoothing produces optimal forecasts for several

Table 2.6 Fitted values ($\hat{x}_t$) given by simple exponential smoothing of the cod time series (first three columns returned by R upon prompting) and the cod time series.

Time	$\hat{x}_t$	level ($\hat{\mu}_t$)	x_t(observed)
1979:1	initial =362.000		
1979:2	$\hat{x}_2(1) = 362.0000$	$\hat{\mu}_2(1) = 362.0000$	381
1979:3	$\hat{x}_3(2) = 362.8792$	$\hat{\mu}_3(2) = 0.0463(381) + 0.9537(362) \approx 362.8792$	317
1979:4	$\hat{x}_4(3) = 360.7562$	$\hat{\mu}_3(2) = 0.0463(317) + 0.9537(362.8) \approx 360.7562$	297
1979:5	357.8059	357.8059	399
1979:6	359.7122	359.7122	402
1979:7	361.6690	361.6690	375
1979:8	362.2859	362.2859	349
1979:9	361.6711	361.6711	386
1979:10	362.7969	362.7969	328
1979:11	361.1867	361.1867	389
1979:12	362.4737	362.4737	343
1980:1	361.5726	361.5726	276
1980:2	357.6128	357.6128	334
1980:3	356.5201	356.5201	394
1980:4	358.2545	358.2545	334
1980:5	357.1321	357.1321	384
1980:6	358.3754	358.3754	314
1980:7	356.3220	356.3220	344
1980:8	355.7518	355.7518	337
1980:9	354.8841	354.8841	345
1980:10	354.4267	354.4267	362
1980:11	354.7772	354.7772	314
1980:12	352.8902	352.8902	365

underlying models, including ARIMA(0,1,1) and the random-walk-plus-noise state space model [80].

2.6.2 Exercises

Exercise 2.17 Suppose $\alpha = 0.2$ and the value of the last three observations are $y_{23} = 362, y_{24} = 314, y_{25} = 365$.

If R gives $a_{23} = 321$, what are the exponential smoothing predictions for the next two periods? What is the out-of-sample forecast value for $t = 26$? Write the equation.

Exercise 2.18 With a simple exponential smoothing, we are looking for a model consisting of a simple smoother. Such a smoother was fitted to a data set of 24 observations that had no trend or seasonals in it. The last three values of the series are $x_{22} = 343, x_{23} = 362, x_{24} = 245$.

Software gave the following estimates for the parameters of the model:

- optimal exponential coefficient $\hat{\alpha}$ = 0.0344
- sum-of-squared residuals (SSE) = 28089.141

• root mean-squared error = 34.21083

Use this output to find a point forecast of the data for $t = 25, t = 26$, and $t = 27$.

Exercise 2.19   We can try to see what happens if we change α manually, that is, we use an alpha that is not optimal. Do the following and write a table with the results.

(a) Fit five models to the cod data set *cod.txt* studied in this section, using $\alpha = 0.1, \alpha = 0.3, \alpha = 0.5, \alpha = 0.7, \alpha = 0.95$. Find the SSE for each model. The only thing new in the fitting command is that now the value of alpha must be input in the program, since we are simulating the models. So now we use

```
HoltWinters(variablename, alpha=value, beta=FALSE,
                    gamma=FALSE)
```

where alpha is made equal to one of the values given earlier, and variablename is the name of the cod time series.

   Create a table with α in one column and SSE in another. Add a line to that table containing the optimal α and its RMSE.

(b) Do five plots with the data and the fitted values, with title in each plot specifying the alpha value used in the fitting command. Use the `par(mfrow(c(3,2))` command to set up the graphs all in one plot. Notice that if the Holt–Winters output is called `cod.hw2`, the plot with the raw data and fitted value can be obtained in each case with function

```
plot(cod.hw2)
```

(c) Discuss what is happening as α increases and why it may not be a good idea to use a value of alpha different from the optimal one.

2.6.3   Additive Holt–Winters Trend-Corrected Exponential Smoothing

This method is most appropriate in smoothing and forecasting a time series that can be modeled as a linear trend that varies slope and intercept over time. The model is

$$x_t = \gamma_{0t} + \gamma_{1t} + w_t, \qquad (2.9)$$

where γ_{0t} and γ_{1t} change slowly over time according to the smoothing equations

$$\hat{\gamma}_{0t} = \hat{\gamma}_{0,t-1} + \alpha(x_{t-1} - \hat{\gamma}_{0,t-1}) + \hat{\gamma}_{1,t-1},$$
$$\hat{\gamma}_{1t} = \beta(\hat{\gamma}_{0t} - \hat{\gamma}_{0,t-1}) + (1 - \beta)\hat{\gamma}_{1,t-1}.$$

The estimates of α and β are the values that minimize the one-step-ahead sum of square errors (SSE) in the training sample, as defined earlier.

   Notice that in R's software documentation the estimate of γ_{0t} is called a_t or `level` and the estimate of γ_{1t} is called b_t or `trend` (where the subscript t is added to a and b here to emphasize their change over time. Thus `level` and `trend` represent the changing parameter values over time represented in Equation 2.9.

> The R command to do Holt–Winters trend-corrected exponential smoothing of a
> `ts()` object (i.e., without any pretransformation) is
>
> `HoltWinters(variablename,   gamma=FALSE)`
>
> where `variablename` is the name of the `ts()` object.

Holt–Winters two-parameter double exponential smoothing is a smoothing approach for forecasting such a time series that employs two smoothing constants. A forecast of a future value of y_t is given by adding those two components (the level or intercept and the trend component).

Example 2.5 We look at weekly thermostat sales in 1972 now. The data set is *thermostat.txt*. The time series has an upward trend in the latter weeks. We can think of the trend as a set of lines that change slope and intercept as time advances. Not sure whether there is seasonality, we might be tempted to tentatively fit a Holt–Winters trend-corrected exponential smoother. We will interpret the output as if trend-corrected was the correct smoother.

We will let R do the Holt-Winters trend-corrected exponential smoothing automatically for us, that is, R will find the optimal α and β that will minimize the SSE and will choose an initial value for the series.

We will use the R code provided in program *trend-exp-smoother.R* to fit the Holt-Winters smoother to the thermostat data, and to obtain the SSE and the forecast and understand how the forecast is obtained. To start with, we look at the output obtained:

```
###OUTPUT
# Holt-Winters exponential
# smoothing with trend and
# without seasonal component.

Call:
HoltWinters(x = therm.ts, gamma = FALSE)

Smoothing parameters:
alpha:  0.6856771
beta :  0.1943463
gamma:  0

Coefficients:
[,1]
a 327.175169  # first out of sample value of a (gamma_0)
b 7.542246  # second out of sample value of b (gamma_1)
```

Table 2.7 Fitted values given by trend-corrected exponential smoothing of the 1972 weekly thermostat sales time series (first four columns returned by R upon prompting) and the thermostat time series.

Time	$\hat{x}_t$	level ($\hat{\gamma}_{0t}$)	trend ($\hat{\gamma}_{1t}$)	x_t (observed)
week 1				206
week 2				245
week 3	284.0000	245.0000	39.00000000	185
week 4	241.9253	216.1180	25.80737970	169
week 5	208.0115	191.9221	16.08943631	162
week 6	186.4205	176.4625	9.95799391	177
week 7	188.6637	179.9611	8.70263286	207
...	...	...	...	...
...	...	...	...	...
week 46	265.7686	266.0972	−0.32857466	312
week 47	303.3006	297.4684	5.83216687	296
week 48	303.1540	298.2947	4.85930078	307
week 49	311.1629	305.7911	5.37180886	281
week 50	291.8332	290.4809	1.35233256	308
week 51	306.4251	302.9184	3.50669631	280
week 52	288.2913	288.3060	−0.01468211	345

After fitting the HW model, we obtain the following smoothing equations:

$$\hat{\gamma}_{0t} = \hat{\gamma}_{0,t-1} + 0.68567(x_{t-1} - \hat{\gamma}_{0,t-1}) + \hat{\gamma}_{1,t-1},$$
$$\hat{\gamma}_{1t} = 0.19434(\hat{\gamma}_{0t} - \hat{\gamma}_{0,t-1}) + (1 - 0.194)\hat{\gamma}_{1,t-1}.$$

The SSE = 61273.09, and the RMSE of the fit is 34.32679.

The first forecast (out-of-sample predicted) values of γ_0 and γ_1 are 327.175169 and 7.542246, respectively.

Table 2.7 shows the level, trend, fitted values and the actual value of the thermostat time series. We present only a few of the first and last observations.

The last values in Table 2.7 help us see how we got the first out-of-sample forecast:

$$327.175169 = 288.3060 + 0.68567(345 - 288.3060) - 0.01468211,$$
$$7.542246 = 0.19434(327.175169 - 288.3060) + (1 - 0.194)(-0.01468211).$$

So, the forecast for the first week of 1973 is

$$\hat{x}_t = 327.175169 + 7.542246 = 334.7174.$$

Notice that we do not update the coefficients alpha and beta as we forecast.

Notice also how R initializes the estimation. The first two values of the time series are used to construct initial $\hat{\gamma}_{0t} = 405$ and initial $\hat{\gamma}_{1t} = 245 - 206 = 39$. These are the values showing for week 3 in Table 2.7.

We learned in this example how to interpret the output of trend-corrected exponential smoothing. ☐

2.6.4 Exercises

Exercise 2.20 Use the estimated values of γ_0 and γ_1 in the example of Section 2.6.3 to show how we manually obtain the one-step-ahead prediction of $\hat{x}_5$ indicated in Table 2.7. Repeat the exercise to predict $\hat{x}_6$.

Exercise 2.21 The data set `LakeHuron` is one of the data sets in Base R's `datasets` library. It contains annual measurements of the level, in feet, of Lake Huron from 1875 to 1972. Annual measurements do not present seasonality. The time series is a `ts()` object.

(a) Type `LakeHuron` to see the source from where Base R got the data.
(b) Do a time plot of the time series and then subset it using the `window()` function to contain only annual observations for 1875 to 1967. This will be the training set or in-sample data.
(c) Use trend-corrected exponential smoothing to predict the values in the training set.
(d) Forecast the value for 1972 and compare it with the actual value of the data in 1972.
(e) Plot the whole data set, the training set prediction and the forecast, using different colors for each.

2.6.5 Additive Holt–Winters Trend and Seasonal Exponential Smoothing

When the data has a trend and a seasonal, we want to use the additive version of Holt–Winters exponential smoothing that will fit such data. The method is most appropriate in smoothing and forecasting a time series that can be modeled as having a trend and a seasonal or just a seasonal. The difference with the Holt–Winters nonseasonal smoother seen in Section 2.6.3 is that we add a third term and a third equation to express the updating of the seasonal:

$$x_t = \gamma_{0t} + \gamma_{1t} + \gamma_{2t} + w_t,$$

where γ_{0t}, γ_{1t} and γ_{2t} change slowly over time according to the smoothing equations:

$$\widehat{\gamma}_{0t} = \alpha(x_{t-1} - \hat{\gamma}_{2,t-p}) + (1-\alpha)(\hat{\gamma}_{0,t-1} + \hat{\gamma}_{1,t-1}),$$
$$\widehat{\gamma}_{1t} = \beta(\hat{\gamma}_{0t} - \hat{\gamma}_{0,t-1}) + (1-\beta)\hat{\gamma}_{1,t-1},$$
$$\widehat{\gamma}_{2t} = \gamma(x_t - \gamma_{0t}) + (1-\gamma)\gamma_{2,t-p},$$

where, if the time series is monthly, $p - 12$; if the time series is quarterly, then $p = 4$, and so on.

The function will find the values of α, β and γ that minimize the one-step-ahead prediction error.

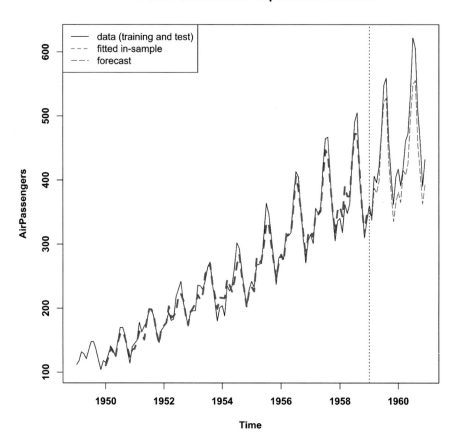

Figure 2.12 The Holt-Winters (HW) trend and seasonal exponential smoother gives the one-step-ahead prediction that minimizes the sum of square errors and is plotted superimposed to the `AirPassengers` time series. The forecast of 24 months out-of-sample is shown after the dotted vertical line.

The R command to do Holt–Winters trend and seasonal exponential smoothing of a `ts()` object (i.e., without any pretransformation) is

`HoltWinters(variablename)`

where `variablename` is the name of the `ts()` object.

Figure 2.12 shows a plot of the `AirPassengers` time series and 24-step-ahead forecasts produced by exponential smoothing. The test set is the last two years of the data. The reader will be asked in an exercise to calculate the RMSE of the forecast.

Separately, Figure 2.13 shows the decomposition of the time series done by exponential smoothing into trend, seasonal and level (average). We notice that exponential smoothing does not produce a random component like classical decomposition

**HW decomposition of AirPassengers
after trend and seasonal HW exponential smoothing**

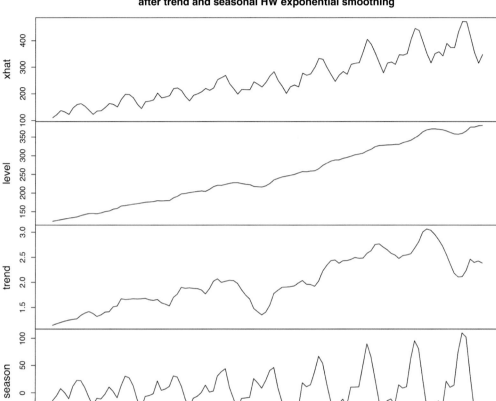

Figure 2.13 Holt-Winters (HW) Trend and seasonal exponential smoothing decomposition of the AirPassengers time series, each plot representing, in order, from top to bottom, the fitted values $\widehat{\gamma}_{0t} + \widehat{\gamma}_{1t} + \widehat{\gamma}_{2t}$, the update of the mean (the level) $\widehat{\gamma}_{0t}$, the trend estimate $\widehat{\gamma}_{1t}$, and the seasonal estimate $\widehat{\gamma}_{2t}$.

did. Notice also how the trend estimate is more sophisticated than the simple moving average smoother that we used in classical decomposition. The seasonal estimate is also different from the seasonal effects that we calculated in classical decomposition.

The program to replicate what we did in this section is *ch2trend-exp-smoother.R*, bottom section.

2.6.6 Exercises

Exercise 2.22 For this exercise, use the AirPassengers data of Base R, but make it a numeric variable. To convert it from ts() object to numeric, use this command:

```
myairpassengers=as.numeric{AirPassengers}
```

Go to the following website:

www.uea.ac.uk/~gj/book/exponentialsmooth.code

Scroll down until you find the function with name `sex4`. Select it all. Copy it and then paste into R. Highlight it and run it. This will compile the function.

Repeat for the function `eforecast5`. Copy and paste into R also. Highlight and run. This will compile this function. Notice that this function will need the sex4 function already compiled. Describe the findings.

Finally, copy and paste into R the `cyc` function and compile it.

Now, in R, type the following code:

```
eforecast5(myairpassengers,12, n)
# where it says n put the length of your time series.
```

This will give a plot with the fitted values (+) and the data (dots and lines). The first line of the output is the three coefficients needed for the trend and seasonal Holt–Winters model. Write the model with these coefficients and forecast the next 12 out-of-sample values. Compare the values of your forecast with the values that you would obtain using R's Holt–Winters function.

We would like to have the RMSE of the fit. So now we could change the R function `eforecast5`. Change the line xhat at the end into two lines:

```
mse=(sum(yseries-fseries)**2)/k)
sqrt(mse)
```

And then copy and paste the new function into R and compile.

Do again in R:

```
eforecast5(myairpassengers,12, n)
```

and now you will get the coefficients of the model and the root mean square error.

To obtain the last 10 seasonals fitted, we will again change the `eforecast5` function. In the place of the two mse lines, put

```
bigc
```

This will give you again the coefficients of the model (alpha, beta, gamma) and the last 12 seasonals snt(T).

Finally, you would like to get the last line of the a(T) and b(T) because those will be your initial values for the forecast. Again, instead of the bigc command, now you can enter

```
m
```

and run. Repeat, putting in the following instead of m:

```
Tr
```

These will give you, respectively, a long list of constant updates (a(T)) and a long list of b(T) updates. Choose the last one in each list. These will be your initial values for a(T) and b(T) for forecasting.

Use the model to forecast the next 12 out-of-sample values. Compare again with the ones you would have obtained with the HoltWinters() function in R as described in Section 2.6.7.

Exercise 2.23 The Bay City Seafood Company owns a fleet of fishing trawlers and operates a fish processing plant. In order to forecast its minimum and maximum possible revenues from cod sales and to plan operations of its fish processing plant, the company desires to make both point forecasts and prediction interval forecasts of its monthly cod catch (measured in tons). The company has recorded the monthly cod catch for the previous two years, 1979 and 1980. The data is in file *cod.txt*.

(a) What is the sample mean of cod?
(b) Describe time features of the data.
(c) Write the exponential smoothing model and the RMSE of the fit and $a_{1980,12}$.
(d) What are the values of the forecast, four steps ahead? Where is it coming from?
(e) Compared with the optimal model, what will the fit with $\alpha = 0.3$ look like? Write the R commands that we would need to type in order to see the graph of the fit.

Exercise 2.24 Split the JohnsonJohnson time series into a training set containing data up to quarter 4 of 1977. Keep the remaining data for the test set.

(a) Apply trend and seasonal Holt–Winters exponential smoothing to the Johnson-Johnson time series. Program *ch2trend-exp-smoother.R* is a good way to start, focusing on the bottom portion of the program. Forecast three years ahead (12 quarters).
(b) Produce a plot like that in 2.12 and calculate the RMSE of the forecast. Notice that the forecast is not the fitted values.
(c) Separately, provide the decomposition into the trend, level and season, as in Figure 2.13, and comment on its features.
(d) Fit an appropriate regression model to the training set of data, like the example in the Regression Smoothers section of this chapter. Forecast 12 quarters ahead and compare the RMSE of the forecast obtained with the regression with that obtained with the trend and seasonal exponential smoothing. Which one produced the lowest RMSE? Notice that it is very common practice in forecasting to fit a gallery of models to the same time series and then compare the forecast errors.
(e) Another practice common in forecasting is to obtain the consensus forecast. This is the average of the forecasts obtained with different models. Then the RMSE of the average forecast is calculated and compared with the individual RMSE. Do that. Does the average forecast give a lower RMSE than the individual model forecasts? Provide the values of the RMSE of each.

Exercise 2.25 Hyndman and Athanasopoulos [76] have a whole chapter on exponential smoothing and interpret the output of trend and seasonal Holt–Winters exponential smoothing. Conduct the analysis they conduct in that chapter to familiarize yourself with how exponential smoothing can be done when using the package `fable` [138]. Notice that the authors include prediction intervals around the forecast to account for uncertainty in the forecast. That is a common practice among forecasters. The HW code we have used does not produce those intervals, but the statistical interpretation of exponential smoothing methods does [80].

2.6.7 Holt–Winters Multiplicative Decomposition

As in the decomposition of a time series seen in Sections 2.3 and 2.4, there is also a multiplicative version of Holt–Winters exponential smoothing. As indicated in Section 2.4, for a time series long enough for us to appreciate the increasing variability of the seasonal swings, such as, for example, Base R's `AirPassengers` data set, multiplicative decomposition models, which account for the increasing variability in the seasonal swing, are more appropriate. An alternative to using multiplicative models is to log the time series (if the increase in variability is proportional to the trend) and then apply additive methods. We choose the former alternative, the Holt–Winters version that we would use for a time series like R's `AirPassengers` using

```
AP.hw1=HoltWinters(AirPassengers, seasonal="mult")
```

The default of the `HoltWinters()` function is to do additive. Therefore, we would need to specify `seasonal="mult"` if we wanted to do multiplicative Holt–Winters. We will leave this approach as an exercise.

2.7 Prophet

Classical decomposition and regression smoothing have a lot in common with functional models used nowadays to forecast at scale in many organizations. Exponential smoothing allows for the fitting of a more sophisticated trend component, and modern approaches fit sophisticated trend functions sometimes. *Prophet* is one such modern forecasting model, designed to additively account for the features that we studied with decomposition: trend and seasonals, with holidays adding to that. The Prophet model is [186]:

$$y_t = g_t + s_t + h_t + \epsilon_t,$$

where g_t is a trend function (the equivalent of the moving average, or the regression smoother), s_t represents the seasonality, and h_t represents the effects of holidays. The ϵ_t is the random term. It is possible to add other variables, as in regression.

Several linear and nonlinear trend functions are accommodated by the Prophet model, functions that, like regression smoothers, could be used for forecasting. Seasonality is modeled additively, so in order to model multiplicative seasonality, log has to be taken. Multiple seasonalities can be accommodated, and different trend assumptions can be made. The observations do not have to be consecutive. So the model is very flexible.

Forecasting with Prophet is a curve-fitting exercise, and the method is not considered a time series model. However, this method is very popular. The fact that there is a package in R that makes the task of using it semiautomatic has a lot to do with its popularity [187]. The package `prophet` is easy to use for beginners; it requires an understanding of the components of a time series that we have discussed in this chapter and a familiarity with time series data like that gained in Chapter 1.

We will use Prophet later in this book, in particular in the case study in Section 6.11. For now, the reader is referred to some examples of its application in the Exercises.

2.7.1   Exercises

Exercise 2.26   Read the article by Taylor and Letham [186] to get acquainted with the language used in forecasting at scale with functional models. Then, in R, execute the following commands:

```
install.packages("prophet")
```

```
library(prophet)
```

```
?prophet()
```

The function contains many arguments (called "parameters" by the ML community). The main arguments, "growth" (for trend) and the seasonality ones, are enough to get started producing some fitting and forecast. Read the documentation.

Direct application of the function requires the data to be entered in a particular format.

2.8   Problems

Problem 2.1   Complete Table 2.8 and do a plot of the original data and the fitted additive decomposition model. Comment on the plot.

Do also a classical decomposition using R's `decompose()` function. Compare the smoothed trend and the seasonal effects. Is R using a four-point moving average to estimate the trend and to obtain the seasonal effects? When done, provide a plot of the data without the seasonal effect, that is, seasonally adjust the data. Separately, provide a plot of the data without the trend, that is, detrend the data, keeping the seasonal effect.

Problem 2.2   The data `JohnsonJohnson` is a quarterly time series that comes with R. We talked about this time series in Section 1.5.

(a) Write R code to access it and check what type of R object it is, where it starts, where it ends, whether it is seasonal, whether the time plot shows trend, and do additive and multiplicative decomposition. Make a training data set that excludes the last year's four quarters. Write a brief summary of what is found.

Table 2.8 Data. Complete this table for Problem 2.1.

Time	y_t	ma4	$\widehat{T}$	$d = y_t - \widehat{T}$	$\widehat{S}$	$\hat{y}_t = \widehat{T} + \widehat{S}$
1989q1	293					
1989q2	392					
1989q3	221					
1989q4	147					
1990q1	388					
1990q2	512					
1990q3	287					
1990q4	184					
1991q1	479					
1991q2	640					
1991q3	347					
1991q4	223					
1992q1	581					
1992q2	755					
1992q3	410					
1992q4	266					

(b) Use R to do additive decomposition and determine whether the seasonal effect given by R's decomposition is the same that we would obtain, more or less, if we did the moving averaging and the determination of the seasonal effect by hand, as in Section 2.3.1. Illustrate, with R code, how to do the manual calculations.

Problem 2.3 Suppose we did additive decomposition of the AirPassengers data set that comes with R. Comment on what the random term looks like. What does the appearance of the random term indicate about the appropriateness of using additive decomposition for this time series? What should we do, based on the conclusions?

Problem 2.4 The data set nhtemp is a ts() class data set in R's datasets library. Inspect the data and decide whether a simple exponential smoothing or a trend-corrected exponential smoothing is needed to fit the data and forecast.

(a) Apply the type of exponential smoothing chosen to a training set of 1912–70. Write the updating equation, specifying the updating parameters.

(b) Obtain the forecasted value for 1971 and compare with the actual value of the data.

Problem 2.5 The article "Forecasting Computer Usage" [62] analyzed monthly data on computer usage by Best Buy Co., Inc (NYSE:BBY) (Millions of Instructions Per Second,MIPS) and total number of Best Buy stores from August 1996 to July 2000. The data set was used to compare time-series forecasting with trend and seasonality components via exponential smoothing and causal forecasting based on simple linear regression. The goal was to forecast the MIPS needed for December 2020 and December 2021 using the data set. Information on the planned number of stores through December 2001 is also available.

Best Buy is a retail store, which had been growing steadily during that period, so it is not difficult to understand that there is seasonality and trend reflecting growth. For example, as holiday seasons require more computer power. It is important for the company to purchase enough mainframe MUPS to guarantee steady surveillance of billing, inventory, and sales.

Columns 1 to 3 contain the date (first day of the month), the MIPS used that month and the number of stores.

Read the data into R and make the data an R time series object and view it by using the following commands:

```
www="http://jse.amstat.org/datasets/bestbuy.dat.txt"
bb.data=read.table(www,fill=T)
head(bb.data)
dim(bb.data)
# make an mts object
bb=ts(bb.data[,2:3],start=c(1996,8))
colnames(bb)=c("MIPS","stores")
head(bb)
plot.ts(bb, main="Best Buy computer usage")
mips=bb[,1]
stores=bb[,2]
```

Fit an appropriate exponential smoothing model for this data set and forecast the required out-of-sample interval.

2.9   Quiz

Question 2.1 The basic idea behind an additive decomposition model is that (choose all that apply)

(a) the seasonal effect is constant for each season over time.

(b) the seasonal effect is proportional to the long-term trend.

(c) there is no random term.

(d) after removing the seasonal, there is always just noise left in the time series.

Question 2.2   The basic idea behind a multiplicative decomposition model is that

(a) the seasonal effect is not constant.

(b) the seasonal effect must be multiplied by the time series to obtain white noise as a result.

(c) the seasonal effect is proportional to the long-term trend.

(d) it is easier to forecast the future with it.

Question 2.3   When we have an additive decomposition model, we adjust the trend value to incorporate the seasonal effect at each time point by

(a) subtracting the seasonal effect.

(b) adding the seasonal effect.

(c) multiplying the seasonal effect.

(d) adding the seasonal swing value at each time point.

Question 2.4   After doing multiplicative decomposition of the quarterly UKgas time series housed in Base R, we extract the trend component or centered four-point moving average, the seasonal effect and the random term, and the data. What was the value of the seasonal swing in quarter 3 of 1984?

(a) −486.05

(b) 0.5584441

(c) 619.75

(d) 32.18

Question 2.5   The seasonal effect is

(a) an average value of the seasonal at each season.

(b) the value of the seasonal swing at each time point.

(c) an estimate of the seasonal swing.

(d) only present if there is seasonality in the time series.

Question 2.6   A time series has the following as the first five quarterly observations: 3.3602, −3.1769, 0.3484, 7.469, 4.4963. What is the first value of $\widehat{T}$ obtained using a centered four-term moving average?

(a) 2.1422

(b) 3.0187

(c) 6.0262

(d) 5.494

Question 2.7   A regression smoother that uses dummy variables to account for the seasonal effect in a monthly time series will have

(a) 12 dummy variables among the explanatory variables.

(b) one dummy variable among the explanatory variables.

(c) 11 dummy variables among the explanatory variables.

(d) 13 dummy variables among the explanatory variables.

Question 2.8 Detrending a time series to which we have fitted a classical additive decomposition model based on moving averages like that produced by the R code decompose requires

(a) multiplying the data by the moving average.

(b) subtracting the seasonal effect from the time series.

(c) adding the seasonal effect to the time series.

(d) subtracting the value of the moving average from the data.

Question 2.9 What is the main difference between classical additive and classical multiplicative decompositions?

(a) In additive decomposition the trend, the seasonal effect and the random term are added, and in multiplicative decomposition they are multiplied, in both cases to recover the value of the time series.

(b) In additive decomposition the trend, the seasonal swing and the random term are added, and in multiplicative decomposition they are multiplied, in both cases to recover the actual value of the time series.

(c) In additive decomposition the trend and the seasonal effect are added, and in multiplicative decomposition they are multiplied, in both cases to recover the actual value of the time series.

(d) Additive decomposition is used when the seasonal swing increases over time, and multiplicative decomposition is used when the seasonal swing is constant over time.

Question 2.10 In which of the following approaches to modeling a time series by decomposing it into its components are we unable to forecast with the model? Select all that apply.

(a) In simple exponential smoothing

(b) In regression smoothers

(c) When using moving average smoothers

(d) In Prophet modeling

2.10 Case Study: LOWESS Smoothing with stl()

Decomposition of a time series into several components has gained a lot of momentum with the growing amount of complex data at our disposal, the increasing demand for automated forecasting of thousands of time series at a time, and the lack of enough personnel trained in time series that could do a less automated analysis. Most automated in-house methods of multiple time series prediction used by high-tech companies are based on decomposition of the time series into components already introduced in this chapter and others such as special holidays. Prophet, Facebook's forecasting method, is one of the modern forecasting methods using some form of decomposition. Prophet is the most modern approach discussed in this chapter, in Section 2.7. Modern methods have grown out of the necessity to be able to forecast time series data representing different frequencies (seasonalities, such as daily, weekly, annual), that could have missing values and could be not recorded at equally spaced intervals.

Decomposition methods research has evolved in order to address multiple seasonalities in very granular data, changing seasonality frequency over time, and incomplete data, which classical decomposition methods do not handle well. This case study uses *Locally Weighted Scatterplot Smoothing* (LOWESS) decomposition. In order to do that, we use `stl()`, which decomposes a time series into its components using LOESS, a different form of smoothing mentioned in Section 2.3.7.

2.10.1   `stl()`

The `stl()` function in R uses local regression to create a smooth trend for the seasonal component, which is then removed to obtain a smooth curve for the long-term trend. More information about this decomposition method can be found in a paper by Cleveland [27].

Although `stl()` is suitable for multiple and varying seasonalities, the data for this case study only has annual seasonality, so the analysis will look a lot, in appearance, like the classical decomposition of annual time series. The data for this case study also has a few missing values, and in order to work with the whole data, we must use a way of making the data an R's time series object that can handle that. It is for this reason that the R's package `zoo` is used [209]. This allows the imputation of values and also allows plotting the incomplete data and time indexes of many types of frequencies. Thus, the `zoo` class, although a wrapper for `ts()`, has wider applicability than `ts()`.

2.10.2   Climate Change and Drinking Water

The case study presented in this section is based on an article written by Wildman [207].

Environmental chemistry and water management are concerned with the harmful effect of chemicals present in water. Lake Mead is the main source of water for Southern Nevada, USA. A small fraction of the water in Lake Mead and therefore taken by Southern Nevada comes from highly diluted former wastewater. Wastewater-derived organic chemicals are carefully monitored by the Southern Nevada Water Authority (SNWA). These chemicals range from pharmaceuticals to personal care products to pesticides ingested in food. Too high a concentration of those endrocrine-disrupting chemicals in the water is considered dangerous.

Two competing research theories regarding the chemicals are:

- Climate change is responsible for the seasonal variation in water level and consequently for the seasonal increase in concentration of harmful treated wastewater chemicals in the water of Lake Mead when water level decreases.
- Specific conductance is the reason for seasonal treated wastewater chemical concentrations in Lake Mead.

According to Wildman [207], "the SNWA practice of discharging treated wastewater to its drinking water source creates the possibility that concentrations of waste-water derived organic chemicals in Lake Mead could increase over time despite considerable dilution because chemicals could be added to water more rapidly than they are degraded as water makes successive passes through the urban system." Thus, [207] concluded that concentration is not due to climate change necessarily. This was in stark contrast with the conclusion reached by other authors, who, according to [207], did not use the right methods to analyze the data.

To reach the conclusion, Wildman first considered the decomposition of four variables: (a) *meprobamate*, which measures concentration of the wastewater chemical of the same name, in nanograms per liter; (b) *elevation* of the water surface of Lake Mead, in meters above sea level; (c) *vol* or volume of Lake Mead expressed in cubic centimeters; (d) *SC* or specific conductance, expressed in microsiemens per centimeter. The data set(s) provided by [207] also include a *date*, a numeric variable. This case study is conducted in program *myprogram.R*. The names of the data sets are very large but they are written in the program. These data sets may be downloaded from the source where Wildman's paper is published.

The author used LOWESS decomposition (`stl()`) of the time series. This is seasonal and trend decomposition using LOWESS. Like [207] we will conduct decomposition of the data and observe the relation.

Wildman decomposes the volume and the meprobamate concentration to compare their long-term trends. Separately, the author decomposes the SC and the waste chemical concentration and compares their seasonal trends with that of volume. The author considers that taking into account long-term trends is only part of the story.

We show in Figure 2.14 the three time series considered in this case study: volume, meprobamate and SC. Notice the break in the meprobamate concentration data. Those are missing values, and the fact that we made the data set a `zoo` data set allows us to visualize where the missing values are in the time series.

Figures 2.15, 2.16 and 2.17 show the LOWESS decomposition of the three time series, respectively. As we can see, meprobamate concentration has a slight nonlinear trend upwards, volume (vol) a nonlinear trend downwards, and specific conductance (SC) also has a nonlinear trend upwards and then downwards.

2.10.3   Comparing the Long-Term Trends

After LOWESS decomposition we compare the long-term trend components of volume and the waste chemical concentration first. We can see in Figure 2.16 the LOWESS decomposition of meprobamate, obtained with `stl` and using default imputation of the missing values by this function [126]. Notice the difference between Figure 2.14 and Figure 2.16. The former has gaps in the top line plot due to the missing values, whereas the latter doesn't due to the software having imputed those values. As we can see, there is a strong seasonal component and a nonlinear trend in meprobamate. On the other hand, we can see the decomposition of Lake Mead water volume

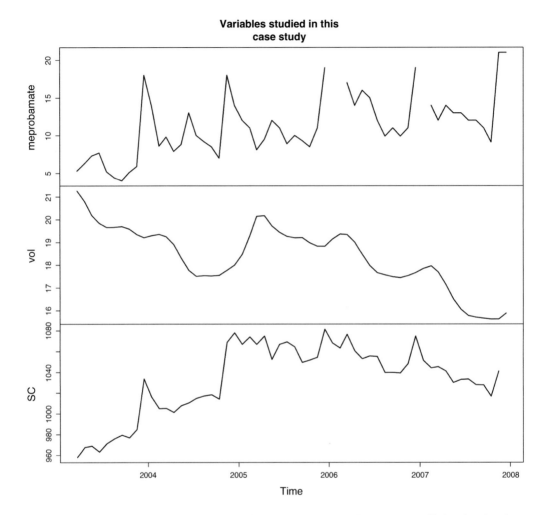

Figure 2.14 Variables studied in Wildman's paper [207] to discuss two conflicting theories about the increase in concentration in Lake Mead.

in Figure 2.15. Relative to the trend values, the seasonality in volume is much less pronounced than the seasonality in concentration.

Figure 2.17 shows the decomposition of the SC variable. There is a strong seasonal component in this variable.

The long-term trends of meprobamate and SC go in opposite directions past 2005. The seasonal effect for meprobamate concentration lead (occur before) those of conductance. The seasonal swings in conductance occur when the seasonal swings of concentration start going down.

The reader will read the article by Wildman [207] to see how the author interprets these findings and incorporates them in a model. This paper presents an example

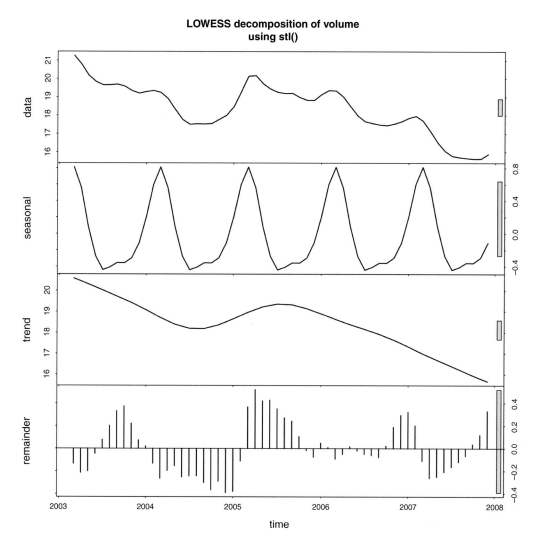

Figure 2.15 A variable studied in Wildman's paper [207]: volume of water in Lake Mead. LOWESS decomposition.The remainder is the equivalent of the random term in classical decomposition. `stl()` plots it as a `type="h"` plot instead of `type=l"`, the latter being used by classical decomposition.

of how the components of time series can be incorporated into a model to try to understand what is going on in a hydrology scenario.

Here are some observations: more chemicals, more conductance (ability to transfer electricity). Even when the chemicals are not being pushed into water, conductance is increasing, because the water is not being pushed.

The program to obtain the plots displayed in this section can be found in program *lakemead.R*.

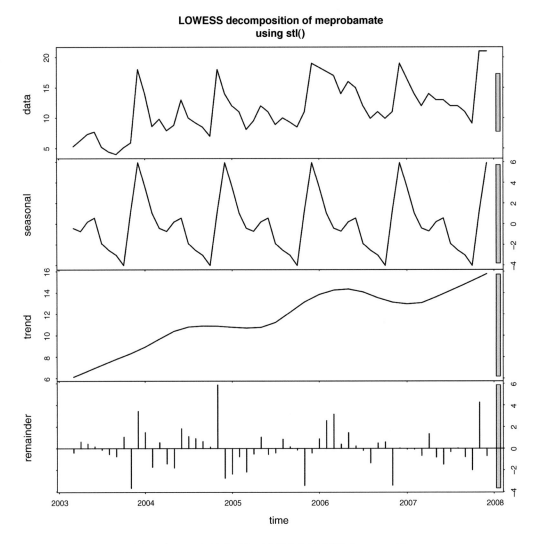

Figure 2.16 Meprobamate in Lake Mead [207]. LOWESS decomposition.

2.10.4   Exercises

Exercise 2.27   Do a LOWESS decomposition of the elevation time series and compare its long-term trend and seasonality with those of the concentration, SC and volume.

Exercise 2.28   Extract the long-term trends from the LOWESS decompositions of all four variables and plot them in one single graph, one plot below the other. Compare the trends. Do the relations observed make sense in the context in which these time series are being studied?

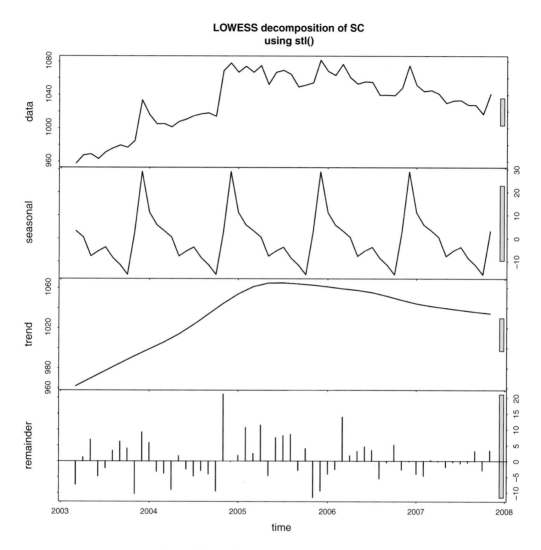

**LOWESS decomposition of SC
using stl()**

Figure 2.17 The SC variable in Wildman's paper [207] has a strong seasonal component, like the meprobamate does.

Exercise 2.29 Extract the seasonal short-term trends from the LOWESS decompositions of all four variables and plot them in one single graph, one plot below the other. Compare them. Do the relations observed make sense in the context in which these time series are being studied? Do they support the conclusions in Wildman's paper? [207]

3 Summary Statistics of Stationary Time Series

3.1 Introduction

The models that we explored in Chapter 2 are curve-fitting models that rely on the components obtained from the decomposition of the time series without explicitly accounting for the temporal dependence structure in the data. Classical time series models rely on that temporal dependence, and that makes it necessary to learn how to measure it. So after having discovered the internal structure of time series data through decomposition in Chapter 2, we will focus in this chapter on the description and summary statistics of the *stationary* random term of a time series in order to learn about the stochastic process or model generating it.[1] We introduce some of the most important descriptive summary statistics of a stationary time series: the sample mean, $\bar{X}$, the sample variance-covariance matrix S^2, which gives the variance of each variable and the auto-covariances of the time series with itself, and the sample auto-correlation matrix.

Autocorrelation analysis of a stationary time series is what helps us identify classical time series models. It is important that the reader remembers what the sample mean, sample variance and standard deviation and correlation between two random variables are. Example 3.12 in the appendix to this chapter will help the reader remember some of those concepts.

> The summary statistics studied in this chapter are helpful to understand the model generating the observed time series only if the time series is stationary. After studying the theory of time series in Chapter 5, we will know more precisely what the

---

[1] In Mathematical Statistics or Introductory Statistics classes, a summary statistic is defined as a function of one or more random variables that does not depend upon any *unknown* parameter. The summary statistic is also called an estimator. Observed values of that function are called *estimates of a parameter* or observed summary statistics. Until it is observed, the summary statistic is a random variable and as such subject to a probability law that we call the *sampling distribution* of the summary statistic. This distribution tells us what values the summary statistic could take under repeated sampling just by chance, if the model assumptions made hold. With the sampling distribution, we conduct classical confidence interval estimation and hypothesis testing. Using that sampling distribution, classical Mathematical Statistics studies the properties of the estimators, such as unbiasedness, efficiency, consistency and asymptotic counterparts. Many classical time series books [13, 12, 60, 179, 20] present the small sample and large sample properties of the estimators seen in this chapter. The reader may consult those sources for more detail.

mathematical expression of "stationary" is. The intuitive meaning was briefly presented in Chapter 2 as a random term that fluctuates around a constant without indication of patterns in the variance or the mean.

Section 3.2 defines the mean and variance of an observed time series. Section 3.3 defines the Sample Autocorrelation Function (ACF), shows how to calculate it, defines the standard error of the autocorrelations, provides interpretation of the tests for autocorrelation at any lag k, introduces the t-test statistic for the significance of individual autocorrelations and discusses the correspondence between lag plots and ACF plots. Base R's acf() and lag() functions are also discussed to make the reader proficient in their use and the interpretation of their output. Section 3.4 describes the Ljung–Box test (or white noise test), a test of several autocorrelations at once, and how to interpret its results. It helps confirm the information obtained in a sample ACF if it shows no significant autocorrelations. Section 3.5 introduces the sample Partial Autocorrelation Function (PACF), also known as a partial correlogram. We learn in that section to calculate the partial correlogram, the standard error, how to conduct tests of hypotheses for partial autocorrelation and how to interpret the partial autocorrelations. Although this book does not focus on the study of time series in the frequency domain, it is interesting to see the spectrum of a time series, how it decomposes the time series into different frequencies (components), and areas of inquiry where the frequency domain is the most important approach to the study of time series. Section 3.6 introduces the spectrum and the sample periodogram to that end, emphasizing the difference between a spectrum that shows a signal in the data and the spectrum of white noise. After that section, Section 3.7 brings to the reader's attention the typical ACF and spectrum of a nonstationary time series, to emphasize why we need to remove the nonstationarity if we want to discover hidden signals in the time series. Section 7.1 warns the reader that there are other, more sophisticated aspects of time series analysis that would require more than the summary statistics studied in this chapter. The chapter continues in Section 3.8 with summaries of the features of a time series studied in this chapter in order to add them to the "features bag" that we are constructing in our ML "features" journey.

Figure 3.1 shows that the ACF and PACF are different for each of the stochastic processes shown. The PACF is studied in Section 3.5.

3.2 The Mean and Variance of a Stationary Time Series

First, we introduce the mean of a stationary time series. The expected value of a stationary time series, μ, is estimated with the *sample mean* of the stationary time series. If y_t denotes the observed stationary time series, for example the stationary random term of a decomposition, then the sample mean is denoted by

$$\bar{y} = \frac{\sum_{t=1}^{n} y_t}{n},$$

where n is the length of the time series. In a stationary stochastic process, the expected value is assumed constant over time. Thus it is expected that the observed time series, if split in several intervals, would have similar sample means across all intervals.

The variance of the stochastic process generating the stationary time series, σ^2, is estimated by the *sample variance*,

$$s^2 = \frac{\sum_{t=1}^{n}(y_t - \bar{y})^2}{n-1}.$$

> **Constant Variance Assumption**
>
> The variance of a time series must be constant before we study its ACF. If an observed time series has changing variability, we have to either transform the original time series to make it have constant variance (via log or square root or other transformation) before removing the trend and the seasonal components, or decompose the time series using multiplicative decomposition and study the ACF of the random term resulting from it.

The reader will notice that we have used the same standard notation for sample mean and sample variance as introductory statistics textbooks. In time series analysis, the sample variance is also denoted by $\hat{\gamma}_0$, denoting the covariance of the variable with itself, and is calculated as follows:

$$\hat{\gamma}_0 = s^2 * \frac{n-1}{n}.$$

More on this subject will be discussed in Chapter 5.

3.3   The Correlogram or Sample Autocorrelation Function (ACF)

The ACF of a time series together with the Partial Autocorrelation Function (PACF) help us discriminate among different stochastic processes, and hence help us select a time series model for the sampled time series.

In Figure 3.1 we see in the first row a window of each of three different stochastic processes, starting at time $t = 2000$ and ending at $t = 2100$. The stochastic processes from which we generate the time series are Autoregressive of order 1, AR(1), ($y_t = 0.8y_{t-1} + w_t$); Moving Average of order 1, MA(1), ($x_t = -0.8w_{t-1} + w_t$); and white noise ($z_t = w_t$), respectively. The second row displays the corresponding ACF of each of the time series and the third row displays the PACF. The autocorrelation that we see at value 1 on the vertical axis must be ignored. It appears in all the ACF and it does not help in modeling.

We simulated only 100 values from each of these processes, hence it is expected that we are seeing an approximation, not a perfect image, of the true ACF and PACF of the models. That it is just an approximation is noticeable, because we have a lot of little vertical lines within the parallel horizontal bands, distracting us from the important information, which is at around the first few lags of the PACF ($k = 1, 2, 3$) for an AR(p) or the first few lags of the ACF for an MA(q). The distraction is an artifact of finite sampling. In Chapter 5, we will derive the pure mathematical functions for some of these stochastic processes and will appreciate seeing the autocorrelation of the true stochastic process without that distraction. Without getting distracted by those numerous little lines, we can see that even though the series might appear to the naked eye as being very similar (as happens with the MA and white noise ones), the ACF and the PACF are very different for each. The ACF and PACF of white noise are indicating to us that there is nothing to model, no signal is available in this data set, because all the little vertical lines (except the one at $k = 0$, which should be ignored) are within the horizontal bands. All autocorrelations are zero. On the contrary, the ACF and PACF of the MA process are indicating that there is a signal, and are suggesting an MA model for it. Again, the names of these models, MA and AR, will make more sense after studying Chapter 5. Figure 3.1 is shown to convince the reader that studying the ACF and PACF is useful to identify a generating model.

Getting used to associating the ACF and PACF plots with the right stochastic process model takes a lot of practice and is hard for the beginner. But it is very important to practice it to avoid doing time series analysis in a vacuum.

It is the correlation structure of what remains after we have accounted for trends, seasonality and other regularly occurring phenomena that gives rise to classical time series theory.

If there are no autocorrelations left after accounting for trends, seasonality and other recurring phenomena (that is, if the random term is white noise), then only the trend, seasonality and indicators of those phenomena should be components of a model. But if there are autocorrelations left, those must be added to the model. Distinguishing between the ACF of white noise and the ACF of a stationary time series that is not white noise, as we did in Figure 3.1, is very important in time series analysis.

As in the study of trends, seasonality and cycles, the temporal order of the observations must be maintained to study autocorrelation.

We obtained Figure 3.1 using program *ch3Section1sim.R*, and the reader should run that program many times to realize that the generated data from each of the three models is different each time.

In this section, we investigate the ACF. By doing so we are estimating, albeit indirectly, the parameters of the stochastic process that we believe is generating the time series, because the autocorrelations are related to the model parameters, as we will show in Chapter 5.

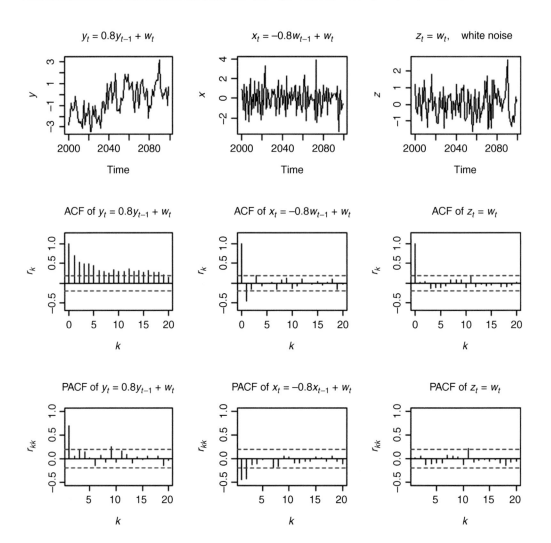

Figure 3.1 Simulated examples of stochastic processes generating time series data and their ACF and PACF.

3.3.1   Calculating Autocorrelations

The sample autocorrelation function or correlogram of a stationary time series is calculated as follows:

$$r_k = \frac{\sum_{t=k+1}^{n}[(y_t - \bar{y})(y_{t-k} - \bar{y})]}{\sum_{t=1}^{n}(y_t - \bar{y})^2}, \qquad k = 0, 1, 2, \ldots, m, \tag{3.1}$$

where m is denoted as lag. Alternatively, this may be written as

$$r_k = \frac{\sum_{t=1}^{n-k}[(y_t - \bar{y})(y_{t+k} - \bar{y})]}{\sum_{t=1}^{n}(y_t - \bar{y})^2}, \qquad k = 0, 1, 2, \ldots, m.$$

The correlogram at lag k measures the linear relationship between y_t and y_{t-k}, separated by a lag of k time units. The r_k is always a number between -1 and 1. It is an estimator for the autocorrelation parameters ρ_k of the unknown model generating the stationary time series. In general, there is no point in computing r_k for $k > n/4$, where n is the length of the time series.

- $r_0 = 1$, because it is the autocorrelation of the time series with itself.
- r_1 is the sample autocorrelation between the observed y_t and the observed y_{t-1}.
- r_k is the autcorrelation between observed y_t and observed y_{t-k}.

Autocorrelation

If $Y_1, Y_2, ..., Y_n$ are random variables representing time series Y_t at each time t, then the correlation between any two Y_t, Y_{t-k} is called autocorrelation at lag k. It makes sense to study the autocorrelation if the time series has constant mean and constant variance. We just call it autocorrelation because both Y_t and Y_{t-k} refer to (in our minds) the same variable, although in time series theory they are two different random variables.

The r_k is an estimator, and as such it is believed to be different for each random time series that the generating process can create. This means that it is random and has a sampling distribution. We see now how to calculate the r_k in Example 3.1.

Example 3.1   Given nine observations of a time series y_t, shown in Table 3.1, we will calculate the autocorrelation coefficients from lags 0 to 3.

The secret to calculating the sample ACF when we have just one observation of each of the variables $Y_1, Y_2, ..., Y_t$ is to rewrite the time series as a rectangular data set with m variables and a varying number of observations for each variable. We will do that now with the time series in this example. First, we create a variable called Y_{t-1}, another called Y_{t-2} and another called Y_{t-3}. The observed values of these variables can be seen in Table 3.1.

In Table 3.1, it is clear that at time $t = 3$, the corresponding observed value of y_{t-1} is 10.5, which is the value of y_t at time $t = 2$. It is this artifact of generating data this way that allows us to use the concept of correlation between two variables learned in introductory statistics to calculate the autocorrelation between the terms of a single time series. The difference with the introductory statistics class is that the variables are generated from one single time series.

To calculate the autocorrelations based on the generated data in Table 3.1, we first calculate the sample mean of the observed time series, $y = \frac{\sum_{t=1}^{9} y_t}{9} = 10.944$, calculated with numbers from column y_t. We then calculate the denominator of the formula for the r_k: $\sum_{t=1}^{9}(y_t - \bar{y})^2 = 5.362$. Notice that all the observations are used for the sample mean and sample variance. Notice also that r_0 is always 1.

Table 3.1 The time series at each time t and new time series representing the same series but at $t-1$; $t-2$; $t-3$. With this artifact, we make the time series look like a rectangular data set similar to those that we are used to studying in Statistics: n rows representing observations, and several columns of variables, which we correlate pairwise to obtain the autocorrelations.

t	y_t	y_{t-1}	y_{t-2}	y_{t-3}	$y_t - \bar{y}$	$y_{t-1} - \bar{y}$
1	10				−0.945	
2	10.5	10			−0.445	−0.945
3	11	10.5	10		0.056	−0.445
4	11.5	11	10.5	10	0.56	0.056
5	12	11.5	11	10.5	1.056	0.56
6	12.1	12	11.5	11	1.156	1.056
7	11.2	12.1	12	11.5	0.256	1.156
8	10	11.2	12.1	12	−0.945	0.256
9	10.2	10	11.2	12.1	−0.745	−0.945

Using Table 3.1, we calculate r_1, the correlation between y_t and y_{t-1}, using the column for y_t from row 2 to 9 and the column for y_{t-1} from row 2 to 9, as follows:

$$r_1 = \frac{\sum_{t=2}^{9}(y_t - \bar{y})(y_{t-1} - \bar{y})}{\sum_{i=1}^{n}(y_t - \bar{y})^2} = \frac{2.989136}{5.36222} = 0.5574435.$$

Notice that we display the elements of the numerator in the last two columns of Table 3.1. This tells us that values at any time t are positively correlated with values at time $t - 1$. There is inertia in the series. Higher values yesterday, for example, are accompanied by higher values today.

We could do this manually in R as follows:

```
y.t=c(10,10.5,11,11.5,12,12.1,11.2,10,10.2)
mean=mean(y.t)
a=sum((y.t[2:9]-mean)*(y.t[1:8]-mean))
b=sum((y.t-mean)^2)
r.1= a/b
```

Or, we could just use R's acf() function as follows:

```
acf(y.t) # to calculate the autocorrelations at many lags
acf(y.t$acf[2]) # to extract the autocorrelation at lag 1
```

Notice that the value we get is not the same value that we would get using the correlation between the two series $y_t[2 : 9]$, $y_t[1 : 8]$ using the *cor*() function in R, because when you use the *cor*() function, the means used are the ones for the shorter series. But the sample correlogram follows the formula given and uses the mean of the whole series, from $t = 1$ to $t = 9$.

We calculate the other autocorrelations using R.

$$r_2 = \frac{\sum_{t=3}^{9}(y_t - \bar{y})(y_{t-2} - \bar{y})}{\sum_{t=1}^{n}(y_t - \bar{y})^2} = -0.113$$

is the autocorrelation between $y_t[3 : 9]$ (that is, y_t from rows 3 to 9) and y_{t-2} from rows 3 to 9 in Table 3.1, or equivalently, $y_t[3 : 9]$ and $y_t[1 : 7]$.

$$r_3 = \frac{\sum_{t=4}^{9}(y_t - \bar{y})(y_{t-3} - \bar{y})}{\sum_{t=1}^{n}(y_t - \bar{y})^2} = -0.493$$

is the autocorrelation between the column for $y_t[4 : 9]$ and the column for $y_{t-3}[4 : 9]$ in Table 3.1.

A more direct way to get all the autocorrelations is to use the $acf()$ function in R,

```
acf(y.t)$acf
```

which gives

```
          [,1]
[1,]    1.00000000
[2,]    0.55744348
[3,]   -0.11387392
[4,]   -0.49323111
[5,]   -0.52343786
[6,]   -0.31162223
[7,]    0.02555602
[8,]    0.22804715
[9,]    0.13111848
```

Notice that our manual calculations resulted in the same numbers up to $t - 3$, which in R's output is represented by the fourth line. Line $[1,]$ in the output corresponds to $t = 0$, the autocorrelation of y_t with itself, which is always 1.[2] $r_1 = 0.55744348, r_2 = -0.11387392, r_3 = -0.49323111$.

We will use directly the $acf()$ function of R in the next example.   □

3.3.2   Exercises

Exercise 3.1   Hand calculate the r_4 corresponding to the time series y_t in Table 3.1. Create a y_{t-4} column.

Exercise 3.2   Complete Table 3.2 and hand calculate the sample autocorrelation coefficients r_1, r_2 showing your work. Compare with the ones obtained using the $acf()$ function in R.

---

[2]Some authors of packages use $acf()$ in the background but then instruct their program to remove the r_0. We prefer to keep it as Base R gives it because the $r_0 = 1$ acts as a point of reference for all the other sample autocorrelations.

Table 3.2 Complete the table

Year	t	x_t	x_{t-1}	x_{t-2}
1970	1	0.94		
1971	2	−0.43		
1972	3	0.62		
1973	4	−1.06		
1974	5	0.8		
1975	6	2.17		

Table 3.3
Annual sales (in
millions),
1961–1968

t	y_t
1961	76
1962	70
1963	66
1964	60
1965	70
1966	72
1967	76
1968	80

Exercise 3.3 Table 3.3 contains a short time series of annual sales (y_t) in millions of dollars. Hand calculate the mean, variance and the autocorrelations at lags $k = 0, 1, 2, 3$. Repeat, but using the acf() function of the ts() object.

3.3.3 The Standard Error of r_k

The reader must have noticed that the ACFs plotted in Figure 3.1 have horizontal bands around the zero line. Those bars refer to two standard errors to each side of 0. If a spike in the ACF is past those bands, and that spike occurs at lower lags and/or at the seasonal lags we say that the autocorrelation at that lag is statistically significant.

The r_k are not the ρ_k, which is the autocorrelation parameter at lag k. Rather, they are just estimates of ρ_k. Therefore, the r_k are subject to sampling error. The standard error of an r_k is defined as [8]

$$s_{r_k} = \frac{\left(1 + 2\sum_{j=1}^{k-1} r_j^2\right)^{\frac{1}{2}}}{(n)^{\frac{1}{2}}}. \tag{3.2}$$

We would use this standard error to compute confidence intervals for ρ_k and to test hypotheses about ρ_k.

Some books, for example [30], give the asymptotic standard error formula

$$s_{r_k} = \frac{1}{\sqrt{n}},$$

which is constant for all lags, unlike 3.2. R uses this last formula. That is why the ACFs seen in Figure 3.1 display two parallel bars around 0. Those bars represent two standard errors below and two standard errors above 0.

The sampling distribution of an r_k is the distribution of all possible values that r_k could take if different time series were drawn from the stochastic process generating y_t, that is, from the ensemble. The sampling distribution of the r_k, under the assumption that the autocorrelation is 0 at all lags k, is centered at $r_k = 0$ approximately. The standard error is that given in the s_{r_k} formula.

3.3.4 Testing Hypotheses about ρ_k

When a value of r_k is statistically significant, what does that mean? The following is the standard test to do to answer that:

$$H_o : \rho_k = 0 \quad \texttt{Autocorrelation at lag k is 0}$$
$$H_a : \rho_k \neq 0 \quad \texttt{Autocorrelation at lag k is not 0.}$$

Under the null hypothesis, r_k has a sampling distribution that is normal with mean $-\frac{1}{n} \approx 0$ and standard error $1/\sqrt{n}$. The test statistic used to do the test is

$$t = \frac{r_k - 0}{se_{r_k}},$$

which follows a t distribution with $n - 1$ degrees of freedom. Any r_k that is beyond two standard errors from the null of zero then is statistically significant, meaning that the result is real and not just due to chance, and that therefore we may reject the null hypothesis that $\rho_k = 0$. In the plot of R's $\texttt{acf()}$ this means that the vertical lines go beyond the dotted parallel horizontal bars around 0.

r_k within the two standard errors bars means that r_k is not statistically significant, and we do not reject the null hypothesis that $\rho_k = 0$. Another way to say this is that if the $p - value < 0.05$, we reject the null hypothesis and say that there is statistically significant autocorrelation at lag k.

The reader should be aware that, as in any hypothesis testing, we expect about 5 percent of the r'_k to be statistically significant even though they are not. In a correlogram, those will be random spikes beyond the two standard error bars, scattered randomly throughout the ACF plot, not meaningfully.

Notice also that we do one test for each ρ_k, resulting in multiple tests.

> It does not make sense to interpret anything in the correlogram if the time series is not variance stationary and it has not had the regular trend and regular seasonal components removed. Mean and variance must be approximately the same across the time series for the ACF and PACF to make sense. We usually focus on the lower lags of the autocorrelation function and on the lags at or around the seasonal ones (the latter if there is seasonality in the time series).

3.3.5   Understanding Base R's Correlogram Plots

Now that we know what r_k measures and what it helps us test, we are in a better position to understand the sample ACF or correlogram plot that Base R produces. Base R's correlogram plot displays r_k against nonnegative values of k. The following summarizes what each part of the plot displays:

- Horizontal axis: the positive lag k.
- Vertical axis: the r_k, $k = 0, 1, 2, 3....$
- All correlograms will show $r_0 = 1$, which should be ignored. The r_0 does not add any information, but since 1 is the maximum that an autocorrelation can be, it helps for comparison with the other r_k, $k > 0$, which will all be less than 1.[3]
- Symmetric parallel bars around 0: two standard errors from 0.
- For lag $k > 0$, a line that falls outside the parallel bars means a significant autocorrelation. The bars are drawn at $-\frac{1}{n} \pm \frac{1}{\sqrt{n}}$.
- About 5 percent of the r'_k will be significant and must be ignored. These will usually be scattered all over the ACF and will not be at the relevant r_k at the lower lags and the seasonal lags.

Example 3.2   We plot in Figure 3.2 the correlogram plot corresponding to the y_t in Table 3.1. The reader will see some vertical lines. Those are the r_k. The plot shows r_k, $k = 0, 1, \ldots, 8$. The dotted lines in the correlogram represent two standard errors of the estimate. If the vertical lines (other than the one at $k = 0$) are above the dotted horizontal lines, then we conclude that the autocorrelation is, statistically speaking, significantly different from 0. If the vertical lines (other than the one at $k = 0$) are within the dotted lines, then we conclude that the r_k is not significantly different from 0. None of the r_k in Figure 3.2 are statistically significant (recall that the one at $k = 0$ must be ignored), hence the time series is white noise. The reader may reproduce Figure 3.2 using program *ch3acfsimpleexample.R*.   □

Stationary time series will usually have autocorrelation at small k's, if there is no seasonality left in them. But they could also have persistent autocorrelations for quite a few lower lags if the model generating them is AR(p). Attention for now should be placed in the lower k, recognizing that if the process is AR, the ACF could have a lot of lower lag r_k that are significant. The 5 percent that appear by chance are usually not at the lower k.

Example 3.3   Contrast the correlogram of the time series in Table 3.1 with the correlogram of the random term of the Johnson & Johnson quarterly earnings per share data (1960:1–1980:4) that comes with R, after a multiplicative decomposition. We looked at this data set already in Chapter 1, Example 1.11. The reader will find the correlogram in Figure 3.3. As we can see, at 1, 3, 4, 8, the r_k are statistically significant. The sample ACF was produced with the code in program *ch3acfsimpleexample.R*, in the lower part of the program.

---

[3] Some R packages do the plot of the correlogram without the r_0.

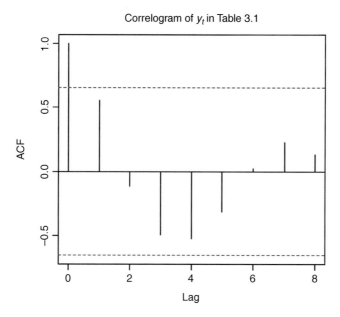

Figure 3.2 The correlogram of the time series y_t in Table 3.1 displays all the autocorrelations in the time series, up to a lag k that defaults to what software decides or that the reader can set, usually at least 25 percent of the length of the series. The autocorrelation at lag $k = 0$ is always 1, and therefore it is not saying anything about autocorrelation in the series. Some R packages remove that. The ACF displayed here is the ACF of white noise. None of the r_k are statistically significant.

The horizontal axis in Figure 3.3 requires explanation. We notice that there is a number 1 in $k = 4$, and a number 2 in $k = 8$. This has to do with the way R processes time series objects. The JohnsonJohnson data set is a quarterly ts() that starts in quarter 1 of 1960. R knows that the full cycle is a year and there are four observations in a year. The 1 represents year 1 and is implicitly lag $k = 4$. The 2 represents year 2 and $k = 8$. We interpret the significant autocorrelation where the 1 is on the horizontal axis as saying that there is some stationary seasonality present in the data because the value of the time series in a given quarter is correlated with the value of that quarter in the following year. But this seasonality is random, not the regular short-term seasonal trend. It could also mean that the decomposition was not successful at removing the regular short-term seasonal trend. □

3.3.6   Autocorrelation Viewed with Scatterplots

The reader is perhaps familiar with scatterplots and correlation and the correlation coefficient, at the level of an introductory statistics course, to visualize and compute the intensity and direction of the linear relation between two variables. Scatterplots and correlation coefficients corresponding to them can be used to obtain an alternative interpretation of the correlogram. For example, consider the random component of the

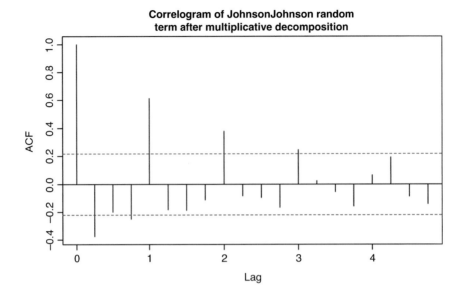

Figure 3.3 A correlogram of the random component of the `JohnsonJohnson` time series after multiplicative decomposition. Because the time series is a `ts()` object of frequency $= 4$, R knows that the full cycle is a year and there are four quarters in it. Thus, $k = 1$ is marking the first year. Before $k = 1$, we see r_0, r_1, r_2, r_3. At $k = 4$ we have r_4.

multiplicative decomposition of the `JohnsonJohnson` time series. A convenient way to visualize many autocorrelations with scatterplots is to use the `lag.plot()` function in R.

Figure 3.4 shows the scatterplots obtained with the `lag.plot()` function. With the naked eye, it is obvious that there is a small negative autocorrelation between y_t and y_{t-1}, between y_t and y_{t-2} and y_t and y_{t-3}, but we can see that there is high autocorrelation between y_t and y_{t-4}. Since we are dealing with a quarterly time series with some seasonality, we are not surprised. Nonregular seasonality will reflect itself in the random component after we have removed the regular seasonality and the overall trend. We calculate the autocorrelations with the `acf()` function. They are $r_1 = -0.374$, $r_2 = -0.2017$, $r_3 = -0.2504$, $r_4 = 0.6140$. The code in program *ch3acfsimpleexample.R* does the plots and the calculations.

3.3.7   Exercises

Exercise 3.4   Do a multiplicative decomposition of the `AirPassengers` data set that comes with R and obtain a lag plot of the random term of the decomposition of up to 12 lags. Comment on the corresponding scatterplots and then obtain the values of the autocorrelation coefficients corresponding to each lag plot. Compare your interpretation of the scatterplots with the autocorrelation coefficients, calculated with the function `acf()`.

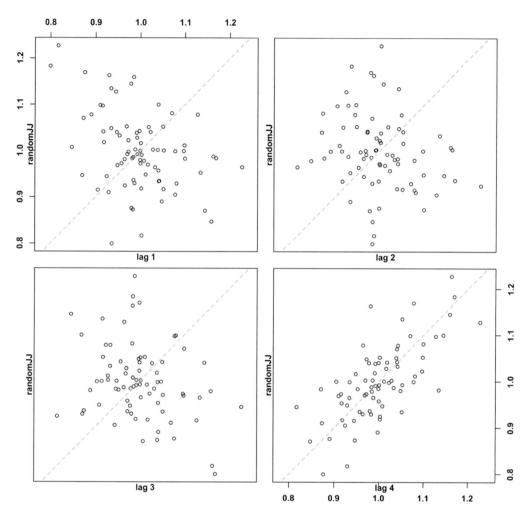

Figure 3.4 From left to right, scatterplot of y_t (randomjj) and y_{t-1} lag(1), y_t and y_{t-2}, and y_t, y_{t-3}, and y_t and y_{t-4}. These are scatterplots to display the relation between two variables, like those in introductory statistics courses, and should be interpreted the same way. The lag 1, lag 2 and lag 3 plots are not showing much correlation between the pairs of time series. But the lag 4 plot is indicating a strong positive linear correlation between y_t and y_{t-4}, which we saw in the ACF.

Exercise 3.5 Figure 3.5 shows the time plot and the ACF of two time series. For each of the time series, sketch the lag plot up to lag $k = 14$. Discuss the difference between the two lag plots and how they reflect the differences between the ACF.

Exercise 3.6 Figure 3.6 contains the lag plots of time series y_t, of length $n = 98$. Guess what the values of the autocorrelation coefficients are. Sketch the correlogram up to lag $k = 14$, including the two-standard-error band. Determine which autocorrelations are statistically significant by just looking at the scatterplots. Explain.

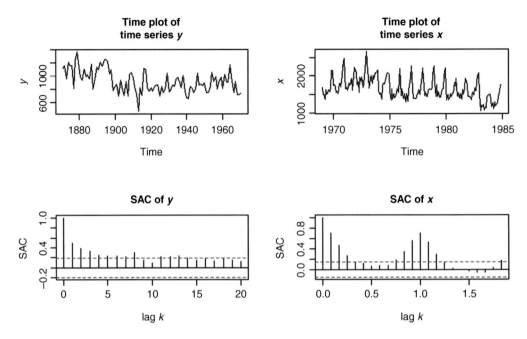

Figure 3.5 Time plots and ACF of observed time series y_t and x_t to be used for Exercise 3.5.

3.4   The Ljung–Box Test for White Noise

The tests that we do with the sample ACF consist of individual tests for each lag k. This is the same idea as in multiple regression, where we test individual regression coefficients with the individual t-tests, one test for each coefficient at a time. When doing multiple regression, we have a test for overall fit, the F-test. Time series analysis also has a test of that nature, which tests whether all autocorrelations at once, up to lag k, are significantly different from zero. This test is known as the Ljung–Box test or white noise test, and it is not F distributed, but chi-square distributed. The null and alternative hypotheses for this test are:

$$H_0: \rho_1 = \rho_2 = \cdots = \rho_k = 0, \quad \text{white noise hypothesis}$$
$$H_a: \text{at least one } \rho_k \neq 0.$$

A test statistic used to conduct this test is

$$Q = N(N+2) \sum_{i=1}^{k} (N-k)^{-1} r_k^2,$$

which is chi-square distributed with k degrees of freedom. This is called the Ljung–Box test statistic. The p-value is the probability that Q is higher than the observed Q if the null is true. If we reject the null (p-value < 0.05 or less), we conclude that there is significant autocorrelation at some lag between 1 and k. If we do not reject the null, then we conclude that the time series represents a white noise stochastic process (a set of independent random variables).

Lag plots

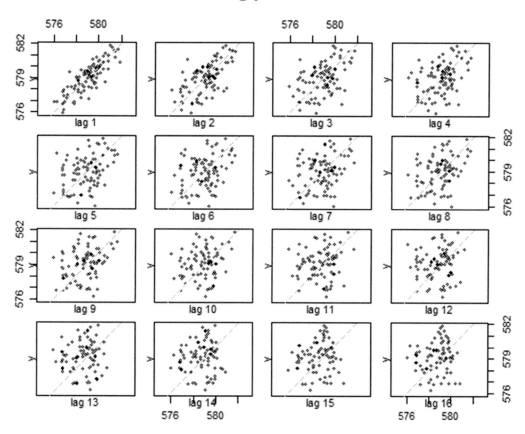

Figure 3.6 Lag plot of time series y_t up to lag $k = 12$. To be used for Exercise 3.6.

It is customary to do the test up to several values of k. For example, first do it for $k = 6$, then for $k = 12$, then for $k = 18$, and so on. If we find statistical significance, then we look at the sample ACF to determine which individual r_k are statistically significant.

Example 3.4   A Ljung–Box test of the random term of R's AirPassengers data set after multiplicative decomposition, up to lag $k = 20$, gave the following results:

$$Q = 131.5329, df = 20, \text{p-value} \approx 0.$$

So we reject the null hypothesis that the random term, after multiplicative decomposition, of the AirPassengers data set is white noise. Therefore, the random term of the AirPassengers data set obtained with classical multiplicative decomposition is autocorrelated and we can proceed to identify a classical time series model for it if it is stationary.   ☐

In R, the Ljung–Box test is calculated as follows:

```
Box.test(timeseriesname,lag=k,type="Ljung")
```

Example 3.5   The time series JohnsonJohnson comes with R. We applied the Ljung–Box test to the random term of the time series obtained with classical decomposition using the following code:

```
y= decompose(JohnsonJohnson,type="mult")$random
Box.test(y,lag=10,type="Ljung")
```

We obtained the following output:

```
Box-Ljung test

data:  y
X-squared = 74.605, df = 10, p-value =
5.68e-12
```

In this output, X-squared is Q. Based on the results, we reject the null hypothesis that the data is white noise.

Of course, rejecting the null hypothesis is only telling us that there are autocorrelations in the series. We need to investigate the sample ACF to determine those autocorrelations the same way as when we do regression: an F test that tells us that there is significant relation between the independent and dependent variables and requires that we go back to the regression to figure out which of the independent variables are significant.

As we mentioned earlier, the important autocorrelations are the ones closer to $k = 0$ and the ones around or at the seasonal components, unless the time series is AR(p), which will sometimes show several autocorrelations significant for quite a few lags.   □

3.4.1   Exercises

Exercise 3.7   A Ljung–Box test gives a p-value larger than 0.05 for all lags k of the autocorrelation function. What does that mean?

Exercise 3.8   A time series y_t has the following autocorrelations given in the format provided by the statistical software SAS when asked to do a Ljung–Box test.

(a) Write the null and alternative hypotheses to determine whether the autocorrelations up to lag 12 are statistically significant according to the Ljung–Box test and determine the the p-value of the test missing in the table. State the conclusion of the test up to lag $k = 12$.
(b) What conclusion can you reach, after looking at all the autocorrelations, regarding the process generating this time series? Say at least two major things that this table implies regarding the time series.
(c) Sketch the ACF of this process, including the two standard error bands.

To lag	Q	DF	P-value	Autocorrelations r_k $k = 1$ on left					
6	4.17	6	0.654	−0.04	−0.037	−0.041	−0.026	0.121	−0.015
12	6.13	12		−0.036	−0.037	0.013	0.06	−0.011	0.052
18	9.06	18	0.958	0.023	−0.079	0.072	−0.017	−0.008	−0.033
24	12.57	24	0.972	0.021	0.035	−0.050	−0.069	0.023	0.078
30	22.78	30	0.824	−0.031	−0.059	−0.067	0.096	0.106	−0.118
36	28.18	36	0.820	−0.04	−0.04	0.104	0.027	−0.076	−0.042
42	33.3	42	0.829	−0.026	−0.003	0.1162	−0.062	0.049	0.007
48	39.06	48	0.817	−0.014	0.075	−0.061	−0.108	0.009	0.03

Exercise 3.9   Using a combination of the acf(), extracting the r_k, and the Ljung–Box test command in R, obtain the information needed to construct a table like that of Exercise 3.8, but for the JohnsonJohnson time series. Repeat Exercise 3.8 but for this time series.

Exercise 3.10   When we run the Ljung–Box test for the random component of the *rooms* time series (used in Chapter 2) up to lag = 12, what is the conclusion of the test? Investigate the autocorrelations if the test says that the time series is not white noise. Use the acf() for that investigation. Compare the Ljung–Box test results obtained with those obtained applying the test to the raw time series (the series without removing the trend and seasonal component).

3.5   The Partial Correlogram of a Time Series

The partial correlogram or Sample Partial Autocorrelation Function (PACF) is only a useful summary statistics of the time series if the sample ACF of the time series is not the ACF of white noise and it is not the ACF of a nonstationary time series. In other words, we will use the sample PACF only if the time series is stationary but not a white noise stationary time series.

To understand the purpose of the sample PACF, it is helpful to remember Simpson's paradox and correlation as introduced in introductory statistics courses. The correlation between two random variables is often due only to the fact that both variables are correlated with the same third variable. This is what we learn in introductory statistics under the name of Simpson's paradox, calling the third variable the confounder. For example, we will find a strong correlation between electricity production and chocolate production time series, but the only reason for that is that the population is growing and both are affected by the population increase. Population is the confounder.

In the time series context, a large portion of the correlation between y_t and y_{t-k} can be due to the correlation that these variables have, individually, with $y_{t-1}, y_{t-2}, \ldots, y_{t-k-1}$, because of the way the autocorrelations of a time series are calculated. To adjust for this possibility, we can calculate the partial autocorrelations,

Table 3.4 First seven autocorrelations for a time series.

k	r_k
0	1.000
1	0.841
2	0.683
3	0.584
4	0.515
5	0.457
6	0.427
7	0.405

which will provide the autocorrelations between y_t and y_{t-k} without the effect of the intermediate correlations.

> The PACF, like the sample ACF, conveys vital information regarding the dependence structure of the time series. It gives, for each lag k, the correlation between y_t and y_{t-k} after removing the effect of $y_{t-1}, \ldots, y_{t-k-1}$.

3.5.1   Calculating Partial Autocorrelations

The partial autocorrelations of a time series r_{kk} are defined as follows:

$$r_{11} = r_1, \quad k = 1,$$

$$r_{kk} = \frac{r_k - \sum_{j=1}^{k-1} \left(r_{k-1,j} r_{k-j} \right)}{1 - \sum_{j=1}^{k-1} \left(r_{k-1,j} r_j \right)}, \quad k = 2, 3, \ldots.$$

In this formula, r_k comes from the sample correlogram, and

$$r_{kj} = r_{k-1,j} - r_{kk} r_{k-1,k-j}, \quad j = 1, 2, \ldots, k - 1.$$

Example 3.6   Consider the sample correlogram for a time series y_t, $\quad t = 1, 2, \ldots, 159$ up to $k = 7$ given in Table 3.4.

We will use the formulas to compute the partial correlations: r_{22}, r_{11}. We will also determine which of the r_{kk} are statistically significant.

if $k = 1$, then $r_{11} = 0.841$

if $k = 2$, $r_{22} = \frac{r_2 - r_1^2}{1 - r_1^2} = \frac{0.683 - 0.841^2}{1 - 0.841^2} = -0.083$ (rounded)

$r_{21} = r_{11} - r_{22} r_{11} = 0.841 - (-0.083 * 0.841) = 0.911$

if $k = 3$,

$$r_{33} = \frac{r_3 - \sum_{j=1}^{2} r_{2j} r_{3-j}}{1 - \sum_{j=1}^{2} r_{2j} r_j}$$

$$= \frac{r_3 - (r_{21}r_2 + r_{22}r_1)}{1 - (r_{21}r_1 + r_{22}r_2)}$$

$$= \frac{0.584 - (0.911(0.683) - 0.083(0.841))}{1 - (0.911(0.841) - 0.083(0.683))}$$

$$= 0.111$$

$r_{31} = r_{21} - r_{33}r_{22} = 0.913$

$r_{32} = r_{22} - r_{33}r_{21} = -0.112$

if $k = 4$, $r_{44} = 0.036$

if $k = 5$, $r_{55} = 0.018$

if $k = 6$, $r_{66} = 0.091$

if $k = 7$, $r_{77} = 0.025$ ☐

The way to obtain the partial correlogram of a time series in R is with the command

```
pacf(timeseriesname, max.lag=k,plot)
```

where `max.lag` should be equal to at least one fourth of the length of the time series, for example:

```
y= decompose(JohnsonJohnson,type="mult")$random
```

```
pacf(y, max.lag=50,plot)
```

It must be noted that, unlike the output of the sample ACF in R, the output of the `pacf` starts at $kk = 1$. That is, the first spike in the PACF does not have vertical coordinate equal to one, nor lag $k = 0$ like the sample ACF does.

3.5.2 The Standard Error of the Sample Partial Autocorrelation

The asymptotic standard error of r_{kk} under the null that $\rho_{kk} = 0$ is given by this formula:

$$s_{r_{kk}} = \frac{1}{\sqrt{n}}.$$

Example 3.7 Consider y_t, $t = 1, 2, ..., 159$ and partial sample autocorrelations up to $k = 7$ given in Exercise 3.6. The asymptotic standard error of r_{kk} is
$$s = 1/\sqrt{159} = 0.079; \quad 2s = 0.1586.$$ ☐

3.5.3 Testing Hypotheses about ρ_{kk}

As in the sample ACF, at each lag k we test whether the partial autocorrelations at lag k are 0 or not. If the t-statistics are greater than two standard errors, we reject the null hypothesis of 0 partial autocorrelation at lag k. That means that there will be a spike beyond the two-standard-error band in the sample partial autocorrelation function at lag k. The hypotheses tested are:

$$H_o : \rho_{kk} = 0$$

$$H_a : \rho_{kk} \neq 0.$$

Rejecting the null, namely, concluding that a partial autocorrelation at some lag k is different from 0, is achieved if the t-test statistic lies outside 2, or equivalently if the r_{kk} lies beyond two standard errors or outside the two-standard-error band. There are many partial autocorrelation coefficients, and therefore there are many tests.

Example 3.8   Consider the sample correlogram for a time series y_t, $t = 1, 2, ..., 159$, up to $k = 7$, which is shown in Table 3.4. As we saw in Example 3.7, the asymptotic standard error of r_{kk} is

$$s = 1/\sqrt{159} = 0.079; \quad 2s = 0.1586.$$

According to these results, the only r_{kk} calculated in Example 3.6 that is statistically speaking significant is $r_{11} = 0.841$ because it is the only one that is beyond the two-standard-errors mark of 0.1586. This means that the autocorrelation of y_t with $y_{t-k}, k > 1$ is zero after we take into account the correlation of y_t with y_{t-k-1}. For example, the autocorrelation of y_t and y_{t-4} is really 0 after we take into account the autocorrelation of y_t with y_{t-2} and with y_{t-3}, even though $r_4 = 0.515$.   □

> The PACF, like the sample ACF, should not be used with time series that are not stationary.

According to the theory of stationary stochastic processes that we will see in Chapter 5, the ACF and the PACF together will help us identify potential stationary models for the time series. As we saw in Figure 3.1, each stationary stochastic process displayed had a different PACF. Thus, in practice, we usually look at both the sample ACF and the PACF to decide what first tentative model to fit to a stationary time series.

> Partial autocorrelations do not help us determine whether a time series is stationary or not. They are only useful to us once we have a stationary series and we want to identify a model, in which case we use both the sample ACF and the sample PACF for identification. They are particularly useful to identify the order of an AR process, because the ACF of an AR process only helps us determine that it is just that, an AR process of some order (the order being determined by the PACF). This will make more sense after Chapter 5.

3.5.4   Interpretation of the Partial Autocorrelations

As we saw in the formula, the PACF is a function of the sample ACF (the partial correlogram of the sample is a function of the correlogram of the sample).

The first r_{11} is the coefficient of the following regression:

$$r_t = r_{11} r_{t-1} + w_{1t}.$$

The second r_{22} is the corresponding coefficient in the regression

$$r_t = r_{12} r_{t-1} + r_{22} r_{t-2} + w_{2t}.$$

The partial autocorrelation at lag k is the partial regression coefficient r_{kk} in the following regression:

$$r_t = r_{1k}r_{t-1} + r_{2k}r_{t-2} + \cdots + r_{kk}r_{t-k} + w_{kt}.$$

The r_{kk} measures the relation between r_t and r_{t-k} after adjustments have been made for intermediate autocorrelations $r_{t-1}, r_{t-2}, ..., r_{t-k-1}$.

Example 3.9 Consider the sample correlogram for a time series y_t, $t = 1, 2, ..., 159$ up to $k = 7$ given in Table 3.4. The only r_{kk} that is statistically significant is $r_{11} = 0.841$ because it is the only one that is beyond the two-standard-errors mark of 0.1586. This means that we cannot reject the null hypothesis that ρ_{kk}, the partial autocorrelation of y_t with $y_{t-k}, k > 1$, is zero after accounting for the correlation of y_t with y_{t-k-1}. For example, the autocorrelation of y_t and y_{t-4} cannot be assumed to be different from 0 after we take into account the autocorrelation of y_t with y_{t-2} and with y_{t-3}, even though $r_4 = 0.515$. □

3.5.5 Exercises

Exercise 3.11 The USA Environmental Protection Agency makes data publicly available at the following site:

www.epa.gov/outdoor-air-quality-data/download-daily-data

This exercise is asking that you download the daily ozone concentration for CA-Los Angeles in the years 2010–20.

(a) Upload the data into R and aggregate the data to monthly average ozone concentration before you make the data a ts() object.
(b) Make the time series a ts() object and provide a time plot of the monthly ozone concentration.
(c) Provide two plots, one of the sample ACF and the other of the PACF, both in a single image, with the PACF plot stacked below the ACF plot.
(d) Calculate r_k for $k = 0, ..., 11$ and then compute the partial autocorrelations $r_{11}, r_{22}, r_{33}, r_{44}$. You may use R to do the calculations, but illustrate manually how the r_{11} and the r_{22} were obtained.
(e) The APA has an Air Quality System (AQS) API which is the primary place to obtain raw data. Sign up to obtain a key to use the API and familiarize yourself with it. Extract some data to practice. There is an R package to access the API from R as well. Information about the AQS API may be found at https://aqs.epa .gov/aqsweb/documents/data_api.html in the EPA website.

3.6 Time Series in the Frequency Domain

Consider audio restoration, used to recover the original sound from parts of a damaged signal. Audio signals are naturally represented by a time series, and audio restoration removes defects and degradation from music recordings by using the methods of analysis of time series learned in this book.

Music is a continuous waveform over time. It may be sampled at discrete time intervals, and the sampling frequency is 44,100 Hz. This gives 44,100 measures of the waveform for one second of music. Having sampled a recording of music, we get a discrete waveform that can be analyzed using time series techniques. But before we get into that, we should visualize what the waveform is.

Music waveforms can be modeled with sine and cosine terms. This differs from the regular sawtooth patterns that we mentioned earlier in seasonal time series, which would not be well represented by sine and cosine waves. Music is made up of different harmonics or frequencies. These harmonics together describe the pitch, tone and timbre of the music. Harmonics are also expressed as pure tones, described by sinusoidal functions of time, also called harmonic components. These functions are characterized by the period, frequency and phase of the sinusoid. The period of a harmonic waveform is the time for one cycle to be completed, or length of the cycle. The absolute frequency is the reciprocal of the period, and the (angular) frequency is this number multiplied by 2π. The phase is the fraction of the cycle completed at $t = 0$, where time is represented by t [166].

Once the audio signal has been represented as a discrete time series, the statistical methods used will depend on the type of defect. Sometimes several methods will be applied in a given order because there could be more than one type of defect. Addressing the defects in the proper order guarantees success in the restoration.

A convenient way to do descriptive analysis of audio signals before modeling them is spectral analysis. In Mathematics, this term is also known as Fourier analysis or harmonic analysis. Spectral analysis is a method that consists of calculating the sample periodogram or sample spectrum of the time series to estimate the true spectrum [170] [169].

Spectral methods are widely used in Physics, Engineering, and the medical sciences, as those fields produce many types of signals, such as, for example, EEG, ECG, radar signals and radio signals. The spectral methods that we describe in this section were also widely used in Economics in the late 1960s and the 1970s [54]. But economic time series do not have exactly periodic behavior like music or radio signals. Time series produced by the economy and other areas involving human activity do not lend themselves to pure sinusoidal wave modeling but rather to linear combinations of many wave signals to capture the random nature of the signal (except for perhaps the regular seasonality due to holidays or natural seasons).

3.6.1   The Periodogram or Sample Spectrum

The sample spectrum is the function of interest in spectral analysis of a time series; it is a decomposition of the variance of a sample into frequency components, from the point of view of a statistician, and it is the Fourier transform of the stationary time series, from the point of view of a mathematician. The periodogram or sample spectrum is an estimator of the true spectrum of the stationary stochastic process generating the time series. As an estimator, its statistical properties are

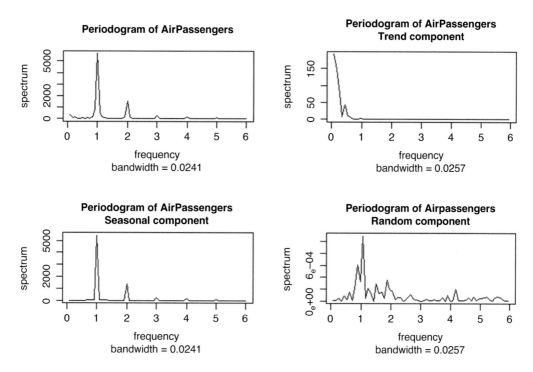

Figure 3.7 Spectrum of R's `AirPassengers` data, and spectrum of each of its components. Each component concentrates on a particular area of the frequencies.

relevant [170]. In fact, the stochastic stationary models that we study in Chapter 5 can all be characterized by their spectrum as well.

The periodogram of a time series y_t at frequency w is calculated using the following expression:

$$C(w) = \frac{1}{n} \left\| \sum_{t=1}^{n} y_t e^{-iwt} \right\|^2 = \sum_{k=0}^{K} \gamma_k e^{-iwk}.$$

Averaged versions of these formulas exist. Notice that the last expression hints at a relation between the sample autocovariance and the periodogram, hence the correspondence between stochastic stationary models and the spectrum.

Figure 3.7 shows the spectrum of the `AirPassengers` data set that comes with R. The spectrum is dominated by the seasonal (at frequency 1 representing a cycle of a year), saying that most of the variability in that time series is due to the seasonal effect. When we decompose the time series and look at the spectrum of the trend component only, we notice how the spectrum has most of the frequency near 0. Time series with long-term trends and cycles extending beyond a year tend to have this type of spectrum. The spectrum of the seasonal component indicates that the largest signal is indeed the seasonal at frequency 1, representing a year cycle, which is large enough to overwhelm the signal in the long-term trend in the spectrum of the original time series. The spectrum of the random component, the one that really interests us because it represents the stationary remaining signal, tells us that there is still some signal left in the data. A small random seasonality that is not regular, other small signals and noise are left. Some stochastic process generating the data has this spectrum. This confirms the information we obtained with the ACF of the time series.

Audio restoration uses analysis of the periodogram of audio signals to diagnose and identify where the signal needs repair.

3.6.2   Synthetic White Noise

Synthetic means simulated. It is important to insist on what the sample ACF of white noise looks like because beginners tend to quickly adopt the misconception that we detrend and seasonally adjust in order to obtain a random term that is white noise. That is not the case, as we have said several times throughout the chapters. White noise is the finish line of a time series analysis. If we have succeeded in modeling all the features of a time series, the only thing left in the residuals of our models will be white noise. If not, we have not succeeded. It is what we want to end up with after we have modeled the signal in a stationary time series.

Another aspect adding to the misconception of beginners is that the correlogram of the sample is not going to look exactly like the correlogram of the model generating the time series, because we are just using a sample. There will be about 5 percent of the autocorrelations in the sample that are significant when they should not be, and that happens also in the sample ACF of data representing a white noise process. In the case of white noise, no r_k should have a spike past the two standard errors, but we may get about 5 percent that have it, just by chance.

Relevant to this section is that the name *white noise* was coined in an article on heat radiation published in *Nature* in April 1922, where it was used to refer to series that contained all frequencies in equal proportions, analogous to white light.

> The term purely random is sometimes incorrectly used for white noise series. Don't use that term. White noise is a zero-mean, constant-variance uncorrelated process. Most stochastic (random) time series processes are autocorrelated.

We simulate 100 observations of a Gaussian white noise process with variance equal to one as follows:

```
set.seed(1)
w = rnorm(100)
```

We saw in Figure 3.1 what the sample ACF of white noise looks like. In theory, the sample ACF should have absolutely no vertical lines branching out from the 0 horizontal line, except for the one at lag 0, which, as we have said, is the correlation of the time series at time t with itself, always 1. White noise means absolutely no significant autocorrelations and no significant partial autocorrelations. However, because of the 5 percent random significant autocorrelation due to sampling, you can see that both in the sample ACF and the PACF there are some autocorrelations that appear borderline

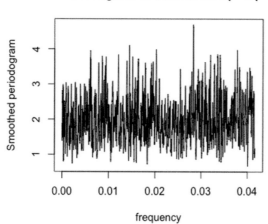

Figure 3.8 shown with title "Periodogram of white noise $z_t = w_t$", y-axis "Smoothed periodogram", x-axis "frequency".

Figure 3.8 The spectrum of white noise does not favor any particular frequency.

significant. As you can see, they do not happen at lower lag values k. They appear randomly throughout the whole plot. Those must be ignored.

As we can see in Figure 3.8, the spectrum of a white noise series also does not favor any particular frequency. All frequencies are uniformly important. Contrast that with the spectrum of the random term in Figure 3.7 where we clearly had signals in the data waiting to be modeled.

The only parameter for a white noise series is the variance σ^2, which is estimated by the white noise series' variance, adjusted by degrees of freedom.

> If your analysis begins on data that are already approximately white noise, then only σ^2 needs to be estimated.

3.6.3 Exercises

Exercise 3.12   Generate several white noise series, and for each of them plot the correlogram, the partial correlogram and the periodogram. What are the main features of the correlogram? What are the main features of the partial correlogram? Describe them. Describe also the periodograms of the generated series.

3.7 The Typical ACF, PACF and Spectrum of a Nonstationary Time Series

The standard calculations of the sample ACF and PACF assume variance and mean stationarity, that is, they assume that we are just looking at the random term of a decomposed time series or a time series of that nature. This implies that the series does not contain trends or seasonality. Nevertheless, computer programs are used to calculate sample ACFs and spectra for any series, irrespective of whether the stationarity condition is satisfied or not [124]. Thus it is very important that a practitioner of time series can discern quickly whether a time series is nonstationary not only by looking at the time plot but also by looking at the ACF and the periodogram.

> The correlogram and spectrum of time series that have long-term persistent trends or very regular cyclical or seasonal components are helpful only to indicate that we need to remove the trend and the regular seasonal component if we want to see the signal in the random component.

If the spikes of a correlogram go beyond the two-standard-errors bands for a continuous sequence of r_k from $k = 0$ to 50 or more, whether exponentially or sinusoidally, then the correlogram reveals that the time series is not a stationary time series, either because of long-term trend or because of very regular seasonal trend. In the correlogram of such a series, most of the periodogram will be concentrated at frequencies very close to 0.

We will see several examples of how that can happen. We must point out that it is possible that a signal is autoregressive, which looks like a time series with heavy trends, but it is stationary. That can be confirmed by looking at the PACF also.

Example 3.10   Consider the `AirPassengers` data set that comes with R. We will consider the decomposition, and the ACF of each part of the decomposition. Program *typicalsac.R* created Figure 3.9, where the typical shape of the sample ACF of the logged `AirPassengers` (Log AP) data set is shown. As we can see, the trend and the seasonal are so dominating that the sample ACF hides any signal that is not trend or seasonal. For the observed time series, the sample ACF goes all the way to almost four years of lags (or almost 48 months). The same happens for the trend component with the typical shape of the sample ACF of a time series with trend.   □

> The typical shape of the sample ACF of a time series with trends or cycles of more than a year's duration will be one of significant slowly decaying autocorrelations up to a large number of lags, 40 or more. The typical shape of the sample ACF of a time series with a very regular periodic component will be that of slowly decaying autocorrelations on both sides of the seasonal lag and multiples of the seasonal lags.
>
> Because all time series with trends and regular seasonality share the same shape, their sample ACF does not let us see the hidden signal in the random component. Their ACF helps us realize that we must remove the trend and the regular seasonal component to see that hidden signal.
>
> In a periodogram, the typical spectral shape of a nonstationary time series is like that seen for the AirPassengers' trend in Figure 3.7.

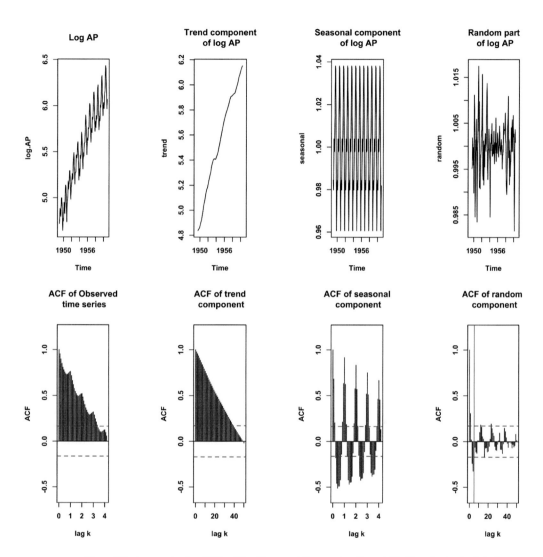

Figure 3.9 As seen in the ACF of the Observed time series (Log AP), if we just calculate the ACF of the raw time series, we may not see the hidden signal in the random component because it is hidden by the ACF of the trend and the seasonal.

3.7.1   Exercises

Exercise 3.13   Consider the plots of the sample ACF in Figure 3.10. Each plot shows 50 autocorrelations, from lag $k = 1$ to $k = 50$. Lag 0 with $r_o = 1$ is included and may be ignored. As we have said repeatedly, it is the autocorrelation of the series with itself, which always is 1. The figure is obtained with program *typicalsac-exercise.R*. For each of the sample ACFs, tentatively assess and answer the following:

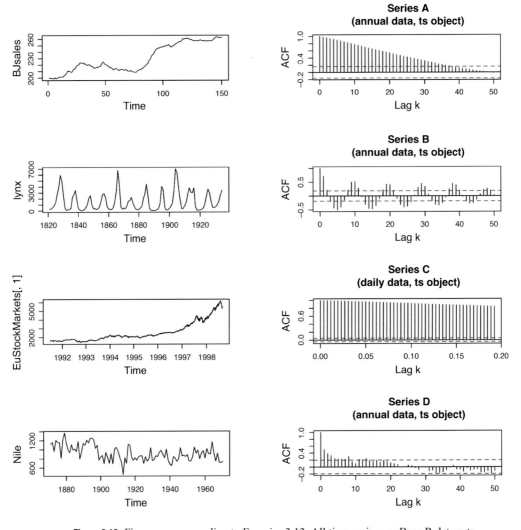

Figure 3.10 Figures corresponding to Exercise 3.13. All time series are Base R data sets.

(a) Is the time series stationary or nonstationary?

(b) If the series is nonstationary, assess where the nonstationary may be coming from: trend, seasonal, or both? Realize that it is hard to tell without looking at the time plot, so suggest tentative possibilities.

(c) If the series is considered stationary, assess whether it is white noise or some stochastic process.

(d) If the series is stationary, identify from the plot the autocorrelation coefficients at lags that are meaningful for a stationary process, namely at lags 1 to about 10, more or less, for a nonseasonal time series.

(e) In each of the plots, use the information in the title to state what the value of the `freq` argument of the `ts` object is and think about the impact it has on how R displays the sample ACF plot.

3.8   Applications in Unsupervised Machine Learning

Suppose we have a lot of hourly time series representing the blood glucose levels of many Type I diabetic individuals, as recorded by a glucose monitor, one time series for each individual. It would be interesting to know whether we can cluster the individuals intro groups characterized by their time series characteristics. Perhaps during work days some individuals have high glucose levels in the middle of the day, while other individuals have high levels at night. Different nutritional advice can then be given to these individuals depending on the cluster in which they fall. With so many individuals and therefore so many time series, it is important to first do descriptive analysis of the autocorrelations and other features of each time series to see what features of the time series are relevant across individuals. Sanchez [174] summarized some of the basic features of one patient's time series window.

In a study like this, the autocorrelation coefficients and the partial autocorrelation coefficients as "features" of the time series can aid in the clustering, in addition to features used in Section 1.8, Example 1.14. But can they? How do we know they will work?

Maharaj and colleagues [112] used simulated data to show that indeed the sample ACF was helpful in discriminating among stochastic processes. That is, stochastic processes that were AR(p) were clustered together and separate from those which were MA(q). The reader will study these classical time series models in Chapter 5. The success in the clustering depended on the method used for clustering. We do a similar kind of analysis here, but using both the sample ACF and the PACF to show that indeed we could add their information to our bag of features in order to improve the clustering of many time series.

Example 3.11   We simulated four time series from an AR process of order 1 (model coefficient 0.4), four from an MA process of order 1 (model coefficient -0.5) and four white noise series. We extracted the first 15 sample autocorrelations for each of them and the first 15 sample partial autocorrelations for each of them, for a total of 30 variables and 12 observations in the features data set. Then we applied k-means to cluster the observations (that is, the time series) according to their sample ACF and sample PACF values. Figure 3.11 shows the clustering of 12 time series with the label of cluster number. All MA processes were placed in the same cluster, all white noise series were put in a separate cluster, but the AR time series had one member placed with the cluster of the white noise. The reader may run program *ch3features.R* and will probably obtain slightly different results, since the simulated time series are random realizations of the model assumed, and each time that the program is run the resulting time series are different. But the conclusions will be very similar.

Had we used as additional features all those used in Section 1.8, Example 1.14 (for these simulated time series), we would have a larger data set of features, and probably more precise clustering.

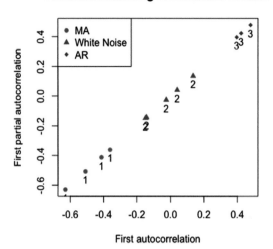

Figure 3.11 Clustering of the stationary time series produced by simulation of AR and MA models and white noise based on the sample autocorrelations and partial autocorrelations only.

The important message is that we can rely on the sample ACF and PACF to do machine learning for large data sets, even if we have just discontinuous fragments of the data. In reality, we often have millions of time series, and working with a data set of features that contains the sample ACF and PACF helps do better unsupervised learning. ☐

3.9　Problems

Problem 3.1　Suppose that the sample autocorrelations of a time series of $n = 89$ observations at $k = 1, 2, 3, 4$ are $r_1 = 0.64279$, $r_2 = 0.32124$, $r_3 = 0.24558$, $r_4 = 0.23751$.

(a) Hand calculate the standard error of r_3, s_{r_3}, and the t-statistic t_{r_3}.
(b) Hand calculate r_{11}, r_{22}, r_{33} and $s_{r_{22}}, t_{r_{22}}$.
(c) Draw conclusions about the statistical significance of r_3 and r_{22}.

Problem 3.2　Consider the following autocorrelations for a time series:
$r_1 = 0.990, r_2 = 0.976, r_3 = 0.958, r_4 = 0.937, r_5 = 0.914,$
$r_6 = 0.887, r_7 = 0.859, r_8 = 0.827, r_9 = 0.794, r_{10} = 0.759,$
$r_{11} = 0.723, r_{12} = 0.685, r_{13} = 0.648, r_{14} = 0.612,$
$r_{15} = 0.577, r_{16} = 0.544, r_{17} = 0.511, r_{18} = 0.479.$
For the following tests, determine whether the null hypothesis would be rejected or not, according to the Ljung–Box test, and give the value of the test statistic, the degrees of freedom and the p-value of the test.

(a)

$$H_o: \rho_1 = \rho_2 = \cdots = \rho_6 = 0$$
$$H_a: \text{at least one } \rho_k \neq 0, \quad k = 1,\ldots,6$$

(b)

$$H_o: \rho_1 = \rho_2 = \cdots = \rho_{12} = 0$$
$$H_a: \text{at least one } \rho_k \neq 0, \quad k = 1,\ldots,12$$

(c)

$$H_o: \rho_1 = \rho_2 = \cdots = \rho_{18} = 0$$
$$H_a: \text{at least one } \rho_k \neq 0, \quad k = 1,\ldots,18$$

Problem 3.3 Given below are the lag k, the sample covariance and the sample auto-correlations r_k and standard errors of the r_k for a time series that has $n = 168$ monthly observations. The sample mean is $\bar{x} = 5.166$, the sample standard deviation is $s = 0.2484$.

Lag k	Covariance $\hat{\gamma}_k$	autocorrelation r_k	standard error of r_k
0	0.081742	1	0
1	0.048996	0.79356	0.077152
2	0.0365	0.59118	0.115971
3	0.023432	0.37952	0.132702
4	0.016210	0.26254	0.139013
5	0.0081180	0.19148	0.141994
6	0.0076632	0.12412	0.142657
7	...	...	...

(a) Is ρ_1 significantly different from 0? Write the null and alternative hypotheses, provide the test statistic and its value, and justify your conclusions. Repeat for ρ_3.
(b) Do a Ljung–Box test to determine whether the time series is white noise up to lag $k = 6$. Provide the null and alternative hypotheses, compute the test statistic and justify your conclusion.

Problem 3.4 Complete Table 3.1 to obtain the information needed to calculate r_4, r_5, r_6 in that example. Show the work done leading to the final result.

Problem 3.5 Access and download the following time series using the Quandl API (see Case Study 1.14).

• Unemployment Rate: College graduates, bachelor's degree and higher, 25 years and over (Series LNU04027662), not seasonally adjusted. January 2009–February 2013.

- Unemployment Rate: College graduates, bachelor's degree and higher, 25 years and over, race Asian (Series LNU04032300), not seasonally adjusted. January 2009–February 2013.

Obtain the sample ACF of the random terms of these series after classical decomposition and interpret their differences.

Problem 3.6   A time series y_t was analyzed with the software R. The autocorrelations of the random component of that series were obtained typing the command

```
acf(y)$acf
```

giving the following output:

```
[1,]   1.00000
[2,]   0.507564127
[3,]   0.105961748
[4,]  -0.131372737
[5,]  -0.035309921
[6,]   0.005037421
[7,]  -0.019030257
[8,]  -0.233208597
[9,]  -0.371457814
[10,]  -0.330918570
[11,]  -0.082869571
```

(a) Sketch the correlogram.
(b) Sketch scatterplots of y_t versus y_{t-1}, then of y_t versus y_{t-2}, then of y_t versus y_{t-3}.
(c) What is the value of r_5? Write code to extract only this value using R.

Problem 3.7   Consider a realization of a time series y_t. The data is entered into R as follows:

```
data=c(362, 381, 317, 297, 399, 402, 375, 349, 386, 328, 389,
  343, 276, 334, 394, 334, 384, 314, 344, 337, 345, 362, 314,
  365)
```

Make this a ts() object called y.t that starts in the first quarter of January and with frequency 4. Calculate the correlation between y_t and y_{t-1} using the acf() function in R. Separately, obtain a time series that contains only y_{t-1}. Calculate manually using R's cor() function the correlation between the two series. If there is a difference, indicate why.

Problem 3.8   A monthly time series y_t, $t = 1, \cdots \cdots 145$ with mean $\bar{y} = 82440.55$ and standard deviation $s = 17131.25$ is recording the total public construction put in place beween January 1980 and December 1991. After some transformation, the sample size is 144. The autocorrelations and Ljung–Box statistic up to the given lags, and the degrees of freedom and p-values of the tests are given in the table.

To lag	Q	DF	P-value	Autocorrelations r_k $k = 1$ on left					
6	774.76	6	< 0.0001	0.974	0.9545	0.937	0.918	0.902	0.888
12	1420.17	12		0.871	0.857	0.841	0.820	0.799	0.788
18	1929.46	18	< 0.0001	0.771	0.752	0.731	0.709	0.687	0.669
24	2292.91	24	< 0.0001	0.646	0.627	0.608	0.586	0.560	0.534
30	2522.80	30	< 0.0001	0.507	0.489	0.471	0.450	0.431	0.411
36	2658.55	36	< 0.0001	0.390	0.371	0.350	0.335	0.319	0.298
42	2719.48	42	< 0.0001	0.274	0.252	0.233	0.212	0.193	0.170
48	2732.33	48	< 0.0001	0.149	0.129	0.107	0.082	0.063	0.046

(a) Write the null and alternative hypotheses to determine whether the autocorrelations up to lag 12 are statistically significant according to the Ljung–Box test and determine the p-value of the test that is missing. State the conclusion of the test up to lag $k = 12$.

(b) What conclusion can you reach, after looking at all the autocorrelations, regarding the process generating this time series? Say at least two major things that this table implies regarding the time series.

3.10   Quiz

Question 3.1   The analysis of a time series y_t gives a correlogram with $r_1 = 0.7$. The standard error of this estimate is 0.25. Which of the following statements is correct?

(a) ρ_1 is significantly different from 0.
(b) r_1 is significantly different from 0.
(c) The correlation between y_t and y_{t-1} is not significantly different from 0.
(d) A scatterplot of y_t against y_{t-1} would show no slope at all.

Question 3.2   Consider time series y_t. Suppose that we want to determine whether there is statistically significant autocorrelation between observations at time t and $t-5$. The null hypothesis to determine this is:

(a) $H_o: r_5 = 0$.
(b) $H_o: r_5 > 0$.
(c) $H_o: \rho_5 = 0$.
(d) $H_o: \rho_5 \neq 0$.

Question 3.3   When we calculate the Ljung–Box test for lag $= 6$ for the rooms data without any transformation, the test statistic has value

(a) 0.000000000
(b) 191.66
(c) 6
(d) 4.1678

Question 3.4   The data set in Base R called `AirPassengers` is monthly data. Run the following R code.

```
new.decomp=decompose(AP,type="mult")  # AP=AirPassengers
 # we try multiplicative decomposition
plot(new.decomp)
trend=new.decomp$trend
seasonal=new.decomp$seasonal
random=new.decomp$random
plot(trend,main="Trend",ylab="average")
plot(seasonal,main="Seasonal component of AP",ylab="AP
-Trend")
plot(random,main="Random component of AP", ylab="AP-trend
-seasonal")
```

The ACF of the trend component is

(a) useful to determine whether there is a signal in the random term.
(b) only useful to determine whether we must remove the trend.
(c) a sample ACF with only lag $k = 0$ significant.
(d) a sample with r_k monotonically increasing as lag k increases.

Question 3.5   In the Ljung–Box test of the residual of the model fit to the transformed rooms data, for a lag 6 test, we obtained:
   chi-square statistic $= 3.4223$; df $= 6$, p-value $= 0.7543$.
   At level of significance 0.05, what would be the conclusion of this test? Check all that apply.

(a) The residuals are white noise.
(b) We reject the null hypothesis.
(c) We can see that the data are nonstationary.
(d) The residuals of the model are not white noise.
(e) We do not reject the null hypothesis.

Question 3.6   Access the time series *bangladesh.txt*.
   According to the conclusions of the Ljung–Box test, the time series is

(a) white noise.
(b) not white noise, and nonstationary.
(c) not white noise, but stationary.
(d) white noise but nonstationary.

Question 3.7   A time series y_t is such that its correlogram is the following up to lag $k = 7$:

k	1	2	3	4	5	6	7
r_k	0.765	0.543	0.584	0.515	0.457	0.342	0.123

The partial autocorrelation at lag $k = 1, r_{11}$, is then

(a) 0.841
(b) −0.083
(c) 0.913
(d) 0.765

Question 3.8   The partial autocorrelation r_{kk} of a time series y_t is best defined as

(a) the correlation of y_t and y_{t-k} after removing the effect of the correlation between y_t and $y_{t-1}, y_{t-2}, \ldots, y_{t-k+1}$.
(b) the autocorrelation of y_t with itself.
(c) the autocorrelation coefficient minus the expected value of the series.
(d) the variance of the time series.

Question 3.9   The Ljung–Box test done of a time series of ozone concentration (a time series without seasonal components) gave the following values for the following test

$$H_0: \rho_1 = \rho_2 = \rho_3 = \rho_4 = \rho_5 = \rho_6$$
$$H_a: \text{at least one of those } \rho_k \neq 0$$

The test statistic $Q = 799.88$
The p-value = 0.0001
What is the conclusion of the test?

(a) The time series is white noise.
(b) The time series is stationary.
(c) The time series is nonstationary.
(d) The time series is not white noise.

Question 3.10   In trying to build features of multiple time series in order to do unsupervised machine learning, particularly when we have multiple and perhaps incomplete time series, we have found that

(a) the sample autocorrelations do not help discriminate among time series.
(b) the sample autocorrelations help discriminate among time series perfectly.
(c) the sample autocorrelations help discriminate among time series but probably there will be some misclassification.
(d) there is no point in using the sample partial autocorrelations.

3.11   Case Study: Multiple Seasonalities in Kaggle Competitions

Some time series have many frequencies: hourly, daily, weekly, monthly, annual, in decreasing level of frequency. In such data, it is not rare to find multiple seasonalities. A typical example of multiple seasonality is the hourly demand for electricity, which could have daily seasonality (use of appliances at different times of the day), weekly seasonality (different use of appliances in weekends and during the week), or yearly seasonality (different use of power in summer and winter, for example). This is in stark contrast with time series of lower frequency (monthly, quarterly), which are more likely to have only one dominant seasonal pattern, as we have seen in several examples in the book.

The Base R `ts()` object cannot handle multiple seasonalities, as we indicated in Chapter 1. Some people use `msts()` from the `forecast` package [77].

Nowadays there is an increasing demand for forecasts of high-frequency time series, due to the wide production of sensor and IoT data. Forecasting competitions have challenged many data scientists to propose solutions to the problem of forecasting time series with multiple seasonalities. These competitions usually challenge by offering a large number of time series, in some cases as many as 145,000 time series [88]. The time series themselves may not be very long, but there are many of them. Methods are sought that forecast well as many of those time series as possible. Two competitions that stand out are the M competitions [5] and the Kaggle competitions [86]. The data set used in this case study is a Kaggle competition data set.

Kaggle is an online data science platform that hosts data science competitions in the areas of solving real business problems, recruitment and academic research. Several of Kaggle's competitions have been forecasting competitions that address real-life, high-frequency business forecasting problems with access to external information, but the forecasting community has largely overlooked the results obtained, according to Bojer and Medgaar [4]. These authors identified six forecasting competitions among the ones run by Kaggle that featured daily or weekly time series with access to external information. They noted that the Kaggle data sets needed initial preprocessing (cleaning). These competitions, however, have the advantage that a large community of data scientists from a variety of backgrounds may compete in the competitions, post their solutions, make beginners aware of problems that may arise just because of the way the data is collected, and participate in the discussion forums by sharing knowledge and discussing potential strategies. In the business-problem-focused competitions, companies provide a data set for a prediction task and typically offer a cash prize to the top performers. There are many solutions, which are ranked from top to bottom. These competitions typically differ from academic competitions because they focus on solving a problem rather than learning why and when a particular method works [4].

3.11.1   Hourly Energy Consumption

In this case study, we are going to look at a particular time series data set from Kaggle: "Hourly Energy Consumption," [127] version 3, which presented hourly energy consumption in megawatts (MW) for different eastern regions of the United States as collected from the PJM website (pjm.com). There are different time series, one for each of the regions, and each of them contains the megawatt energy consumption and a field indicating the date and time. Each of them is given in an individual .csv data file.

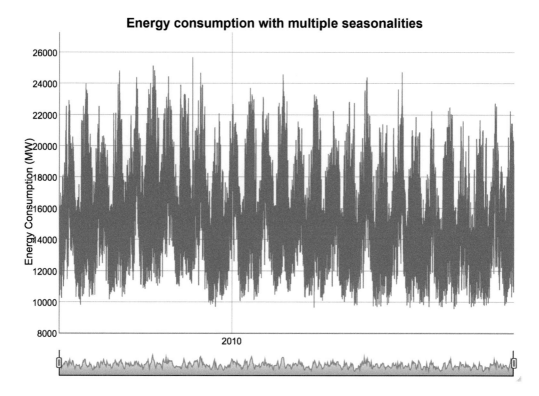

Figure 3.12 With so many observations, it is hard to discern the multiple seasonalities, but we can see that indeed the energy consumption follows a seasonal pattern. Studying the seasonality at different levels of granularity will help us to discover which seasonality dominates the movement of the series.

For example, AEP_hourly is the name of the data set that contains the estimated energy consumption for the American Electric Power Co., Inc. in a variable labeled AEP_MW and a date variable called datetime. The energy consumption estimates are reported from 2004 to 2018. In this case study, we will study the AEP_hourly data set. There is also a data file that combines all the individual time series. All the discussion that follows can be replicated by running program *chapter3case.R*. We recommend that the reader runs in particular the beginning plots in Figures 3.12 and 3.13. By playing with the range selector at the bottom of the image obtained by running the time plot code, the reader can observe the data at different levels of granularity and appreciate daily, weekly, monthly and annual seasonalities. All the reader has to do once the graph appears in R is to put the cursor in the range selector and move it right or left.

3.11.2   Data Inspection

All time series data analysis should start with a very careful inspection of the data. In the PJM data set, all time series are complete, meaning that there are no missing

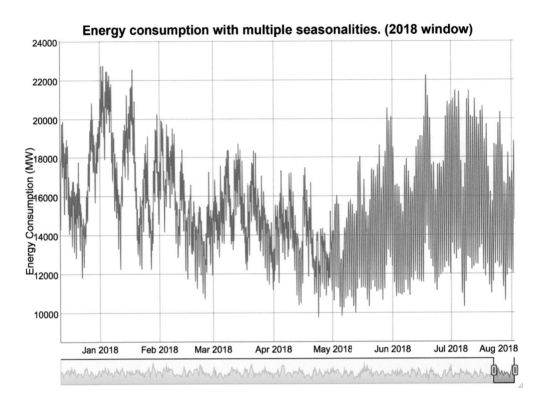

Figure 3.13 A window of the image in Figure 3.12 lets us see more closely the multiple seasonalities. We can see smaller windows of data by just dragging the range selector at the bottom of the graph (the reader would have to be inside R to do that interactively).

values. However, one data file, which aggregates them all, has a lot of missing values due to the fact that the regions have changed over the years, so data may only appear for certain dates per region when combined together in one single file. One thing that is clear upon inspection of the data sets is that lengths of the series are different and the first time stamp and the last time stamp listed in the data sets are also different across regions. Although the last time stamp corresponded to January 2, 2018, for all of them, the first time stamp varies from 1998 to 2011. Thus, the number of records per series varies from 32,896 hours to 145,366 hours.

We will concentrate on the time series for American Electric Power Co., Inc. (AEP) and look at it a little closer.

The Temporal Index Is Not in Order
At first glance, the AEP_hourly has the first observation on December 31, 2004, and the last observation on January 2, 2018. It therefore contains 121,273 hourly observations. However, upon further inspection and ordering in the right chronological order by converting the data frame to a xts() class object, which has the time stamp as index and the observed hourly energy consumption, we notice that the first observation is on October 1, 2004, at 1:00:00 AM and the last observation is on August 3, 2018,

at the 00:00 hour. The xts() class can be used if the R package xts is installed [167].

Thus, we need to make sure, for future analysis, that the data has been sorted properly.

> It is very important that the data is sorted in the right date and time order. Many time series data sets are not sorted that way, and it is hard to notice at first glance when the number of observations is very large. Converting the time series to an object that will respect the actual order is very important.

Another thing that we noted, by accident, is that, in 14 instances, different years and months, there is no observation or time index collected at 3:00:00 AM. Without further information, it is hard to know why that is the case. Other participants in the Kaggle competition noticed other irregularities. The daylight-savings time changes are a possible explanation. That would be something to keep track of.

Since our task in this case study is to illustrate the analysis of multiple seasonalities in the data set and their reflections in the sample ACF, we will not impute those values like other authors do. The time series object xts recognizes gaps in the time index and allows us to plot descriptive graphs. Similarly, the ts() object allows us to get the ACF of data that has no missing values.

dygraph Interactive Time Plot

We will start by looking at a dygraph of the properly sorted time series. We will play with the interactive aspect of the graph to show how the granularity of the data appears more clearly. With so much data, it is hard to discern patterns unless we look at windows of the data, which dygraph allows us to do interactively. As we can see in Figures 3.12 and 3.13, the data reveals a periodic behavior, confirming our suspicion that there are perhaps multiple seasonalities. This is more clearly seen in Figure 3.13, which shows a window containing only the more than 8,000 observations for 2018. In the latter, we can see the monthly seasonality, with the winter months and the summer months showing an increase in consumption.

Detailed Inspection of Seasonalities

A simple way to look at seasonality that is familiar already to us from Chapter 1 is the seasonal box plot. In order to see the seasonality at the different levels of granularity of the data, we extract from the date variable the month, the weekday and the hour of the day.

Figure 3.14 lets us see the variation throughout the hour of the day, the day of the week, and the month of the year. Figure 3.14, panel (a), tells us that, as we expected, winter and summer months are the months of highest median consumption of energy, but during the summer months there is much more variability, higher interquartile range, than in other months. Panel (b) tells us that, as we expected, there is less consumption of energy during the night hours, but then during the day there is similar median consumption, although the hours of 5 to 8 PM indicate much higher variability

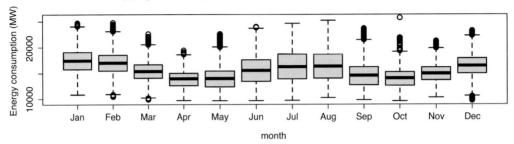

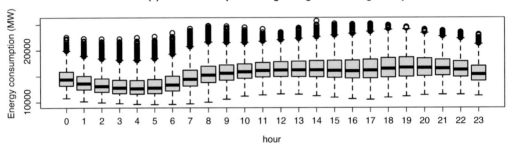

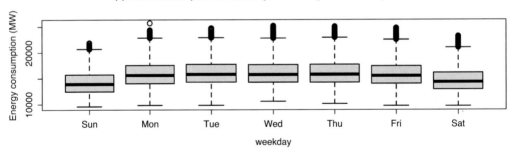

Figure 3.14 Seasonal box plots at different levels of granularity allow us to see in more detail the multiple seasonalities.

than the other hours. And panel (c) tells us that Sunday and Saturday are the days of lowest energy consumption, with similar variability in all the days.

We also notice in the box plots the large amounts of outliers. All the distributions behind each of the boxes have substantial right tails. This requires further exploration.

Autocorrelation

The ACF of the correctly sorted hourly data reveals that the daily seasonality is the largest contributor. However, as we saw in the box plots, there are a lot of outliers, suggesting that the other seasonalities also play a role. For that reason, we proceed to investigate the ACF under different cycle assumptions.

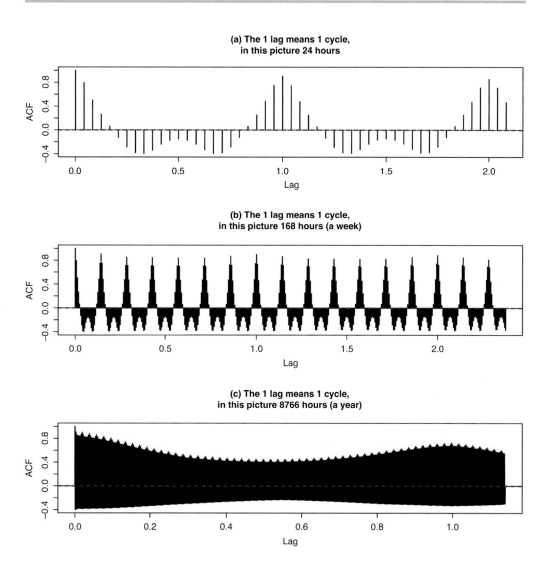

(a) The 1 lag means 1 cycle, in this picture 24 hours

(b) The 1 lag means 1 cycle, in this picture 168 hours (a week)

(c) The 1 lag means 1 cycle, in this picture 8766 hours (a year)

Figure 3.15 Sample autocorrelation interpretations depending on the frequency set for a ts () object.

Figure 3.15 shows three sample ACFs of the change in energy consumption. We differenced because the data seems to have some mild trend to it. Panel (a) is the sample ACF when we have set up a cycle to be the day, and hence the frequency to 24 intervals in a day. Recall that when we set the frequency in the ts () command, the ACF at lag 1 represents the autocorrelation at the lag corresponding to a full cycle. Hence panel (a) is telling us that there is daily seasonality, something that we already knew from the seasonal box plots in Figure 3.14. Panel (b) was calculated setting the cycle to be the week. Correspondingly, the frequency is the number of hours in a week, since our data is hourly. Here too we observe that there is weekly seasonality, illustrated by the peak at lag 1, which, again, represents a week's worth of hours.

Finally, panel (c) is the sample ACF reflecting annual seasonality. The number of hours in a year is very large, but nonetheless that is the frequency that we set. Here too we detect an annual seasonality, which we also knew of.

We conclude therefore that multiple seasonalities are important in their own way in this data set.

3.11.3   What to Do after Discovering Multiple Seasonalities?

After discovering multiple seasonalities, the thing to do is to incorporate them in subsequent analyses of the data. An alternative is to downsample, also known as aggregate, the data in some way to have perhaps just one average measurement per week.

Modern modeling efforts in time series forecasting consist of incorporating those seasonality features, trend features and other variables in a curve-fitting model that looks like a multiple linear regression model, or a Prophet model or some other ML model.

3.11.4   Exercises

Exercise 3.14   Repeat the analysis done in this chapter for the energy data with the daily ozone data from Exercise 3.11. Use the dygraphs to plot and inspect the time series. Then do all the seasonal plots done to investigate daily, weekly or monthly seasonality. Describe all your findings in a report.

Exercise 3.15   Choose a particular time series data set from Kaggle: "Hourly Energy Consumption," [127] version 3, that is not the AEP_hourly. Repeat the data analysis done in this case study but for your chosen time series. Make sure to first inspect the time series and assess whether it is properly sorted and/or whether there are any times without records. Review the case study for hints as to where to start.

3.12   Appendix

This chapter is easier to follow if some concepts from introductory statistics are remembered. The following example is presented as questions to assess whether the concepts are familiar.

Example 3.12   The following data describes the relationship between grades in quiz 1 (denoted by X) and grades in quiz 2 (denoted by Y) in section T of a history class. The sample average of X is $\bar{x} = 4$, the sample standard deviation of X is $s_x = 2$, the sample average of Y is $\bar{y} = 7$ and the standard deviation of Y is $s_y = 4$. After completing the table and calculating Pearson's sample correlation coefficient, answer the questions.

(a)   Is the relation between X and Y a positive association, a negative association, or no association overall? Why? Note that you will have to complete the table to answer this question.

x	y	$\frac{x-\bar{x}}{s_x}$	$\frac{y-\bar{y}}{s_y}$	$\left(\frac{x-\bar{x}}{s_x}\right)\left(\frac{y-\bar{y}}{s_y}\right)$
1	5	−1.5	−0.5	
3	9			−0.25
4	7		0.0	
5	1	0.5	−1.5	
7	13	1.5	1.5	

(b)   Lecture M had a different table, which resulted in a correlation coefficient of −0.8, and lecture N also had a different table, which resulted in a correlation coefficient of −0.2. Sketch the scatterplot of grades X,Y for each lecture, assuming there are no outliers and the relationship is linear.   □

   If the reader feels that the preceding exercise is hard to answer, a review of correlation, sample mean and sample standard deviation is perhaps necessary before embarking into the study of this chapter. Appropriate references for a refresher are [52, 17, 34, 37, 108, 190, 147, 183, 184, 199] among many others. Those summary statistics are also relevant in the study of stationary time series, in particular the correlation coefficient. However, correlation calculations take a peculiar form and interpretation in time series, different from that learned for data with independent observations in introductory statistics. Two variables are needed to calculate correlation, but we have only one time series. This chapter provides the skill set needed to get out of this dilemma.

Part II

Univariate Models of Temporal Dependence

After description of data, the next step in an introductory statistics course is to learn the probability needed to understand statistical inference. Time series analysis is no different. Modeling the temporal dependence of a time series and drawing conclusions about models fitted to time series and forecasting require probabilistic statements. The probabilistic properties of the models fitted to data must be checked against the data. It is for this reason that this part presents the theory of univariate stochastic processes in Chapter 5. These models use the temporal dependence present in a time series, and detected by the autocorrelation function, in order to forecast the yet unknown data. The models contrast with the curve fitting or decomposition-based models introduced in Part I.

This part focuses on the Box–Jenkins methodology for handling trends and seasonals and for modeling the random term with stationary stochastic models. Box–Jenkins methodology has been widely used for decades and is still the benchmark against which new methods are evaluated. The main models in this methodology are ARIMA models, studied in Chapter 6. Understanding these models requires understanding first stationary stochastic models, studied in Chapter 5. The latter require an understanding of the backshift and differencing operations, which are introduced as alternative ways of reducing a time series to its random term in Chapter 4. The theory of stationary stochastic models and the practice of backshifting and differencing meet in Chapter 6, where the most general ARIMA(p,d,q)(P,D,Q) model is introduced and applied to time series data. The practices of pretransforming, differencing, identifying, model fitting, diagnosing and forecasting are introduced with different types of time series models. The concepts of volatility, garch and arch are also introduced briefly in Chapter 6.

Chapter 5 makes generous use of simulation. The latter assumes that we know a model. After we introduce models and their properties, we simulate them to generate synthetic data that can make the reader aware that even in a perfect world, the data never will act exactly as the theory predicts. It is important for the reader to understand that the same process can manifest itself in many different paths. Hence the probabilistic transition to modeling needed.

Together, Chapters 5, 4 and 6 will give the reader a good overview of what has been the most prevalent type of classical time series modeling and forecasting for decades: the Box–Jenkins methodology.

4 The Algebra of Differencing and Backshifting

4.1 Introduction

We have seen in Chapter 2 how we can decompose a time series into trend, seasonal and random terms to form additive or multiplicative decomposition models. By subtracting trend or seasonal or both, we could obtain detrended or seasonally adjusted time series, or both, respectively. Differencing is an alternative way to achieve the same goal: a random term free of long-term trend and short-term regular seasonal trend. The backshift operator translates differencing into an alternative expression that is convenient for expressing the processing done to a time series and for model specification. The ARIMA models of Chapter 6 are a unified way of modeling a time series that particularly relies on the notation of differencing and backshifting to not only achieve the same goal as decomposition but to, at the same time, identify and check the properties of the remaining random term and compactly specify a model.

In the context of ARIMA modeling, a time series X_t with constant variance is *integrated* of order d, denoted as $I(d)$, if the dth difference of X_t is stationary, that is, free of long-term trend and regular seasonality. We difference a time series with constant variance that has trends and/or cycles in order to obtain a stationary random term that has mean and variance that do not depend on time. The stationary random term is then assigned a stationary model (like those studied in Chapter 5), and then when using the model to forecast the future value of the time series, the trend and seasonality that were removed are integrated into the model.

The backshift operator helps us specify the differencing in an alternative form that serves two purposes. First, it helps us see whether the polynomial in the backshift operator has roots outside the unit circle; second, it helps us, after we fit a model, to combine the differencing operations with the model fit to obtain the final forecasting model. All of that will become more clear when we study Chapter 6.

Thus backshift and difference operators must be understood in order to understand ARIMA modeling. They are also convenient to simplify the specification of models. That is why we study them in this chapter. Section 4.2 introduces the difference and backshift operators and their relationship. We distinguish there between regular differencing done to remove the long-term trend, and seasonal differencing to remove the regular seasonal component. In 4.3 we describe what to look for in the ACF of a properly differenced time series. Section 4.4 tells us what to look for in the spectrum after differencing. In Section 4.7, we continue becoming acquainted with the realities of modern, imperfect and complex time series data, and explore again the use of features to do unsupervised machine learning based on compressed data. With data recorded every half hour, differencing presents new challenges, which are explored in the exercises.

Many beginners tend to think that differencing, detrending and seasonal adjustment are done to convert the time series to white noise. However, as we said in past chapters, and the reader will see in this chapter, that is not the goal. A stationary time series is not a white noise series. The latter is a special case of stationarity. Differencing, like detrending and seasonal adjustment, is done to make a time series mean stationary. Prior to doing those operations, we must make sure that the time series is variance stationary, that is, has variance that does not change much over time.

4.2 The Difference Operator and the Backshift Operator

Differencing is an operation done to a time series to either remove the trend, or the seasonal or both. The *difference* operator is the operator that calculates changes in the level of a time series from time $t - k$ to time t. If y_t is the observed time series, then the kth difference of y_t is

$$\nabla_k y_t = y_t - y_{t-k}.$$

Before differencing, we must make sure that the time series is variance stationary, that is, that the variance does not depend on time.

Differencing is done in R using the `diff()` function

`diff(y, lag= k  ,  diff=1)`

where `y` is a `ts()` object containing the observed time series, `lag` is the k and `diff` is how many times we repeat the differencing. If diff>1, the differencing is done in the previously differenced time series. We represent the repeated differencing mathematically by powers of the difference operator:

$$\nabla_k^d y_t = (y_t - y_{t-k})^d.$$

In R,

`diff(y, lag= k  ,  diff=d)`

Example 4.1   Consider observed annual time series y_t, from 1950 to 1958 inclusive, with values 2, 5, 9, 3, 4, 10, 15, 20, 1, where $y_{1950} = 2$ and $y_{1958} = 1$ (see Table 4.1).

$y_t^* = \nabla y_t = y_t - y_{t-1}$, applied to the time series, gives 3, 4, −6, 1, 6, 5, 5, −19, a shorter time series with just eight values.

Table 4.1 Differencing and backshift translation. Notational conventions to use for work. It is good practice to change the name of a time series after a transformation because it helps distinguish the raw time series, the y_t in this case, from the transformed series.

Time	y_t	y_{t-1} $= By_t$	y_t^* $= \nabla y_t$	y_t^{**} $= \nabla^2 y_t$ $= \nabla y_t^*$	y_{t-2} $= B^2 y_t$	y_t^{***} $= y_t - y_{t-2}$ $= \nabla_2 y_t$
1950	2					
1951	5	2	3			
1952	9	5	4	1	2	7
1953	3	9	-6	-10	5	-2
1954	4	3	1	7	9	-5
1955	10	4	6	5	3	7
1956	15	10	5	-1	4	11
1957	20	15	5	0	10	10
1958	1	20	-19	-24	15	-14

$y_t^{***} = \nabla_2 y_t = y_t - y_{t-2}$, when applied to the time series, gives time series $7, -2, -5, 7, 11, 10, -14$, a shorter time series with just seven values.

$y_t^{**} = \nabla^2 y_t = \nabla y_t^* = y_t^* - y_{t-1}^* = (y_t - y_{t-1}) - (y_{t-1} - y_{t-2}) = y_t - 2y_{t-1} + y_{t-2}$ gives the time series $1, -10, 7, 5, -1, 0, -24$.

The operations would be quickly obtained with R using

```
y=ts(c(2,5,9,3,4,10,15,20,1), start=1950, frequency=1)
y.star=diff(y,lag=1,differences=1)
y.3.star=diff(y, lag=2, differences=1)
y.2.star=diff(y,lag=1,differences=2)
```

The *backshift* operator, B^k, shifts the subscript of the time series backward in time by k units. It produces the y_{t-k} value of the time series,

$$B^k y_t = y_{t-k},$$

and it is convenient to represent the differencing operation.

Example 4.2

$$By_{50} = y_{49}.$$

Example 4.3

$$B^{12} y_{50} = y_{50-12} = y_{38}.$$

We can combine the differencing and backshift operators as follows:

$$\nabla_k y_t = y_t - y_{t-k} = y_t - B^k y_t = (1 - B^k) y_t$$

and

$$\nabla_k^d y_t = (1 - B^k)^d y_t.$$

The $(1 - B^k)^d$ is a backshift polynomial.

Example 4.4

$$\nabla x_t = x_t - x_{t-1} = x_t - B x_t = (1 - B) x_t,$$

so $\nabla = (1 - B)$. ☐

Example 4.5

$$\nabla^2 x_t = (1 - B)^2 x_t = (1 - 2B + B^2) x_t = x_t - 2x_{t-1} + x_{t-2}.$$

Thus $\nabla^2 = (1 - B)^2$. ☐

In general, differences of order d are defined as

$$\nabla^d x_t = (1 - B)^d x_t = \left[\sum_{k=0}^{d} \binom{d}{k} 1^k B^{d-k} \right] x_t.$$

Example 4.6

$$\nabla^3 y_t = (1 - B)^3 y_t = (1 - 3B + 3B^2 - B^3) y_t$$
$$= y_t - 3y_{t-1} + 3y_{t-2} - y_{t-3}.$$ ☐

Powers of the backshift polynomial do not always have to be positive, although most of the time they are.

Example 4.7 Let y_t be an observed time series. Then

$$B^{-1} y_t = y_{t+1},$$
$$B^{-k} y_t = y_{t+k},$$
$$\nabla y_t = y_t - y_{t-1} = (1 - B) y_t.$$

The latter is a very intuitive differencing, as it gives us the change of a time series from time $t - 1$ to t. Sometimes, the change value is more interesting than the absolute value at each t.

$$\nabla^{-1} x_t = \sum_{j=0}^{\infty} x_{t-j} = x_t + x_{t-1} + x_{t-2}\ldots\ldots = (1 + B + B^2 + \cdots) x_t = (1 - B)^{-1} x_t.$$

☐

In ARIMA modeling, we do *differencing* in order to remove (or rather account) for the trend and/or the regular seasonal components of an observed time series y_t that has constant variance. The end result should be the stationary random term of the time series, which will often have a hidden signal that we want to extract and model. The backshift polynomial is then used to write in compact form all the parts of the model, as in a decomposition model, but with the backshift polynomial notation instead, to reflect that we differenced.

For the practice of time series analysis of many types of time series, it is convenient to distinguish between regular differencing and seasonal differencing. In the language of decomposition, the equivalent operations are smoothing, detrending and seasonally adjusting.

4.2.1 Exercises

Exercise 4.1   Given nine observations, 10, 10.5, 11, 11.5, 12, 12.1, 11.2, 10, 10.2, on a discrete time series denoted by x_t, find, showing all your work, the value of the new variables and write the expression in backshift polynomial form:

(a) $x_t^* = \nabla x_t$
(b) $x_t^{**} = \nabla_4 x_t$
(c) $x_t^* = \nabla^2 x_t$
(d) $y_t = \nabla^4 x_t$
(e) $w_t = \nabla_4 x_9$
(f) $z_t = \nabla \left(\nabla^4 x_t \right),$

where the names given to the new variables are for notational convenience, to distinguish the original series from the transformed series. For communication purposes, it is very important that reports and answers to exercises assign a different name to the variables that have been obtained from raw data variables by differencing transformations.

4.2.2 Regular Differencing

Regular differencing is the differencing we do to remove the long-term trend of the time series. We use the difference operator to represent the differencing.

> Regular differencing refers to the act of differencing with very small lag k, often $k = 1$, to remove a long-term trend. Given a time series y_t, the *first regular difference* is
> $$y_t^* = \nabla y_t = y_t - y_{t-1} = (1 - B)y_t.$$
> This is always the first differencing done to a time series with a long-term trend or long-term cycles, such as, for example, business cycles of several years. It could be followed with second regular difference, that is, first difference of the first difference, but it is rare that we need to do that.

> If the stationary random term of a time series is obtained right after first regular differencing, we say in the language of ARIMA that the time series is integrated of order $d = 1$, or $I(1)$.

Example 4.8 Suppose we have a time series x_t that has been taken log of in order to make the variance constant. The time series x_t has a long-term trend. Our constant variance series is denoted by y_t. The *first regular difference* converts it to

$$y_t^* = \nabla y_t = y_t - y_{t-1} = (1 - B)y_t.$$

In R, we use

```
y.star= diff(y, lag=1, diff=1)
```
□

Example 4.9 Paper towels sales is an annual time series extending from 1978. The data set name is *towels.txt*. If a time series is annual, there is no need to do seasonal differencing because there is no seasonality. The time series, however, shows long-term trend that first goes down and then goes up, that is, a trend that extends over a year. In order to obtain the random component of the time series, first difference of the time series is conducted. Figure 4.1 shows both the time plot of the towels time series, the time plot of the difference and the correlogram of the difference.

Let y_t denote the towel sales value. Figure 4.1, in the plot of the regular difference, reveals that the regular difference ∇y_t is enough to obtain the stationary random component of the time series. The latter shows a time series fluctuating around a constant mean, with no regular trend in mean. The sample ACF reveals that the ACF dies away quickly at lower lags, another sign of stationarity.

The R code used to obtain the figure and do the calculations is in program *ch4roomsdiffbasic.R*. □

Sometimes, but very rarely, we may need to do regular differencing of y^*, that is, regular difference twice. In difference operator notation, that is

$$y_t^{**} = \nabla^2 y_t = \nabla \nabla y_t = (1 - B)^2 y_t = (1 - 2B + B^2)y_t = y_t - 2y_{t-1} + y_{t-2}.$$

This is like regular difference again converting y_t^* to

$$y_t^{**} = \nabla y_t^* = (1 - B)^2 y_t.$$

Second regular differencing in R may be achieved in two ways:

```
## do it directly on y_t   using diff=2
y.double.star = diff(y,lag=1, diff=2)
##or  do it on y^* using diff=1
y.double.star= diff(y.star, lag=1, diff=1)
```

However, as we said earlier, it is rare to have to regular difference twice. Most often, what remains after the trend in mean has been removed is a very strong seasonal trend

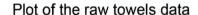

Plot of the raw towels data

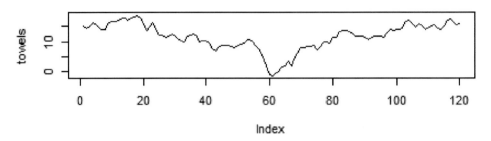

Regular difference of towels

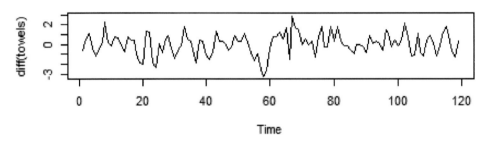

Series diff(towels)

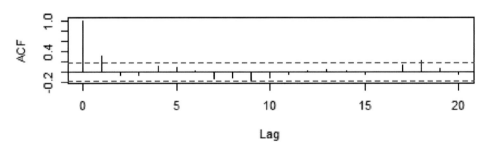

Figure 4.1 Towels, regular difference of towels and correlogram of the latter. The correlogram reveals that the differencing has made the time series stationary because the autocorrelations at lower lags die away quickly.

if there is one in the raw data, or just the random term (if there is no seasonality in the raw data) or perhaps a random seasonality that is part of the random term (if there is regular and random seasonality). The natural step to take after first regular differencing in the event that there is a strong seasonality after regular differencing is to do seasonal differencing, which we discuss next.

4.2.3   Seasonal Differencing

Seasonal differencing is not different from regular differencing. But the lag used to difference will usually be the frequency of the time series stated when we create the time series object.

Example 4.10   Suppose we have a monthly time series that exhibits a regular seasonal trend, that is, each year, during the summer months, there is an overwhelming increase or decrease in the value of the series. But the series does not have a trend. Because it is monthly, the time series is a time series object of frequency 12. We would like to remove the trending seasonal pattern in the series each year, that is, we would like to see just the random term, and perhaps the random seasonal part. In order to do that, we do the following:

$$\nabla_{12}y_t = y_t - y_{t-12} = (1 - B^{12})y_t.$$ □

Example 4.11   If the time series had been quarterly, and exhibited seasonal trend each fourth quarter, then frequency would be 4 and we would have removed the trendy seasonal part as follows:

$$\nabla_4 y_t = y_t - y_{t-4} = (1 - B^4)y_t.$$ □

Example 4.12   We rarely do seasonal differencing twice. Notice that when we do seasonal differencing, we lose observations. Moreover, overdifferencing always causes a lot of problems, such as, for example, rendering the remaining time series useless for the purposes of detecting any signal in the random term.

$$\nabla_{12}^2 y_t = (1 - B^{12})^2 y_t = (1 + B^{24} - 2B^{12})y_t = y_t - 2y_{t-12} + y_{t-24}.$$ □

Example 4.13   The *nottem* time series that comes with R reports the Average Monthly Temperatures at Nottingham, 1920–1939. The series has no long-term trend but it has a very regular seasonality every year. Figure 4.2 shows the effect of performing first seasonal differencing on the average temperature series. See program *ch4nottemdifbasic.R*.

The differencing has removed the short-term seasonal trend, leaving us with just a time series that is stationary, because there are only a few significant autocorrelations at small lags and at the seasonal lag.   □

4.2.4   Seasonal Differencing of the Regular Difference

We want to be economical in our differencing practice because overdifferencing brings patterns in the data that were not there to start with, rendering any analysis useless. Usually, a first regular difference ∇y_t will take care of the long-term trend in mean and then after that, if there is regular seasonality, differencing the remaining series once to remove the regular seasonal trend suffices. Or, sometimes, if the regular seasonal is a very strong component of the time series, a first seasonal difference is enough to give a time series without long-term trend and regular differencing is not needed.

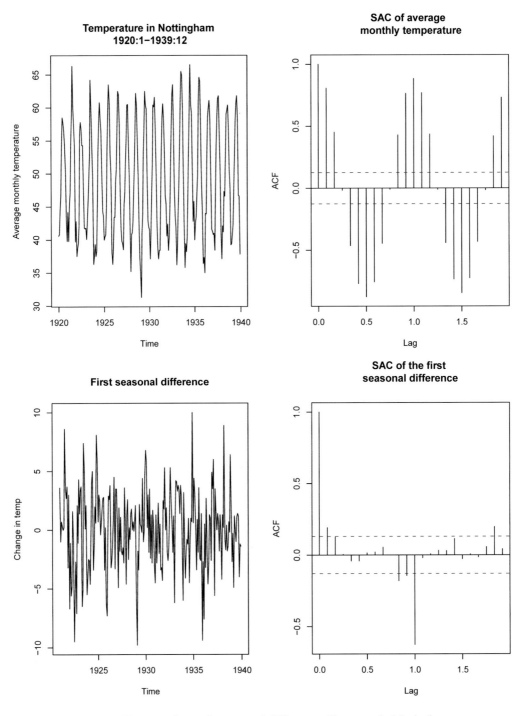

Figure 4.2 We can see how a first seasonal difference will convert the Nottingham average monthly temperature series into a stationary time series. The sample ACF (SAC in the title) of the stationary time series, that is, of the first seasonal difference, is very different from that of the nonstationary series, as we can see. The nonstationarity in the Nottingham time series is due to the seasonality.

Example 4.14 Consider quarterly time series x_t that has trend and that appears to have seasonality. Suppose we take the log and denote the new series by y_t, then take regular first difference at lag $k = 1$ and denote the new series by y_t^*. That did not take care of the regular seasonality in our example, so we need to do seasonal differencing:

$$y_t^{**} = \nabla_4 y_t^* = y_t^* - y_{t-4}^* = (1 - B^4)y_t^* = (1 - B^4)(1 - B)y_t.$$

In R, after doing regular differencing,

```
# seasonal diff of regular diff of log x_t
y.star.star=diff(y.star, lag=4, diff=1)
```

If the time series had been monthly, then `lag=12`. □

It is often not necessary to seasonally difference more than once. Notice that we lose a lot of data for model fitting when we do seasonal differencing. Thus, we always try this as a last resort, after we have tried regular differencing and a strong regular seasonal signal persists.

Sometimes, a time series has no long-term trend in mean, but has a strong seasonal trend. In that case, the only differencing needed is the seasonal differencing of the original time series, appropriate to the frequency of the time series.

And sometimes, even, the time series has long-term trend in mean and, without doing regular differencing, the seasonal differencing removes the long-term trend in mean and the seasonal trend.

Given a time series x_t that has been converted, if needed, to constant variance time series y_t, and presents a trend in mean and seasonality, a good practice is the following:

(i) Regular differencing once, $y_t - y_{t-1}$. This could be enough to get a time series without long-term trend and regular seasonal trend. This represents "change" in y_t, which has meaning to us.

(ii) Separately, seasonal differencing once only, $y_t - y_{t-F}$, where F is the frequency. This, also, at times, results in a a time series without long-term trend and without regular seasonal trend. The result is a series representing change from cycle to cycle, which also has meaning to us.

(iii) If neither of those two work separately, do regular differencing first and then do seasonal differencing on the regular difference. If y_t^* is the regular first difference time series, then do $y_t^* - y_{t-F}^*$ to seasonally adjust. The combined operation does not have a clear meaning. The more we difference, the harder it is to interpret what the resulting time series represents.

It is very common that one of those three combinations does the job of making the time series mean stationary. It is rare to need to do more than one of those operations.

> And we must recall that making the variance approximately constant is required
> before differencing. If that is not the case, the random component obtained after
> differencing will show nonconstant variance.

4.2.5   Notational Conventions to Use When Differencing and Backshifting

For reporting and for the sake of communicating accurately, it is important to keep
a notational convention when doing differencing and backshifting, because when
modeling with ARIMA in Chapter 6 the notation is needed to translate the model
specification into a familiar model form. To that end, it is necessary to keep track
of the operations done to the time series. Indicating with asterisks or different letters
any transformation done helps the person to whom the process is being explained and
the person doing the reports understand the analysis and not make mistakes in the
interpretation.

Example 4.15   As indicated earlier, quite often we regular difference and then after
that we do seasonal differencing. For example, given a time series x_t, which after log
to make the time series stationary we denote by y_t, we may do the following:

$$y_t^{**} = \nabla_4 \nabla y_t = (1 - B^4)(1 - B)y_t = (1 - B - B^4 + B^5)y_t = y_t - y_{t-1} - y_{t-4} + y_{t-5}.$$

In R, we would do this:

```
#### we do this first regular diff
y=log(x)
        y.star= diff(y, lag=1, diff=1)
###   then seasonal diff of reg diff
y.star.star=diff(y.star, lag=4, diff=1)
```

Notice how the polynomial specification of the operations that we did explains what
we did in a simplified and orderly fashion. First, the regular difference polynomial
$(1 - B)$ is written closest to the y_t because it was the first operation done. Second, the
seasonal difference polynomial $(1 - B^4)$, more distant from y_t, is written to the left of
the regular difference one, because it was done later. Writing the operations this way,
conveying the order in which they were done by proximity to the time series letter, and
with the leftmost one representing the last operation, helps translate quickly what the
practitioner did and contributes to reproducibility.   □

The differencing operations that we have talked about can be summarized in the following expression on a constant variance time series y_t, which is called the *polynomial
expression*:

$$\nabla_k^D \nabla^d y_t = (1 - B^k)^D (1 - B)^d y_t \qquad (4.1)$$

where

- d = number of times we do regular differencing. For example, if we regular difference once, $d = 1$, and $(1 - B)^d = (1 - B)y_t = \nabla y_t$. If we do not regular difference,
  then $d = 0$.
- D = number of times we do seasonal differencing $(1 - B^k)y_t = \nabla_k y_t$, k = the
  seasonal cycle or frequency of the time series. For example, if the seasonal differencing is for a monthly time series, and we seasonally difference just once,
  $(1 - B^k)^D = (1 - B^{12})$. If we do not seasonally difference, then $D = 0$.
- y_t is the log or square root transform of the time series x_t if x_t has nonconstant
  variance but $y_t = x_t$ if that is not the case.

The polynomial expression in Equation 4.1 is convenient, because, in just one line,
it summarizes to others what has been done to the time series in order to make it a
mean-stationary series (a series devoid of long-term trend and regular seasonal trend).
This operation will prove very useful in Chapter 6.

4.2.6 Exercises

Exercise 4.2 A time series x_t was logged to make the variance independent of time.
After that, the time series was regularly differenced twice and twice seasonally differenced. This practice is not recommended but just presented here for the sake of
practice. Write the operations done in compact backshift polynomial form and then in
difference operator form, as in Equation 4.1.

Exercise 4.3 Consider the `AirPassengers` time series. Determine first whether
any transformation is needed to make the variance not time dependent. Then decide
the type of differencing needed to make the time series mean stationary. Write the
operations done using the backshift operator and also write them with the difference
operator. Justify the steps taken by providing the ACF that prompted the differencing
and the final ACF.

4.3 The Outcome of Differencing

After differencing a time series that has constant variance, we expect to have accounted
for the trend and the seasonal and to have extracted the random component of the
time series properly. It takes practice to learn to recognize when that happens. The
act of regularly differencing is equivalent to detrending the time series and the act of
seasonally differencing is equivalent to seasonally adjusting the time series. However,
the detrending done using classical decomposition components is of a very different
nature from the detrending done using differencing.

An indication that we have reached stationarity (a time series without long-term
regular trend and without regular seasonal trend) would be a correlogram of a
stationary time series. For example:

- The correlogram of the differenced time series has statistically significant autocorrelations only at lower values of k if the data does not have seasonality, and will die away quickly after that (meaning that the autocorrelations will not be significant after that).
- The correlogram of the differenced time series has significant autocorrelations both at lower values of k and also at two or three significant autocorrelations around the seasonal k (12 if monthly, 4 if quarterly, etc.) if the time series has seasonality.

One common misunderstanding is that we are differencing the series to make it white noise. So people difference until they leave no correlation at all in the data, producing a sample ACF of white noise. That is not what we want. We want to capture the correlations at lower lags or at the seasonal lags, without all the clutter that comes with the nonstationarity caused by long-term trends and strong seasonals.

If there is no autocorrelation, then there is no point in modeling. Time series analysis is about modeling all the autocorrelation present in the data, in order to exploit that autocorrelation to perform better forecasts. We only want to see white noise after we have fitted the best possible model. The residuals of that model are what we want to look like noise.

4.4   The Spectrum of Time Series after Differencing

As indicated in Chapter 3, Section 3.6, the spectrum is the function of interest in spectrum analysis; it is a decomposition of the variance of a time series into frequency components. As such, it provides a clear interpretation for the decomposition of time domain processes into trend, cyclical, seasonal, and irregular components. Trends in mean will give rise to a spectrum that peaks at low frequencies. Seasonal trends will give rise to a spectrum that peaks at the seasonal frequencies. Cycles in a time series of frequency larger than the one of the seasonal will have frequency between that of the seasonal and that of trend in mean, and random irregular components will show other frequencies without regularity. After regular and seasonal differencing of a time series that has constant variance, we would expect that the spectrum of the remaining random term reveals only the true signals in the random term.

Each stationary stochastic process has a specific spectrum. The sample spectrum or periodogram may be regarded as an estimator of the true spectrum of the random process generating the sample. Like the correlogram, the sample spectrum or periodogram is full of artifacts that are not really present in the data. It is important that after regular and/or seasonal differencing, the signals present in the stationary time series are displayed in the periodogram clearly and match those of the stochastic process generating the data. Overdifferencing will render the periodogram useless as an indicator of the stochastic model that it represents. If the signal reflected in the periodogram is distorted because the wrong differencing was done, that periodogram will not reflect

Table 4.2 The reader will complete the table.

Time y_t	$y_{t-1} = By_t$	$y_t^* = \nabla y_t$	$y_t^{**} = \nabla^2 y_t = \nabla y_t^*$	$y_{t-2} = B^2 y_t$	$y_t^{***} = y_t - y_{t-2}$ $= \nabla_2 y_t$
2001 15					
2002 14.4064					
2003 14.9383					
2004 16.0374					
2005 15.6320					
2006 14.3975					

the true signal in the random term. Thus, sample periodogram analysis calls for the same cautions recommended for the interpretation of the sample ACF.

4.5 Problems

Problem 4.1 Complete Table 4.2, showing manual work. Then double-check your answers using R code.

Problem 4.2 Consider a model written as follows:

$$\left(1 - \alpha_1 B - \alpha_2 B^2\right) Y_t = (1 - \beta B^2) W_t,$$

where $\alpha_1, \alpha_2, \beta$ are model parameters, and B is the backshift operator. Write the model in a way that y_t is on the left-hand side of the equal sign, and lagged values of Y_t and W_t are on the right-hand side.

Problem 4.3 Consider the first 20 observations of monthly values of housing starts (not seasonally adjusted total new privately owned housing units started in the United States) given in Table 4.3. You may see a plot of the monthly nonseasonally adjusted time series between January 1959 and November 2019 in https://fred.stlouisfed.org/series/HOUSTNSA. You may also download the data set and read it into R from the directory where it is located by typing the following code in R :

```
houstnsa=read.csv("HOUSTNSA.csv",header=T)
```

or, alternatively, use the instructions given in the case study of Chapter 1 and use the API Quandl.

(a) Hand calculate the last two columns of Table 4.3, showing the calculations.
(b) After completing part (a), read the time series into R, make it a time series object, and obtain a time plot of the time series and another time plot of the seasonally differenced time series. Compare the two plots, and determine whether there are still some long-term trends of many years or a set of years such as business cycles that should be taken care of by regular differencing. If so, then do regular differencing. The following R code could be useful. Comment on your findings.

Table 4.3 The reader will complete the table to show the seasonally differenced values of the time series. Retrieved from FRED, Federal Reserve Bank of St. Louis; http://fred.stlouisfed.org/series/HOUSTNSA. FRED® graphs and images provided courtesy of the Federal Reserve Bank of St. Louis. ©2020 Federal Reserve Bank of St. Louis. All rights reserved. FRED® and the FRED® logo are the registered trademarks of the Federal Reserve Bank of St. Louis and are used with permission.

DATE	HOUSTNSA (y_t)	y_{t-12}	$\nabla_{12} y_t$
1959-01-01	96.2		
1959-02-01	99.0		
1959-03-01	127.7		
1959-04-01	150.8		
1959-05-01	152.5		
1959-06-01	147.8		
1959-07-01	148.1		
1959-08-01	138.2		
1959-09-01	136.4		
1959-10-01	120.0		
1959-11-01	104.7		
1959-12-01	95.6		
1960-01-01	86.0		
1960-02-01	90.7		
1960-03-01	90.5		
1960-04-01	123.0		
1960-05-01	130.2		
1960-06-01	122.8		
1960-07-01	114.3		
1960-08-01	130.3		

```
houstnsa=read.csv("HOUSTNSA.csv",header=T)
houstnsa[1:20,]
y=ts(houstnsa[,2],start=c(1959,1),
end=c(2019, 11),frequency=12)
plot.ts(y)
y.1star=diff(y,lag=12,differences=1)
plot.ts(y.1star,main=expression)
y.2star=diff(y.1star)
plot.ts(y.2star)
```

(c) Should we have taken log of the time series before the time series was differenced? Why? If we should have, do it, take logs or square roots to make the variance constant and repeat the analysis done in (a) and (b).

(d) Show the correlogram of the resulting random term of the time series.

(e) Compare the random term obtained with the differencing transformations and the random term that would result from decomposition of the time series. Decide whether multiplicative or additive decomposition is needed. If it is decided that the data needs variance stabilizing, do the decomposition on the variance-stabilized time series and compare with the variance-stabilized differenced data.

4.6 Quiz

Question 4.1

$$\nabla^3 y_t$$

is equivalent to

(a)

$$(1 - B^3)y_t$$

(b)

$$(1 + B^2 - B^3)y_t$$

(c)

$$(y_t - y_{t-3})$$

(d)

$$(1 + 3B^2 - 3B - B^3)y_t$$

Question 4.2 Consider the time series x_t:

```
x=ts(c(2,5,9,3,4,10,15,20,1),start=1950,frequency=1)

x11=diff(x,lag=1,differences=2)
```

In terms of the backshift operator, what is x11?

(a)

$$(1 - B)^2 x_t$$

(b)

$$(1 - B^2)x_t$$

(c)

$$(1 - B^2)^2 x_t$$

(d)

$$(1 - B)(1 - B_{12})y = x_{t-1}$$

Question 4.3 The AirPassengers data set that comes with R is a time series object. A data scientist did the following to remove the long-term trend of the series (without taking logs first, which perhaps they should have done):

```
dAP= diff(AirPassengers,lag=1,differences=1)
ddAP=diff(dAP,lag=12,differences=1)
```

Is that enough to obtain a time series that has no long-term trend and no regular seasonal trend?

(a) Yes, because the sample ACF of ddAP is the sample ACF of white noise.
(b) No, because the time plot of the ddAP shows increasing trend in mean.
(c) Yes, because the sample ACF of the ddAP shows that ddAP is stationary.
(d) No, because the time plot shows increasing variability with time.

Question 4.4   The `AirPassengers` data set that comes with R is a time series object.
   The following two operations were done:

```
dAP= diff(AirPassengers,lag=1,differences=1)
ddAP=diff(dAP,lag=12,differences=1)
```

Using the backshift and differencing operators, and denoting `AirPassengers` by y_t, those two operations are which of the following?

(a)
$$(1 - B - B^{12} + B^{13})y_t$$

(b)
$$(1 - B)^{12}y_t$$

(c)
$$(1 - B^{12} + B)y_{t-1}$$

(d)
$$(1 - B)(1 - B_{12})y_{t-1}$$

Question 4.5   The data set `AirPassengers`, y_t, is a time series containing the number of air passengers from January 1949 to December 1960.
   Someone typed the following commands in R:

```
AP=AirPassengers
air=AP[1:(length(AP)-12)]
newvar=air[1:(length(air)-1)] -air[2:length(air)]
```

This person did which of the following?

(a) $(1 - B)y_t$
(b) $(1 - B^{12})y_t$
(c) First regular difference of y_t
(d) None of the above

Question 4.6 $\nabla_2^2 y_t$ equals

(a) $y_t - 2y_{t-1} + y_{t-2}$
(b) $y_t - 4y_{t-4} + y_{t-2}$
(c) $y_t - 2y_{t-2} + 4y_{t-4}$
(d) $y_t - 2y_{t-2} + y_{t-4}$

Question 4.7 Consider the following time series:

t	y
1	2
2	5
3	9
4	3
5	4
6	10
7	15
8	20
9	1

An operation is done to the time series in R with the following command:

```
diff(y,lag=2,differences=2)
```

Which of the following is true?

(a) The new time series is $(1 - B^2)^2 y_t$.
(b) The new time series is $(1 - B)^4 y_t$.
(c) The new time series is $(1 - B)y_t^4$.
(d) The new time series is $(1 - B)^2 y_t^2$.

Question 4.8

$$\nabla^4 y_t$$

equals which of the following?

(a) $(1 - B^4)y_t$
(b) $(1 - B)^4 y_t$
(c) $(1 - B)y_{t-4}$
(d) $y_t - y_{t-4}$

Question 4.9 Read the following time series:

```
data=scan("viscosity.txt")
```

It is not a time a series object yet. Differencing is something that we would need to do if the time series has long-term trend or seasonal trend.

Would we need to difference this data set?

(a) Yes, because the sample ACF shows clearly that the time series is nonstationary in mean.
(b) Yes, because the time plot shows a very steep trend in mean.
(c) No, because the sample ACF shows that the time series is white noise.
(d) No, because the sample ACF shows that the data has constant mean.

Question 4.10   The *towels.txt* time series is a quarterly time series studied in Example 4.9. After making the time series an R time series object of class ts(), do the following operations separately.

(i) Seasonal differencing of towels once. Then do a time plot and an ACF plot of the seasonally differenced time series.
(ii) Regular differencing of towels once. Then do a time plot and an ACF plot of the regularly differenced series.
(iii) Seasonal differencing once of the regular differencing done in *(ii)*. Then do a time plot and an ACF plot of the resulting time series.

Which of the differencings resulted in a more stationary random term?

(a) The differencing in *(i)*.
(b) The differencing in *(ii)*.
(c) The differencing in *(iii)*.
(d) None of the differencing approaches tried made the resulting random term stationary.

4.7   Case Study: Smart Cities

With the Internet of Things (IoT), it has become routine to know the state of a city by using sensors to monitor data on traffic, temperatures, pollution levels, car parks' occupancy rates, gas usage, crime, and many other aspects of city life. Continuous monitoring is the main aspect of a smart city. In particular, nowadays it is possible to use sensors to count the number of vehicles entering and leaving an off-street car park and make these data publicly available to help make decisions and predictions based on the data. With the time series produced by those sensors, it is possible to study historical trends and use that to predict parking space availability for each parking alternative and offer this information to users on a web page or their mobile phones. Needless to say, smart cities have brought many new challenges to the practice of storing and analyzing the large amounts of data produced. Large-scale forecasting systems have become the norm in order to help local authorities plan.

The *Parking Birmingham Data Set* is one of the time series data sets of the UCI Machine Learning Repository (UCI ML) [40] [182]. The repository maintains many data sets used by the machine learning community for the empirical analysis of machine learning algorithms. A user-friendly searchable interface allows finding data sets of many types, including over 100 time series data sets [40]. These are flat files, downloadable into a computer. The advantage of having flat files is that, unlike API, users accessing the database at different times will get the same version of the data set, hence making comparisons between methods used by each user easier.

4.7.1   The Data

Solfi and colleagues [182] addressed the study of parking occupancy data published by the Birmingham city council of the United Kingdom, with the aim of testing several prediction strategies and allowing users to consult the occupancy rate forecast to satisfy their parking needs up to one day in advance. The data is a flat file downloaded from the UCI ML Repository that contains occupancy rates updated every 30 minutes (8:00 to 16:30) from 2016/10/04 to 2016/12/19 (11 weeks, consecutive days). We can find four variables: a SystemCodeNumber for the sensor, parking capacity, parking occupancy and the time the occupancy was last updated. As downloaded from the UCI ML repository, the data has 35,717 instances and no indication of missing values (in R's terms, no NA). A variable of interest is the occupancy rate, which can be calculated by dividing the parking occupancy by the parking capacity. The head() and tail() of the data, as provided by the UCI ML, are as indicated in the following lines. As the reader will notice, the data is given in long format (i.e., each parking data set stacked on top of the other):

```
  SystemCodeNumber  Capacity  Occupancy     LastUpdated
1      BHMBCCMKT01        577        61   10/4/16 7:59
2      BHMBCCMKT01        577        64   10/4/16 8:25
3      BHMBCCMKT01        577        80   10/4/16 8:59
4      BHMBCCMKT01        577       107   10/4/16 9:32
5      BHMBCCMKT01        577       150   10/4/16 9:59
6      BHMBCCMKT01        577       177  10/4/16 10:26

    . . . . . . . . . . . . . . . . . . . . . . . . . . . . .

    . . . . . . . . . . . . . . . . . . . . . . . . . . . . .

    . . . . . . . . . . . . . . . . . . . . . . . . . .

35712          Shopping       1920        1521 12/19/16 14:03
35713          Shopping       1920        1517 12/19/16 14:30
35714          Shopping       1920        1487 12/19/16 15:03
35715          Shopping       1920        1432 12/19/16 15:29
35716          Shopping       1920        1321 12/19/16 16:03
35717          Shopping       1920        1180 12/19/16 16:30
```

4.7.2   Inspecting the Quality of the Data

Without metadata or other information, we cannot do much to clean the data. But this data set is accompanied by a paper, which helps confirm our observations. In their paper, Solfi and colleagues [182] mention that the data is not very accurate, as sometimes the sensors are faulty or the whole data set may not be updated in a whole day in some parks. We used the program *Birmingham.R* to confirm what the authors say. The reader is encouraged to run this program while reading this section. With the program, we confirmed that some parkings have average percentage occupancy rates larger than 100 percent, which is obviously an error.

Also, we confirmed that some parkings have negative occupancy rate, which is impossible.

Two parks have very incomplete data, with less than 200 observations.

So those parks were looked at in more detail. There are 30 parks, but two of them have almost no information in them, and one has negative occupancy rates. It is also the case that the number of observations per parking differs. Before doing their analysis, Solfi and colleagues [182] did some adjustments, removed some observations and imputed others.

Still some parkings had less than the expected number of observations, perhaps because the sensor did not work that day. Some days do not have an observation at all (not indicated by NA or anything, just not recorded, because the sensor did not record anything).

Solfi and colleagues [182] did also observe that some parkings, in some days, had a very small standard deviation in the occupancy, and they interpreted that also as a faulty sensor. They also imputed some data by replacing values that were not recorded by the sensor with the previous hour's value or similar conventions. According to the authors, their finalized data set involved occupancy values for 29 car parks over ten days from 8:00 AM to 4:30 PM with measurements every half hour. We show the results of our data summaries for those 29 parks in Table 4.4. The reader will appreciate, after the discussions that we have had in this book so far, how closely we must look at the data in order to know what might have gone wrong and what needs fixing before we embark in any statistical or ML analysis.

With the few features of the parks given in Table 4.4, we proceeded to group the parks into similar clusters. As in Section 3.8, we applied kmeans to the data in the table, except the data for BHMBRTARC01. Looking at the cluster allocation in the table does not give a good idea about how well the clusters separate. So, as in 3.8, we show a plot of the allocations in Figure 4.3, labeled by cluster number and looking at the median occupancy and the maximum occupancy on one plot, and the median occupancy and standard deviation on the other plot. We can see that the parks cluster well, the different types are well separated from each other, and there is little overlap.

4.7.3   Subsetting and Reshaping the Data for App Testing

A vendor is interested in creating a phone app that predicts parkings and plans to test the app using part of the Birmingham data set. The vendor selects six parks of the

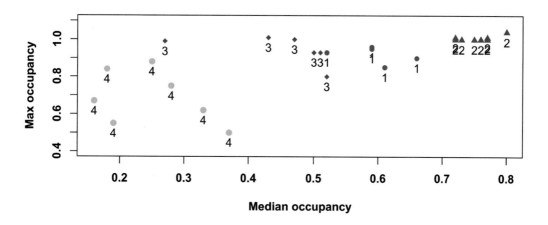

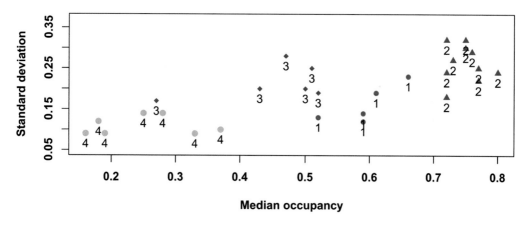

Figure 4.3 Clustering of parking structures based on selected features (summary statistics per half hour).

Birmingham data set, as downloaded from UCI, that had the same number of observations. The lots chosen were "Others-CCCPS135a", "Shopping", "Others-CCCPS8", "Broad Street", "Others-CCCPS98" and "BHMBCCMKT01", which all have 1,312 observations. The structures chosen differ in overall occupancy (some had overall low, others medium and others high occupancy). Figure 4.4 shows the temporal trace of the occupancy rate in each of the selected parks. As we can see, during the same time period, some parks were heavily occupied and others barely occupied. The scale is the same in each graph, so that means that some time series are closer to full rate (1 or 100 percent) than others. Broad is the busiest at all times and CCPS98 the least busy. All

Table 4.4 Summary statistics of the occupancy rates of 29 parking structures in the Birmingham data set (per half hour). This helps see that some occupancy rates are larger than 1, which is impossible; the sensors did not work all the time they should have worked (data length different across parks).

Park ID	mean	sd	max	min	median	length	cluster
BHMBCCPST01	0.43	0.20	1.01	0.00	0.43	1276	1
Others-CCCPS105a	0.56	0.14	0.95	0.22	0.59	1312	1
BHMNCPNST01	0.59	0.12	0.96	0.28	0.59	1312	1
Others-CCCPS135a	0.59	0.23	0.90	0.12	0.66	1312	1
Others-CCCPS8	0.50	0.13	0.93	0.17	0.52	1312	1
Shopping	0.55	0.19	0.85	0.13	0.61	1312	1
BHMBCCTHL01	0.74	0.24	1.04	0.10	0.80	1312	2
BHMEURBRD01	0.64	0.29	1.00	0.06	0.76	1312	2
BHMMBMMBX01	0.69	0.18	1.00	0.25	0.72	1312	2
BHMNCPHST01	0.46	0.19	0.80	0.05	0.52	1312	2
Broad Street	0.6	0.30	1.00	0.07	0.75	1312	2
BHMBRCBRG01	0.64	0.32	1.00	0.00	0.72	1186	2
BHMNCPNHS01	0.71	0.25	1.01	0.04	0.77	1038	2
BHMNCPRAN01	0.64	0.27	1.00	0.08	0.73	1186	2
BHMNCPLDH01	0.70	0.22	1.00	0.22	0.77	1292	2
BHMBCCSNH01	0.66	0.24	1.01	0.08	0.72	1294	2
BHMEURBRD02	0.62	0.32	1.00	0.01	0.75	1276	2
BHMBCCMKT01	0.28	0.17	0.99	0.00	0.27	1312	3
Others-CCCPS133	0.47	0.20	0.93	0.11	0.50	1294	3
BHMBRCBRG02	0.47	0.28	1.00	0.02	0.47	1186	3
Bull Ring.	0.47	0.25	0.93	0.03	0.51	1186	3
NIA Car Parks	0.16	0.09	0.67	0.00	0.16	1204	4
NIA South	0.25	0.14	0.88	0.00	0.25	1204	4
BHMNCPPLS01	0.19	0.12	0.84	0.01	0.18	1291	4
Others-CCCPS119a	0.19	0.09	0.55	0.02	0.19	1312	4
Others-CCCPS202	0.34	0.10	0.50	0.11	0.37	1312	4
BHMBRCBRG03	0.29	0.14	0.75	0.02	0.28	1186	4
Others-CCCPS98	0.33	0.09	0.62	0.13	0.33	1312	4
BHMBRTARC01	0.78	0.04	0.84	0.71	0.78	88	

of the time series exhibit daily seasonality (the working hours of the day have heavier occupancy) and weekly seasonality (the weekend days experience lower occupancy). Because there are barely three months worth of data, we do not have information to comment on annual seasonality (whether some months have higher occupancy rate).

The reader will notice that there is a big decrease during the weekend days. Sundays are the least busy days. The reader will also notice that during the week of October 17 to October 24 there is a strange line. Upon inspection we discovered that the sensor did not function between October 19 at 4:25 PM and October 22 at 7:59. The strange line in all the plots is not just data, but a jump covering those missing observations. Similar

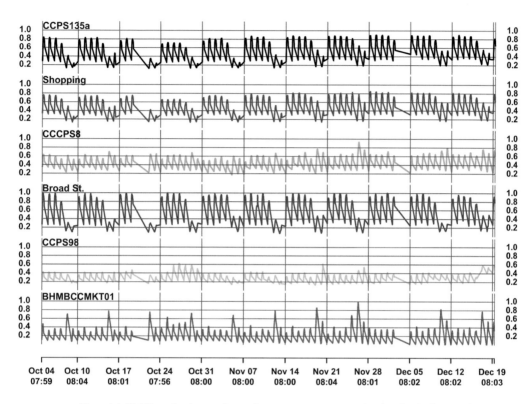

Figure 4.4 Half-hourly observations of percentage occupancy in six Birmingham parking structures during 11 consecutive weeks of 2016 show daily seasonality and weekly seasonality, as well as some anomalies in the data during the weeks of October 17 to October 24 and December 2 to December 5. Notice that the colors in this plot correspond to the colors in Figure 4.5.

phenomena happened during the week of November 28 to December 5. During the 3rd and the 4th of December, the sensor was not working.

A panel of multiple time plots like that in Figure 4.5 offers more detail of the daily and weekly seasonality. Each panel shows the daily trace of occupancy rate for each of the six parking structures studied. By looking at the traces of the time series this way, throughout the day and across the different days of the week, we can appreciate, for example, the differences between the weekend and the working weekdays' pattern.

The reader is encouraged to download the data from the UCI [182] and run R program *Birmingham.R*. The data displayed in Figure 4.4 is a reshaped version of the original data in order to have one time series for each parking lot as follows: one column for each parking lot's percentage occupancy rate. Also, because the time stamp was of type character we had to convert the character date to a `date` object in R. The reader will notice in the R program that we used the `mdy_hm()` function of the `lubridate` package. That function was used because the time stamp is given in `month/day/year hour:minute` format.

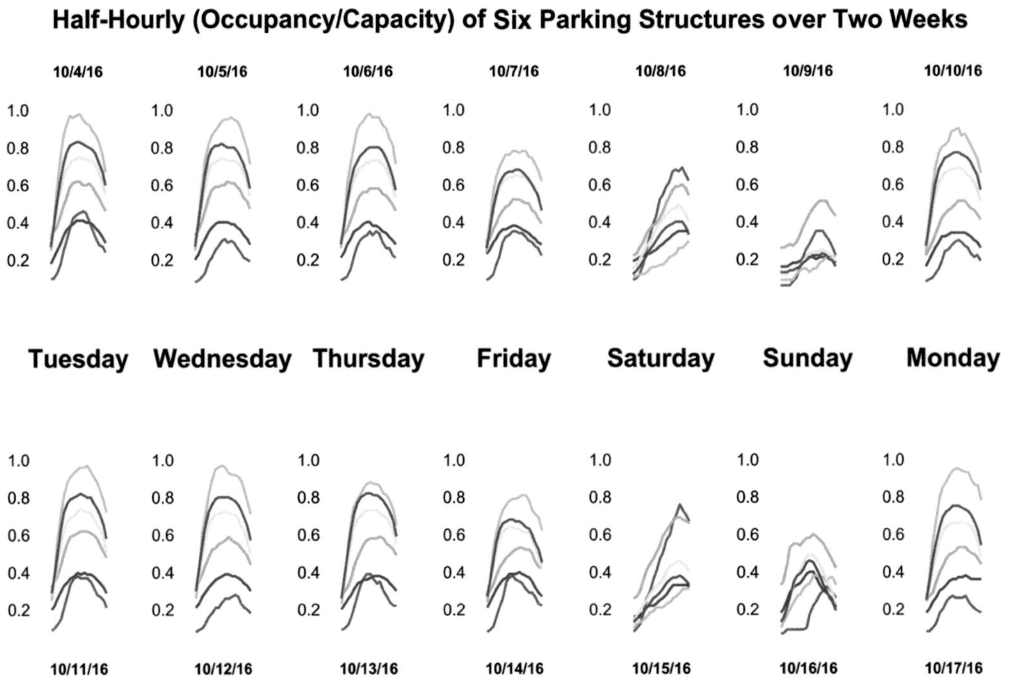

Half-Hourly (Occupancy/Capacity) of Six Parking Structures over Two Weeks

Figure 4.5 In this panel of multivariate time plots of percentage occupancy over two weeks, we observe the daily seasonality and the weekly seasonality in more detail. The Monday-to-Friday pattern is very similar across weeks, but Friday has slightly lower occupancy. The weekend has very different pattern and the lowest occupancy. The maximum percentage occupancy is 100 percent (1) for each park. At any given time in a given day, the city has parking somewhere. Notice that the colors in this plot do not correspond to the colors in Figure 4.4. Can the reader ascertain which parking structure displayed in Figure 4.4 corresponds to each curve color in Figure 4.5?

According to Hyndman [72], the easiest way to incorporate the daily nature of the time series into a model, when there is data only for one year, which is our case, is to use, for a time series denoted by x_t,

```
y=ts(x, frequency=7)
```

But the Birmingham data set is half-hourly data. In that case, Hyndman recommends using the number of half hours per day as one frequency and the total number of half hours recorded in the week as another frequency. So, the seasonal frequencies are set, using the `forecast` package, to 18, and then $18 \times 7 = 126$.

We can now proceed to do the same seasonal and correlation analysis of this subset data set that we did with the energy data set in Chapter 3, but we will leave that as an exercise.

4.7.4 Features That the App Will Need to Take into Account

If the vendor is interested in creating an app to help people in Birmingham find parking without wasting too much time, the vendor's algorithm will have to take into account the things that the exploration of the data set done here reveals:

- a daily seasonality (occupancy increases during the middle of the day) in all the parking structures during the working weekdays);
- a weekly seasonality (weekdays have higher occupancy rate than weekends) in all the parking structures;
- different shapes of the curve representing the occupancy during weekends and weekdays, with Sunday and Saturday showing a peak in occupancy at later hours of the day than during the weekdays (the fact that the peak hours are different during weekdays and weekends might complicate the daily seasonality analysis);
- the location of the parking structure, with some parking structures having much larger hourly occupancy throughout than others;
- if there were calendar holidays or events happening in the city, those too would have to be incorporated in the algorithm.

In just one case study we have found four different features of the time series (the day of the week, the hour of the day, the parking structure, holidays). That help explain the observed occupancy rate and would be important to consider were we to create one of the curve-fitting or functional models that are used nowadays to do forecasting at scale (that is, when for example, many parking structures have to be forecast at once with Prophet or a regression model).

> Let the excursion through the Birmingham data set be another indication to the reader of the large amount of exploration that we must do to understand the quality of a data set before we jump to conclusions about the data.
>
> But let this also be an opportunity to realize that most people that work as data scientists are always working with the same type of data. Facebook knows the typical problems encountered with its data. Google knows what features its data has. Sony knows its data, as well. So these companies have already prepared proprietary software that automatically checks the typical problems encountered with their data.

4.7.5 Exercises

Exercise 4.4 Do a sample ACF plot of the data set for each of the six parking structures selected and determine whether you should difference the data. Use what you learned in Chapter 4 to do an appropriate differencing of the six parking structures data in order to remove the weekly seasonality and the daily seasonality. Determine whether both are needed. Is there some significant autocorrelation in the time series after the differencing in any of the parking structures? (See the case study in Chapter 3).

Exercise 4.5   Some time series analysts approach the problem of missing observations by imputing the missing values with some alternative value. There are several approaches to doing that. Some impute the value before or after the missing one, or an average of those two. Others impute the predicted value given by some regression, or the average of the data for that day and hour in similar days, and so on. Each of these has its dangers. Those are methods for independent observations, and time series data has temporal dependence. How should we impute a value for the missing data of October 17 to October 24 and of December 3 and 4, when the sensor malfunctioned? Investigate the package `imputeTS` [126] and try one of the approaches mentioned in the documentation. Think about the effect that the imputation method could have in future analysis done with the data.[1]

Exercise 4.6   Consider the analysis that we did in the case study of Chapter 3 and displayed in Figure 3.14. Do similar seasonal box plots of the hour of the day and the day of the week for the six parking structures selected in this case study.

---

[1] If imputation is not a technique familiar to the reader in the context of independent observations, it is recommended to first become familiar with imputation in that context. The following consulting website has a very thorough discussion of the implications of imputing when the data are independent observations. Think how the problems mentioned there would extend to the temporally dependent time series data. This website has a brief summary, albeit for another software: https://stats.oarc.ucla.edu/stata/seminars/mi_in_stata_pt1_new/.

5   Stationary Stochastic Processes

5.1   Introduction

We have decomposed time series in Chapter 2 in order to obtain a random term devoid of trends and seasonals; we have distinguished the ACF of time series with and without trends or seasonals in Chapters 3 and 4. We have referred occasionally to the random term as stationary, as we were seeking constant mean and variance. But we have not yet given a precise mathematical definition of what a stationary time series is. The theoretical mathematical concept behind "free of trends and regular seasonals" and "constant variance," which are terms used in past chapters, is that of a *stationary stochastic process*. These processes have expected value, variance and autocorrelation that do not depend on time. They are appropriate for time series that are stationary.

> A stationary stochastic process is a statistical model of a time series that gives a precise statistical meaning to being free of long-term trends or seasonality. The theory of stationary stochastic processes explicitly declares what properties a process must have in order to be labeled as stationary and therefore to be a good candidate to be modeled by a stationary stochastic process. Box–Jenkins models that we will study in Chapter 6 model the random term of time series with stationary stochastic processes.

In competitions such as Kaggle [4] or M5 [5] [81], it is standard practice to compare new methods with benchmark Box–Jenkins models. The latter have been the most widely used forecasting models since the 1970s.

Although there are many stationary stochastic processes, each such model is characterized by its theoretical autocorrelation function and its theoretical partial autocorrelation function. If the time series sample has been generated by a stationary stochastic model, then the sample ACF and sample PACF will look approximately like the theoretical counterparts. Becoming familiar with some theoretical ACF and PACF, like we do in this chapter, will be helpful for the practice of Box–Jenkins modeling in Chapter 6, where we will have to identify a model for the random component of the data given the sample ACF and the sample PACF of the data.

Before the description in Section 5.3 of white noise and MA(q) models for stationary time series, we study in Section 5.2 the probability theory of stationary stochastic processes. The reader immediately interested in the applications may skip Section 5.2. Section 5.4 describes the theory of the AR(p) stochastic process. Section 5.5 describes ARMA(p,q) processes, and Section 5.6 provides practical guidelines for the identification of a stochastic process based on the sample ACF and sample PACF of a sampled time series. Simulation is then used in Section 5.7 to illustrate how to estimate the parameters of the stochastic process when all we have is a finite sample time series from that process. Some reflections on the limitations of stationary stochastic processes follow in Section 5.8. The spectrum is again briefly mentioned in Section 5.9. The case study of this chapter in Section 5.12 is about using simulation to understand future uncertainty, with a special application to COVID-19.

In Section 5.3 we restrict attention to special classes of processes that are particularly useful for modeling many of the time series that are encountered in practice. We shall supplement those with simulations, as other authors before us have done [30] [32], in order to train our eyes to identify the process in its sample ACF and sample PACF, a task that is always hard for beginners.

> The theory of stochastic processes teaches us that significant autocorrelations of stationary stochastic processes to look at in observed correlograms are those that occur usually at small lags $k = 1, 2, 3$ or closely around a seasonal lag k, if there is a seasonal random term. For those reasons, it is good practice to train our eyes to see the signal in the noise with simulated data before embarking on the interpretation of correlograms of real data.

The reader will find most of the code needed to reproduce the images in program *ch5simcode.R*. A warning is necessary. The reader should run simulations several times in order to appreciate the notion of ensemble, namely that there are many different pictures of the same process, as we indicate in Figure 5.1.

5.2   Stationary Stochastic Processes Theory

Many classical time series books start with the theory of stationary stochastic processes and they concentrate throughout the book on the probability theory of such processes [13] [60]. Understanding the probabilistic theory of stochastic processes, and hence those books, is facilitated immensely by having some elementary background in probability. In particular, the ability to use the expectation operator to calculate expectation, variance and covariance helps derive the theoretical autocorrelation function associated with a stationary stochastic process. The generalization of the theory to vector random variables is useful for a more compressed version of the theory. The reader who is not already familiar with probability and random vectors should first read Appendix A and Appendix B in Sections 5.13 and 5.14, respectively, where a concise account of the required background is given. More in-depth background can be obtained in many probability theory sources [173], [17], [175] and online [180], among others, and in the time series sources mentioned previously.

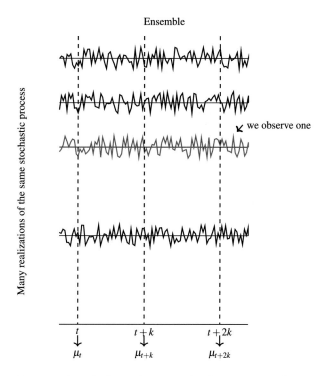

Figure 5.1 A particular stationary stochastic process runs forever and manifests itself in many different ways. Some authors call the many ways the *ensemble population* [30]. Each of the time series in this figure represents a window of one of the ways, and there are many more. In our practice, we observe only one of those ways, for example, the one in red, and we do not observe the whole duration of this way either, but a time window, an interval of an infinite process. At each time t there is a different average of all potential possible values that could occur at that time t, μ_t, and a variance, σ_t^2, of those values. The same stochastic model could be a good model for any of those different-looking time series. This notion informs the practice and the philosophy of many automatic forecasting (at scale) methods prevalent nowadays, such as Facebook's Prophet, Uber's Pyro, Google's and other contemporary best practices for forecasting at scale.

Probability plays a very important role in the study of stationary stochastic processes because these processes are viewed as processes about the joint behavior of the random variables $Y_1, Y_2, ..., Y_n$, also denoted as $\{Y_t\}$, or simply Y_t. The observed time series, y_t, contains only one observation of each of those random variables (the equivalent of the red time series in Figure 5.1). We develop ideal models for the joint distributions of those random variables and assume that the single observed time series is generated by one of those models.

In elementary statistics, we usually need not consider joint behavior of observations because they are assumed to be independent [131], and we usually have more than one observation for each random variable.

5.2.1   The Ensemble

Defining a stationary stochastic process as a sequence of random variables is equivalent to saying that a stationary stochastic process is like an ensemble of time series, as illustrated in Figure 5.1. A random variable can result in many possible values at a particular time t. Thus, a stochastic process could manifest itself in practice in many different sequences. And we, in practice, will just observe one, for example the red sequence, y_t, $t = 0, 1, 2, ...$ in Figure 5.1, and have to draw inferences about the whole process based on the autocorrelation function of that single red one. As Granger [55] once put it, in practice we are in a position such that:

"One can not stop the economy, go back to some starting point, and let it go once more to see if a different pattern would emerge" [55].

Our job as data scientists is to learn about the stochastic process using a single sequence.

Example 5.1   Figure 5.2 displays eight possible looks of the stochastic process

$$Y_t = 0.8Y_{t-1} + W_t,$$

where W_t are unobserved white noise, that is, a sequence of independent and identically distributed random variables, with expected value 0, and variance σ^2. Independence implies correlation between W_t, W_{t-k} being 0 for all k. The parameter $\alpha_1 = 0.8$ is constant. The reader will notice that this model looks, at first glance, like a simple regression model with dependent variable at each time t being Y_t and independent variable the value that Y takes the period before that, Y_{t-1}.

The signal in this model is $Y_t = 0.8Y_{t-1}$. We can see this signal in Figure 5.3, which, like the images in Figure 5.2, was obtained using program *ch5simcode.R*.

Notice that there could be many possible "realizations" of, or time series generated by, this process, because of the randomness of the W_t for given α_1. In the simulations, W_t has $\sigma^2 = 1$. As we can see in Figure 5.2, the eight realizations are different from each other, but the model generating them is the same. If we observe only series 6, for example, then we will have to try to discover the signal, $Y_t = 0.8Y_{t-1}$, using only that series. The reader should notice that because there are many possible time series that could be generated from the model, when the reader runs program *ch5simcode.R* the reader will get different time series. We did not use `set.seed()` to fix the images to avoid the reader getting the impression that those are the only images in the ensemble for the model.

This model implies that Y_t is itself a random variable, and the generation of the random variables that represent a time series sequence proceeds as follows:

$$Y_{t+1} = 0.8Y_t + W_{t+1},$$
$$Y_{t+2} = 0.8Y_{t+1} + W_{t+2},$$
$$\cdots\cdots\cdots .$$

Similarly, past values are generated by the same process as

$$Y_{t-1} = 0.8Y_{t-2} + W_{t-1},$$
$$Y_{t-2} = 0.8Y_{t-1} + W_{t-21}$$

where Y_t, for $t = \cdots - 2, -1, 0, 1, 2....$ are random variables.

Program *ch5simcode.R* will allow the reader to generate an ensemble similar to that of Figure 5.2. The reader should run the program several times to realize that the sequences will be different each time, and in none of them is the signal, seen in Figure 5.3, apparent. Understanding the concept of an ensemble generated by a stochastic process this way is crucial for understanding why we do time series analysis the way we do it, and the uncertainty that accompanies statistical analysis of time series.

A lesson learned from this example is that the same behavior over time, as indicated by the theoretical model, manifests itself in many different ways. Statistical analysis of time series is helpful because it helps us learn the commonality in many different patterns of behavior. ☐

To reinforce this idea of a stationary stochastic process viewed as a set of random variables, Figure 5.4 shows the value y_t for each of the eight series of the ensemble of Figure 5.2 at only a few times, $t = 1, \ldots, 10$. Notice how, at each time t, values of Y_t are spread out around the expected value of 0. The vertical dotted line at $t = 7$ (1998:5), for example, helps emphasize that some values of Y_t at time $t = 7$ could be above the expected value of 0, and some could be below the expected value of 0. The possible values in the ensemble shown are $3.318, 0.68, -0.93, 2.35, -3.07, 0.073, 2.25, 0.40$. In other words, a given stochastic process could generate very different values at any given time t.

The preceding discussion begs the question: if we observe only one of the sequences of the ensemble in Figure 5.2, for example, what other properties must be invoked in order to draw inferences about the model generating all those very different time series of Figure 5.2? With only one sequence, we just have one value of Y at each time t. We were told in introductory statistics that we need large samples for inferences to be reliable. Section 5.2.2 introduces some of those additional properties of stationary stochastic properties that must be invoked to answer that question. We must require the stochastic process to satisfy some assumptions for us to be able to learn the model from one observed series alone. Those properties are what makes the theory of time series sometimes very hard for beginners, particularly if the background in probability theory is not there. We next define what those properties are. At this point, the reader may want to go back to Appendix A and Appendix B to review the necessary probability concepts needed.

5.2.2 Weak-Stationary Stochastic Process

In this section, we present the general theory of stationary stochastic processes, which applies to any vector of random variables.

We say that vector **Y** is a weak-stationary stochastic process if it is first-order weak stationary and second-order weak stationary:

One stochastic process, many time series from it

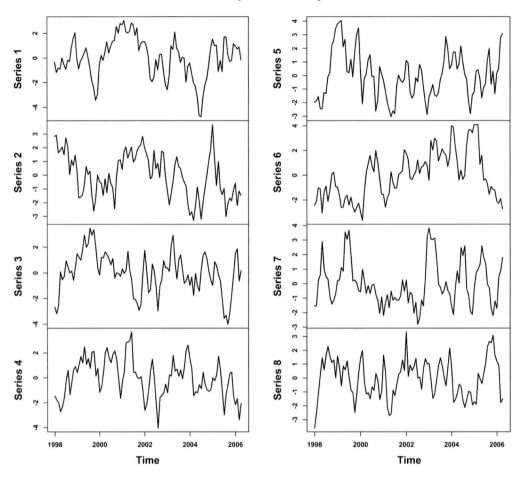

Figure 5.2 A subset of an ensemble. Several realizations of the following AR(1) stochastic process: $Y_t = 0.8Y_{t-1} + W_t$. We can appreciate that the time series look very different. Yet the model generating them is the same model, albeit with white noise added to it that generates the difference in the images. Discovering the signal, the $Y_t = 0.8Y_{t-1}$, hidden in that noise, when we, in practice, observe only one of those series, is the objective of time series analysis.

- First-order weak stationary: each element of the expectation vector is the same constant, μ, namely

$$
E(\mathbf{Y}) =
\begin{bmatrix}
E(Y_1) \\
E(Y_2) \\
\cdots \\
E(Y_i) \\
\cdots \\
E(Y_n)
\end{bmatrix}
=
\begin{bmatrix}
\mu_1 \\
\mu_2 \\
\cdots \\
\mu_i \\
\cdots \\
\mu_n
\end{bmatrix}
=
\begin{bmatrix}
\mu \\
\mu \\
\cdots \\
\mu \\
\cdots \\
\mu
\end{bmatrix}
= \mu.
$$

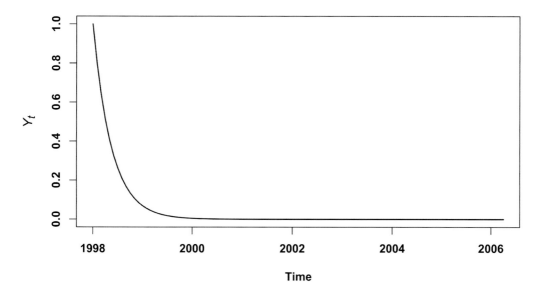

Figure 5.3 The signal hidden in each realization of the process represented in Figure 5.2, $Y_t = 0.8Y_{t-1}$ decays very exponentially and quickly. When considering the noise, one could not imagine that this signal is present in any of the time series in Figure 5.2. That is why we need statistical methods to find the signal in all the noise present in time series.

That is, if you took the mean of all possible values in the ensemble of Figure 5.2 at each time t, you would see that it is the same for all t if the process is weak stationary.

- Second-order weak stationary:

  – In the variance-covariance matrix, all the diagonal elements are the same and constant,

  $$\text{Var}(Y_t) = \sigma^2 \quad \text{for all } t.$$

  If we assume that $\text{Var}(Y_1) = \text{Var}(Y_2) = \cdots = \text{Var}(Y_n) = \sigma^2$, then we will denote it as γ_0, the covariance of the variable with itself.

  – The variance-covariance matrix's off-diagonal elements depend only on the absolute value of the difference between t and $t-k$; that is, the covariances depend only on lag k, the time does not matter, and we denote it by

  $$\text{Cov}(Y_t, Y_{t-k}) = E[(Y_t - \mu)(Y_{t-k} - \mu)] = \gamma_k.$$

  For example, if $\text{Cov}(Y_{10}, Y_{25}) = 3$, then $\text{Cov}(Y_{100}, Y_{115}) = 3$, $\text{Cov}(Y_{31}, Y_{46}) = 3$, and so on. It is for this reason that we can denote the covariances with only the lag subscript k. It is also for this reason that we may write the variance-covariance matrix's general expression as

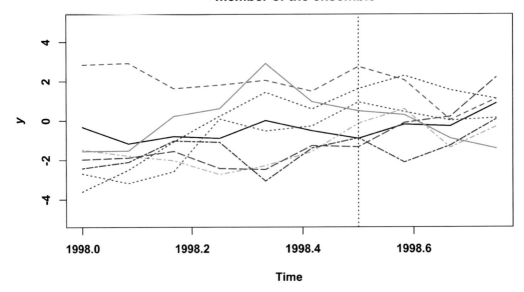

Figure 5.4 The values of the ensemble members of Figure 5.2 at selected times. Moving our eyes along the vertical dotted line, we appreciate that at a given time t, there could be very different values observed, depending on the realization of the process that we are given.

$$\Sigma_{\mathbf{Y}} = \text{VarCov}(Y) = E[(Y - E(Y))(Y - E(Y))^T]$$

$$= \begin{bmatrix} \gamma_0 = \sigma^2 & \gamma_1 & \gamma_2 & \gamma_3 & \cdots & ..\gamma_{n-1}. \\ \gamma_1 & \gamma_0 & \gamma_1 & \gamma_2 & \cdots & \gamma_{n-2} \\ \gamma_2 & \gamma_1 & \gamma_0 & \gamma_1 & \cdots & \gamma_{n-3} \\ \gamma_{k-1} & \gamma_{k-2} & \cdots & \gamma_0 & \cdots & \gamma_{n-4} \\ \cdots & \cdots & \cdots & \cdots & \gamma_0 \cdots & \cdots \\ \cdots & \cdots & \cdots & \cdots & \cdots & \cdots \end{bmatrix},$$

where $\gamma_k = \text{Cov}(Y_t, Y_{t-k})$.

Notice the structure of this matrix. The only parameters in $\Sigma_{\mathbf{Y}}$ are the covariances $\gamma_1, \ldots s, \gamma_n$ and the variance $\sigma^2 = \gamma_0$. The diagonal elements are identical and equal to the variance. The matrix is symmetric. Thus, this matrix is a Toeplitz matrix [56]. As a consequence, the autocorrelation of the stationary stochastic process with lagged values of itself also depends only on k and is also a Toeplitz matrix. The autocorrelation at lag k is defined as

$$\rho_k = \frac{\gamma_k}{\gamma_0} = \frac{\gamma_k}{\sigma^2}, \qquad k = 0, 1, 2, 3, \ldots,$$

and the autocorrelation matrix is given by

$$\rho(Y) = \frac{1}{\sigma^2} E[(\mathbf{Y} - E(\mathbf{Y}))(\mathbf{Y} - E(\mathbf{Y}))^T]$$

$$
= \begin{bmatrix}
1 & \rho_1 & \rho_2 & \rho_3 & \cdots & ..\rho_{n-1} \\
\rho_1 & 1 & \rho_1 & \rho_2 & \cdots & \rho_{n-2} \\
\rho_2 & \rho_1 & 1 & \rho_1 & \cdots & \rho_{n-3} \\
\rho_1 & \rho_2 & \cdots & 1 & \cdots & \rho_{n-i} \\
\cdots & \cdots & \cdots & \cdots & 1\cdots & \cdots \\
\cdots & \cdots & \cdots & \cdots & \cdots & \cdots
\end{bmatrix},
$$

where

$$
\rho_k = \mathrm{Corr}(Y_t, Y_{t-k}) = \frac{\mathrm{Cov}(Y_t, Y_{t-k})}{\sqrt{\sigma_t^2, \sigma_{t-k}^2}} = \frac{\mathrm{Cov}(Y_t, Y_{t-k})}{\sqrt{\sigma^2, \sigma^2}} = \frac{\mathrm{Cov}(Y_t, Y_{t-k})}{\sigma^2} = \frac{\gamma_k}{\gamma_0}.
$$

The diagonals are $\rho_0 = \frac{\mathrm{Cov}(Y_t, Y_t)}{\sigma_t, \sigma_t} = \frac{\sigma_t^2}{\sigma_t^2} = 1$. This explains why we saw in Chapter 3 that the plot of the values of the sample ACF always showed an autocorrelation equal to 1 at lag 0, a consequence of how we define the autocorrelations. The autocorrelation matrix is also symmetric and positive definite.

Notice that:

- ρ_0 is always 1, because it is the autocorrelation of each random element of the Y vector with itself.
- ρ_1 is the autocorrelation between random variable Y_t and Y_{t-1}. For example, $\rho_1 = \mathrm{Corr}(Y_1, Y_2) = \mathrm{Corr}(Y_4, Y_5) = \mathrm{Corr}(Y_{100}, Y_{101}), \ldots$.
- ρ_2 is the autocorrelation between variable Y_t and Y_{t-2}. For example, $\rho_2 = \mathrm{Corr}(Y_1, Y_3) = \mathrm{Corr}(Y_4, Y_6) = \mathrm{Corr}(Y_{100}, Y_{102}), \ldots$, and so on.

Figure 5.5 shows the mapping from the autocorrelation matrix of a stationary stochastic process to the theoretical autocorrelation function (ACF). At each lag k, the theoretical autocorrelation function gives the corresponding ρ_k in the autocorrelation matrix. This explains the appearance of the sample ACF observed in Chapter 3.

Stationarity is a concept that implicitly means "in a state of equilibrium." The behavior of the time series is stable, even though there is change at every time t. The covariance matrix and autocorrelation matrix do not change with time. They are only affected by the lag k. The mean does not change with time if the series is stationary either.

Weak-stationary stochastic process means:

- Constant mean μ for Y_t at all t.
- Constant variance σ^2 at all time t.
- Covariance at lag k that depends only on lag k (not on time t) and on the parameters of the model generating the time series. Notice that the variance is $\sigma^2 = \gamma_0$.
- Correlation at lag k that depends only on k, and on the parameters of the model generating the time series.

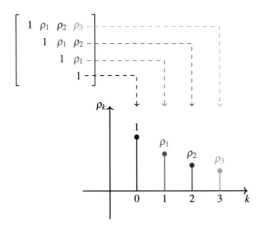

Figure 5.5 The autocorrelations in the autocorrelation matrix are what we see in the ACF, as indicated by this image.

Sometimes it is easier to understand a concept by looking at an example in which the concept does not apply.

Example 5.2   The following process,

$$Y_t = 6 + 10t + W_t,$$

where $\{W_t\}$ is white noise and t is time, $t = 1, 2, \ldots$, is not stationary. To prove it, we have to show that either the expected value, the variance or the covariances depend on time t. The following is a purely theoretical calculation, using the expectation and variance of a linear function of a random variable.

$$E(Y_t) = E(6 + 10t + W_t) = 6 + E(10t) + E(W_t) = 6 + 10t,$$

which is a function of t, and hence depends on t. The mean vector is not a vector of constants, it changes when t changes. Hence we already know that this process is not stationary because it is not stationary in the mean.

$$\mathrm{Var}(Y_t) = \mathrm{Var}(W_t) = \sigma^2,$$

and therefore does not depend on t.

$$\begin{aligned}
\mathrm{Cov}(Y_t, Y_{t-k}) &= E[(6 + 10t + W_t) - (6 + 10t)][(6 + 10(t-k) + W_{t-k}) \\
&\quad - (6 + 10(t-k))] \\
&= E(W_t W_{t-k}) \\
&= 0
\end{aligned}$$

for all t.

Therefore Y_t is NOT stationary because the expected value is a function of t even though the covariance is not.   □

The following is an example of a nonstationary process that can be converted to stationary using the differencing operations learned in Chapter 4.

Example 5.3 Suppose that $Y_t = \beta_0 + \beta_1 t + X_t$ where $\{X_t\}$ is a zero mean stationary series with covariances γ_k, and β_0 and β_1 are constants. Show that $\{Y_t\}$ is not stationary but that $Z_t = \nabla Y_t = Y_t - Y_{t-1}$ is stationary [32].

$$E(Y_t) = E(\beta_0 + \beta_1 t + X_t) = \beta_0 + \beta_1 t,$$

which clearly depends on time t, so this already indicates that Y_t is not stationary.

$$\begin{aligned}
\text{Cov}(Y_t, Y_{t-k}) &= E(\beta_0 + \beta_1 t + X_t - (\beta_0 + \beta_1 t))(\beta_0 + \beta_1 (t - k) + X_{t-k} \\
&\quad - (\beta_0 + \beta_1 (t - k))) \\
&= E(X_t X_{t-k}) \\
&= \gamma_k.
\end{aligned}$$

The covariance does not depend on t, because it is the covariance of the series X_t.

Notice that $Z_t = Y_t - Y_{t-1} = (\beta_0 + \beta_1 t + X_t) - (\beta_0 + \beta_1(t - 1) + X_{t-1}) = \beta_1 + (X_t - X_{t-1})$.

$$E(Z_t) = E(Y_t - Y_{t-1}) = E(Y_t) - E(Y_{t-1}) = \beta_0 + \beta_1 t - (\beta_0 + \beta_1(t - 1)) = \beta_1,$$

which is constant, it does not depend on t.

$$\begin{aligned}
\text{Cov}(Z_t, Z_{t-k}) &= E((\beta_1 + X_t - X_{t-1} - \beta_1)(\beta_1 + X_{t-k} - X_{t-k-1} - \beta_1) \\
&= E((X_t - X_{t-1})(X_{t-k} - X_{t-k-1})) \\
&= E(X_t X_{t-k}) - E(X_t X_{t-k-1}) - E(X_{t-1} X_{t-k}) + E(X_{t-1} X_{t-k-1}) \\
&= \gamma_{k-1} - \gamma_{k+1} + 2\gamma_k.
\end{aligned}$$

We know that none of the γ_k depend on t because X_t, the problem says, is stationary. Thus, neither the expected value of Z_t nor the covariance depend on t. So the Z_t is stationary.

We have shown, then, that although the Y_t is not stationary, the first regular difference of Y_t is stationary. ☐

More examples like those in Example 5.2 and 5.3 can be found in [32].

Example 5.4 A random walk, which is used to model financial time series and the position of particles suspended in fluid, among many other phenomena, is a nonstationary stochastic process, despite having a name that may lead us to think otherwise.

The derivation of the expected value, variance, and covariance of a random walk is found in numerous textbooks [32, 60, 151]. ☐

5.2.3   Ergodicity

Stationarity alone does not justify basing our whole conclusion about a stochastic process on the sample ACF and the sample PACF, the mean and the variance of one single observed time series. For that, we need the additional property of ergodicity.

The property of a time series stochastic process that allows us to estimate all the parameters that characterize an ensemble with just one member time series of the ensemble, the member that we observe in our practice, is *ergodicity*.

Ergodicity is a property of a time series that requires that observations that are sufficiently far apart in time should be almost uncorrelated [60], so that by averaging a series through time one is continually adding new and useful information to the average.

An ergodic process means that the ensemble averages and autocorrelations can be computed from a window of the process.

Ergodicity refers to one type of asymptotic independence [151]. If a time series stochastic process is ergodic, the summary statistics learned in Chapter 3 are consistent estimates of the mean, variance and autocorrelations of the stochastic process. Thus, the time average

$$\bar{y} = \frac{\sum_{t=1}^{n} y_t}{n},$$

n being the length of the observed time series, is an unbiased and consistent estimate of the population mean μ, so that var($\bar{y}$) goes to 0 as n goes to infinity and $E(\bar{y}) = \mu$, for all n. Similarly, the r_k will also be consistent estimates of ρ_k.

Given stationarity and ergodicity, one can form good estimates of the quantities of immediate interest by averaging through time rather than being forced to depend on the ensemble averages across realizations. Unfortunately, it is not possible to test for ergodicity using a single realization, but if we do not have strict cyclical components or strong trends of any kind, we can be confident that it is satisfied. A necessary condition for ergodicity, but not a sufficient one, is that the correlations go to zero for lag k large.

For many models that we study in this book, stationarity implies ergodicity and we will make the assumption that the property is guaranteed. All the procedures learned so far in past chapters will help us do that.

Weak-stationarity and ergodicity imply, for the practice of time series analysis, that

- There is only one mean to estimate from data, the μ. So although we have only one observation for each of $Y_1, Y_2, \ldots$, we could estimate the μ by using the mean of our observed $y_1, y_2, \ldots, y_n$.
- There is only one variance to estimate, so we obtain the observed sample variance calculated from the observed time series.

- There is only one correlation for each lag k, and we can use the observed time series to estimate it, as we explained in detail in Chapter 3, Section 3.3.

By assuming stationarity, we reduce the number of parameters that we have to estimate from the observed time series that we usually start our analysis with. The summary statistics described in Chapter 3 make sense if there is stationarity. That is why, in past chapters, we insisted so much in calculating the summary statistics only of the random term of a time series, and not of time series with trends or very regular seasonals. In Chapter 3, we introduced the $\bar{y}, s^2, r_k$ of a time series free of trends and regular seasonals. They are the estimates of the variance-covariance and mean vector of the ensemble.

The property of stationarity reduces the number of parameters we must estimate.

5.2.4 Mathematical Autocorrelation Function (ACF)

The mathematical ACF of a stationary stochastic process is a graphical display of the upper triangular values in the autocorrelation matrix, as indicated in Figure 5.5. The ACF has very distinctive properties:

- The theoretical autocorrelation function must show the value of ρ_k for lags $k = 0, 1, 2, \ldots$, a large number of lags.
- The ρ_k are functions of the parameters of the model representing the stochastic process. Thus, one can go from the parameters to the ρ_k and vice versa.
- The theoretical autocorrelation function lacks the extraneous features that one often sees in sample correlograms.

Notice that $\gamma_k < \gamma_0$ implies that $|\rho_k| \leq 1$, and notice also that $\rho_k = \rho_{-k}$ because the variance-covariance matrix of the vector corresponding to the time series is a symmetric positive definite matrix. It has a nonnegative determinant.

5.2.5 Strong Stationarity

There is also the concept of strong stationarity, which says that the joint distribution of $Y_1, Y_2, \ldots, Y_k$ is unchanged throughout the duration of the series. This is a very restrictive assumption and it is not used much.

5.2.6 Exercises

Exercise 5.1 Suppose $Y_t = 8 + 3t + X_t$ where X_t is a zero mean stationary series with autocovariance function γ_k.

(a) Find the mean function for Y_t.

(b) Find the autocovariance function for Y_t, starting with γ_0.

(c) Is Y_t stationary? Why or why not?

Exercise 5.2   Suppose $\text{Cov}(X_t, X_{t-k}) = \gamma_k$ is free of t but that $E(X_t) = 3t$ [32].

(a) Is X_t stationary? Show work.

(b) Let $Y_t = 7 - 3t + X_t$. Is $\{Y_t\}$ stationary? Show work.

Exercise 5.3   Suppose that X_t is a random variable with zero mean and constant variance. Define a time series process by $Y_t = (-1)^t X_t$ [32].

(a) Find the mean function for Y_t.

(b) Find the covariance function for Y_t.

(c) Is Y_t stationary?

Exercise 5.4   The following information is given to you about a stochastic process Y_t: $\rho_0 = 1$, $\rho_1 = -0.3796633$, $\rho_k = 0$ for $k > 1$; $\mu = 0$, $\sigma^2 = 1$. Which of the following models could represent this process?

(a) $Y_t = W_t + 0.4683749 W_{t-1}$

(b) $Y_t = W_t - 0.4683749 W_{t-1}$

The W_t is white noise (0 mean and uncorrelated series).

5.3   Models for Stationary Stochastic Processes

In Chapter 6, once we have a stationary time series, we use the sample ACF and the sample PACF studied in Chapter 3 to identify a stationary stochastic model describing the observed stationary time series values (the random component). Two useful types of stationary stochastic models are *autoregressive models* and *moving average models*. Familiarizing ourselves with the mathematical properties of these models helps us determine what it is that we must look at in a sample ACF and sample PACF to identify a model based on the time series data we have.

In the next few sections, we will derive the expected value and variance-covariance and correlation matrices of some common stochastic processes, and we will draw the implied mathematical autocorrelation function (ACF). The intention of doing this is twofold: first, it helps to see that the covariances depend on the model parameters; second, it helps us in the identification of a model to fit to an observed time series by matching the observed sample ACF with the mathematical one, which is something we will do in Chapter 6.

For each of the stochastic process models discussed, we will also show an example of a time series that could have been generated by that model, and that generated series' sample ACF and sample PACF. The purpose of doing this is to make the beginner aware of the fact that an observed time series' properties will not be exactly like those of the stochastic model generating it. Sometimes it will be hard to discern which theoretical ACF is the one matching the observed sample ACF.

Some training is needed to distinguish chance autocorrelation observed in the sample ACF from authentic signal autocorrelation present in the unknown stochastic process generating our time series.

The sample ACF of an observed time series does not exactly look like the mathematical ACF of the stochastic process generating the series.

The stationary stochastic processes that we will be seeing in the next few sections are: white noise, moving average processes (MA), autoregressive processes (AR), and combined autoregressive and moving average processes (ARMA). White noise goes first, because it is part of all the other models.

5.3.1   White Noise Stochastic Process

A white noise stochastic process is the simplest type of time series model, relevant when a time series Y_t consists of uncorrelated random variables having mean 0 and constant variance σ^2. This represents the effect of unknown factors on the value of Y_t.

The name "white noise" was coined in an article on heat radiation published in *Nature* on April 1922, where it was used to refer to series that contained all frequencies in equal proportions, analogous to white light [30].

The time series stochastic model for white noise is

$$Y_t = W_t,$$

where W_t is a sequence of statistically independent and identically distributed variables with

$$E(W_t) = 0, \quad t = \ldots, -3, -2, -1, 0, 1, 2, 3, \ldots,$$

and

$$\text{Var}(W_t) = \sigma^2, \quad t = \ldots, -3, -2, -1, 0, 1, 2, 3, \ldots,$$

the same for each and every time period. The covariance for a white noise process is

$$\gamma_k = E[(W_t - 0)(W_{t-k} - 0)] = 0, \quad k = 1, 2, 3, \ldots,$$

the latter being true because independence implies covariance and correlation equal to 0 (probability theory teaches us).

$$\rho(Y_t, Y_{t-k}) = 0, \quad k = 1, 2, \ldots$$

As we can see, the only parameter to be estimated is the σ^2, which is estimated by the variance of the data, adjusted by degrees of freedom.

With these properties, we can check the stationarity of the process represented by Y_t by using the properties of the expectation operator:

$$E(Y_t) = E(W_t) = 0, \quad t = \cdots, -3, -2, -1, 0, 1, 2, 3, \ldots$$
$$\text{Var}(Y_t) = \text{Var}(W_t) = \sigma^2, \quad t = \cdots, -3, -2, -1, 0, 1, 2, 3, \ldots$$
$$\gamma_k = E[(Y_t - \mu)(Y_{t-k} - \mu)] = E[(Y_t - 0)(Y_{t-k} - 0)] = E[(W_t)(W_{t-k})] = 0$$

for all k. This last property implies that all autocorrelations are 0.

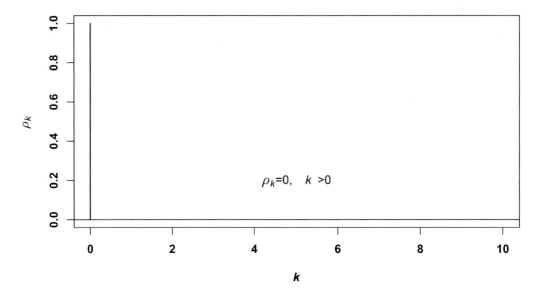

Figure 5.6 The ACF of a white noise process has all autocorrelations $\rho_k = 0, k > 0$. As in all stationary stochastic processes, the autocorrelation of the time series at time t with itself at time t is 1, and therefore that vertical line at lag $k = 0$ should be ignored in all ACFs.

> A white noise stochastic process is the only stationary process that has all autocorrelations equal to 0, except ρ_0, which is 1 for every process, and should be ignored in the ACF. These properties mean that there is nothing left to model.

Thus, white noise is stationary because the expected value, the variance and the ACF do not depend on time t, as we just showed. In fact, a white noise process has autocorrelation 0 at each lag k different from 0. The ACF looks like that in Figure 5.6.

5.3.2   Observed Time Series Generated by a White Noise Process

Two time series generated by two white noise processes are simulated and their sample ACF are shown in Figure 5.7. Program *ch5simcode.R* is used for that.[1] The reader will notice that data that is generated by white noise processes has the properties of the process itself, albeit not as perfectly, since data is never the actual model. There is usually a lot of extraneous details in the sample ACF that are not in the mathematical ACF, shown in Figure 5.6. Those extra details are due only to the fact that the data is

---

[1] To give the reader a more convincing evidence that data generated from stochastic processes can manifest itself in many different ways, as shown in the ensembles displayed in Sections 5.1 and 5.2, we do not use a seed. Thus the reader will obtain a different white noise process when running the program, and consequently a different sample ACF . We recommend running the problem several times and observing what we describe in this section.

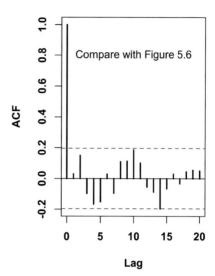

SACF of white noise is not exactly like the ACF

Compare with Figure 5.6

SACF could even have significant chance autocorrelations

Compare with Figure 5.6

Figure 5.7 Two different realizations of a white noise process give very different sample ACF (SACF), and sometimes the observed white noise's sample ACF will show significant autocorrelations which are of course chance autocorrelations. The randomly scattered approximately 5% significant autocorrelations are scattered along the sample ACF, they make no sense and should be ignored.

only a sample, not the whole process. Those details are there by chance. There will be around 5 percent random significant r'_k randomly scattered along the correlogram, as we mentioned in Chapter 3, and should be ignored. Those are the artifacts that make a sample correlogram different from the theoretical autocorrelation function. The reader can appreciate those artifacts in Figure 5.7. Another thing to ignore is the autocorrelation at lag $k = 0$, which, as we have indicated earlier, is 1 for every observed time series.

> The curse of time series analysis in practice is that about 5 percent of the r_k observed in the correlogram of a sampled time series will be statistically significant by chance alone, and there will be a lot of nonzero autocorrelations that are arti- facts of the sampling, not real. Those are randomly scattered along the correlogram and should be ignored.

If an observed time series exhibits behavior of a white noise process (ignoring the 5 percent by chance significant ones), then the Ljung–Box test will indicate so.

The following are examples of observed time series in which we would expect to observe white noise behavior:

- Residuals obtained after fitting a good regression model or one of the time series models learned later in this book to the random term of a time series. Such behavior in residuals obtained this way is an indication that the model fitted to the random term captured all the relevant autocorrelations or patterns in the random term, and there is no room for improvement. If y_t is an observed time series, and $\hat{y}_t$ is the fitted model, then the residual is $y_t - \hat{y}_t$. As the residuals occur in time, they also form a time series. This residual time series being white noise indicates to us that we may stop modeling.
- Some financial time series, for example, stock return or stock volume. These time series, however, are such that perhaps the correlogram of the square series shows relevant statistically significant r_k, indicating that they are conditionally heteroscedastic time series. The variance of these time series will have to be modeled with a model that accounts for its autocorrelation, such as ARCH or GARCH models, which will be mentioned in Chapter 6.
- The difference of a random walk time series is white noise. A random walk process, which is an infinite sum of discrete white noise terms, and is nonstationary, turns into a white noise time series after differencing. That is not the case for most differenced nonstationary time series.

Very rarely, the random term of a time series that has been properly decomposed or differenced and is not a random walk will be found to be white noise.

5.3.3   Exercises

Exercise 5.5   Use the R program used to construct Figure 5.7, which you will find in program *ch5simcode.R*, to generate five different white noise time series of length $n = 100$. Do not use seed to guarantee that all the time series generated are different. For each of them, find the sample ACF and the sample PACF. In how many of them did you find some significant autocorrelation in the sample ACF? Repeat the process again and answer the question again.

5.3.4   Moving Average Models

Moving average stochastic processes of order q, also known as MA(q) processes, represent the time series as a finite linear combination of white noise terms:

$$Y_t = \beta_1 W_{t-1} + \cdots + \beta_q W_{t-q} + W_t,$$

and because they are made of stationary white noise terms, they are stationary. The model is saying that the stationary observed time series equals a random shock corresponding to time period t plus a constant linear combination of past shocks at times $t-1, \ldots, t-q$. The constants β_i are unknown parameters that must be estimated from sample data.

In polynomial form,

$$Y_t = \left(1 + \beta_1 B + \cdots + \beta_q B^q\right) W_t.$$

This process is *invertible* if it can be expressed as an infinite linear combination of past values of Y_t that converges to a finite value. Some conditions must be imposed on the coefficients in order to have invertibility. For example, an MA(1) is invertible if $| \beta | < 1$, which is required for convergence [30]. If the condition is not satisfied, the process is not invertible. Bowerman [8] contains a table indicating what are the parameter conditions for MA of several orders and the reader is invited to check that source.

A faster way to check invertibility is by calculating the roots of the backshift polynomial and checking whether all exceed 1 in absolute value.

Invertibility must be satisfied for us to be able to identify a model.

- All MA(q) stochastic processes are stationary because they are linear combinations of white noise terms.
- MA(q) are invertible if the roots of their backshift polynomial all exceed 1 in absolute value. An MA model cannot be identified if it is not invertible.

Example 5.5   Consider the process:

$$Y_t = W_t - \frac{1}{2}W_{t-1} + \frac{1}{4}W_{t-2}.$$

We rewrite it as

$$Y_t = \left(1 - \frac{1}{2}B + \frac{1}{4}B^2\right)W_t.$$

In R, we find the modulus of the roots of this polynomial in B:

```
Mod(polyroot(c(1, 0.5, 0.25)))
```

which gives roots $2, 2$. The modulus of the roots is larger than 1, therefore the process is invertible. ☐

Being able to derive mathematically the properties of MA(q) processes helps gain an understanding of what stationarity means, the relation between the model parameters and the autocorrelations, and the autocorrelations that we should expect to see as statistically significant in a sample ACF of data generated by an MA process. We will illustrate with models of small order.

5.3.5   The Moving Average of Order 1, MA(1)

Consider an MA(1) process:

$$Y_t = W_t + \beta_1 W_{t-1} = (1 + \beta_1 B)W_t,$$

where the W_t is white noise.

Then

$$\mu_t = E(Y_t) = E[W_t + \beta_1 W_{t-1}]$$

$$= E[W_t] + \beta_1 E[W_{t-1}]$$
$$= 0 - \beta_1(0)$$
$$= 0,$$

which is constant and does not depend on t; and

$$\text{Var}(Y_t) = \gamma_0 = \text{Var}(W_t) + \beta_1^2 \text{Var}(W_{t-1})$$
$$= (1 + \beta_1^2)\sigma^2,$$

which is constant and does not depend on time t; and

$$\gamma_1 = \text{Cov}(Y_t, Y_{t-1}) = E[(Y_t - 0)(Y_{t-1} - 0)]$$
$$= E[W_t + \beta_1 W_{t-1})(W_{t-1} + \beta_1 W_{t-2})]$$
$$= E[(W_t W_{t-1})] + \beta_1 E[(W_{t-1}^2)] + \beta_1 E[(W_t W_{t-2})]$$
$$+ \beta_1^2 E[(W_{t-1} W_{t-2})]$$
$$= \beta_1 E[(W_{t-1}^2)]$$
$$= \beta_1 \sigma^2.$$

So, the autocorrelation for lag $k = 1$ is

$$\rho_1 = \frac{\gamma_1}{\gamma_0}$$
$$= \frac{\beta_1 \sigma^2}{(1 + \beta_1^2)\sigma^2}$$
$$= \frac{\beta_1}{1 + \beta_1^2}.$$

Notice how the autocorrelation is a function of the model parameter β_1. We can show that

$$\gamma_k = \text{Cov}(Y_t, Y_{t-k}) = 0, \qquad k > 1$$

and therefore,

$$\rho_k = 0, \qquad k > 1.$$

We have shown that the mean and variance of an MA(1) process do not depend on t. We have also shown that for any lag k, the autocovariance and therefore the autocorrelation do not depend on t. Therefore, we have shown that an MA(1) process,

$$Y_t = W_t + \beta_1 W_{t-1},$$

is a stationary stochastic process.

Example 5.6   Figure 5.8(a) shows the ACF of an MA(1) with parameter 0.5. In this case, the only autocorrelation different from zero other than at lag 0 is at lag $k = 1$, and its value is 0.4, calculated as

$$\rho_1 = \frac{0.5}{1 + (0.5)^2} = 0.4.$$

□

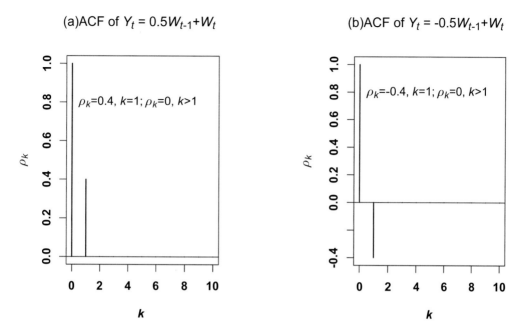

Figure 5.8 The ACF of an MA(1) process has only lag $k = 1$ autocorrelation different from 0. If the β_1 coefficient is positive (as in (a)) autocorrelation will be positive. It will be negative otherwise. If a process is MA, the ACF determines the order of the process.

Example 5.7   What would be the ACF and PACF of an MA(1) with parameter -0.5? In this case, the only autocorrelation is at lag $k = 1$, and its value is -0.4, calculated as

$$\rho_1 = \frac{-0.5}{1 + (-0.5)^2} = -0.4.$$

We can see the theoretical ACF in Figure 5.8(b).   □

Notice that when β_1 is positive, ρ_1 is positive and successive values of Y_t are positively correlated, and so the process will tend to be smoother than the random series W_t. On the other hand, a negative value of β_1 will yield a series that is more irregular than a white noise series, in the sense that positive values of Y_t tend to be followed by negative values and vice versa. This is reflected in the autocorrelation function, as ρ_1 is negative when $\beta_1 < 0$.

The MA(1) stochastic process has these properties:

- $\mu = 0$ for Y_t at all t
- $\mathrm{Var}(Y_t) = (1 + \beta_1^2)\sigma^2$ at all time t
- $\rho_1 = \frac{\beta_1}{1+\beta_1^2}$
- $\rho_k = 0, k > 1$

An MA(1) is a stationary process with those features.

> The autocorrelation function is not completely specified until we have shown what is the autocorrelation at each lag, that is, ρ_k, $k = 0, 1, 2, 3, 4, \ldots$.

5.3.6 Be Aware of Software Conventions

It is important to know which model specification the software is using. R uses the one we are using here, but SAS, for example, expresses the model with negative signs before the coefficients. Even in some R packages created by different authors for their textbooks (not base R though) the specification is with negative signs. For example, the library TSA in R specifies the model with the negative sign. Be aware that the assumption

$$Y_t = W_t + \beta_1 W_{t-1}$$

will give different results than

$$Y_t = W_t - \beta_1 W_{t-1},$$

and the signs in the γ_k will be different.

5.3.7 Observed Time Series from an MA(1)

Figure 5.9 shows the sample ACF of two simulated time series from the MA processes

$$Y_t = 0.5 W_{t-1} + W_t$$

and

$$Y_t = -0.5 W_{t-1} + W_t.$$

The reader should compare the image for the sample autocorrelations with those of the mathematical ACF in Figure 5.8 to appreciate how the ACF of a sample from the stochastic process makes the sample ACF contain a lot of extraneous autocorrelations, the mentioned 5 percent statistically significant.

5.3.8 Exercises

Exercise 5.6 Simulate from an MA(1) with $\beta_1 = 0.4$,

$$Y_t = W_t + 0.4 W_{t-1},$$

where W_t is a white noise process.

Use program *ch5simcode.R* and answer the following questions:

(a) Describe the time plot and the correlogram of this MA(1). Compare the correlogram to the theoretical autocorrelation function.
(b) Are there any statistically significant r'_ks in the correlogram? Which ones? What are their value? Compare to what would be expected if this had been the theoretical autocorrelation function.

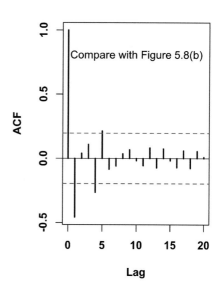

(a) SACF of observed time series not exactly like the ACF

Compare with Figure 5.8(a)

(b) SACF could have more than expected autocorrelations

Compare with Figure 5.8(b)

Figure 5.9 (a) The sample autocorrelation, SACF, from a moving average process $Y_t = 0.5W_{t-1} + W_t$ will not look exactly like the theoretical one. Artifacts showing chance autocorrelations, at least 5 percent of the autocorrelations, should be ignored. (b) The sample autocorrelation from a moving average process $Y_t = -0.5W_{t-1} + W_t$ has the same problem.

(c) Repeat all the code four times to see if you get the same conclusions you reached in the last two questions.

(d) What is the conclusion of the Ljung–Box white noise test?

(e) Plot together the correlogram and the partial correlogram. Describe their main features.

(f) Change the sign of β_1 and make it negative. Repeat everything we did for the MA(1) model since the start. Write your conclusions here. Has anything changed? Explain.

(g) To realize that when the time series is not very long, there will be more garbage in the correlogram, generate a time series of size 100 and go over the questions asked earlier again. What is the difference? What should your eyes be most focused on amidst all the garbage that appears?

Exercise 5.7 Show that ρ_2 and ρ_3 are 0 for the MA process of Example 5.6 .

Exercise 5.8 In an MA(1) model, show that ρ_1 is the same for $\beta_1 = 1/2$ as for $\beta_1 = 1/(1/2)$. If we know that an MA(1) process had $\rho_1 = 0.4$, we still could not tell the precise value of β_1. Relate this problem to the property of invertibility.

Exercise 5.9 An MA(1) stochastic process has autocorrelation coefficient $\rho_1 = 0.4$. Write the formula for the MA(1) model.

Exercise 5.10   Using program *ch5simcode.R*, generate eight realizations of an MA(1) with coefficient $\beta_1 = -0.8$. Get the sample ACF and sample PACF for each. Comment on the differences of the results. What is common to all? What is different? What is important to focus on only?

Exercise 5.11   Consider an MA(1) process with model coefficient $\beta_1 = 0.4$. Simulate this process and then several other processes, with $\beta_1 = 0.6$, another with $\beta_1 = 0.8$, and another with $\beta_1 = 0.9$. What happens to the sample ACF and the sample PACF as the β_1 increases? Describe. Provide also the numerical value of the relevant autocorrelation coefficients corresponding to each case.

Exercise 5.12   (a) Generate 100 observations from an MA(1) with parameter $\beta_1 = -0.8$ and plot the sample ACF and the sample PACF. Interpret what you get (i.e., which lag k is statistically significant?).

(b) Generate 100 observations from an MA(1) with parameter $\beta_1 = +0.8$ and plot the sample ACF and the sample PACF. Interpret what you get (i.e., which lag k is statistically significant?).

Exercise 5.13   Obtain a realization of an MA(1) with $\beta = 0.9$. As we saw, this process will have $\rho_0 = 1$, $\rho_1 = \frac{0.9}{1+0.9^2} = 0.4972376$. Compare the theoretical value of ρ_1 with the value of r_1.

5.3.9   The MA Process of Order 2, MA(2)

An MA process of order 2 follows the model

$$Y_t = \beta_1 W_{t-1} + \beta_2 W_{t-2} + W_t.$$

We can use the same methodology employed in Section 5.3.5 to derive the properties of this stationary stochastic process to find that

$$E(Y_t) = \beta_1 E(W_{t-1}) + \beta_2 E(W_{t-2}) + E(W_t) = 0.$$

Again, we notice that the autocovariances and the autocorrelations depend on the model parameters:

$$\gamma_0 = \text{Var}(Y_t) = \beta_1^2 \text{Var}(W_{t-1}) + \beta_2^2 \text{Var}(W_{t-2}) + \text{Var}(W_t) = \sigma^2(1 + \beta_1^2 + \beta_2^2)$$
$$\gamma_1 = E[(Y_t - 0)(Y_{t-1} - 0)] = \sigma^2(\beta_1 + \beta_2 \beta_1)$$
$$\gamma_2 = \beta_2 \sigma^2$$
$$\gamma_k = 0, \quad \text{for } k > 2.$$
$$\rho_1 = \frac{\sigma^2(\beta_1 + \beta_2 \beta_1)}{\sigma^2(1 + \beta_1^2 + \beta_2^2)} = \frac{(\beta_1 + \beta_2 \beta_1)}{(1 + \beta_1^2 + \beta_2^2)}$$
$$\rho_2 = \frac{\beta_2 \sigma^2}{\sigma^2(1 + \beta_1^2 + \beta_2^2)} = \frac{\beta_2}{(1 + \beta_1^2 + \beta_2^2)}$$
$$\rho_k = 0, \quad k > 0.$$

Example 5.8   Let

$$Y_t = 5 + W_t - \frac{1}{2} W_{t-1} + \frac{1}{4} W_{t-2}$$

$$E(Y_t) = 5$$

$$\gamma_0 = \text{Var}(Y_t) = \sigma^2 \left(1 - \frac{1}{4} + \frac{1}{16} \right) = \left(\frac{21}{16} \right) \sigma^2$$

$$\gamma_1 = E[(Y_t - 5)(Y_{t-1} - 5)]$$

$$= E\left[\left(W_t - \frac{1}{2}W_{t-1} + \frac{1}{4}W_{t-2} \right)\left(W_{t-1} - \frac{1}{2}W_{t-2} + \frac{1}{4}W_{t-3} \right) \right]$$

$$= -\frac{1}{2}E\left[W_{t-1}^2 \right] - \left(\frac{1}{4} \right)\left(\frac{1}{2} \right) W_{t-2}^2$$

$$= -\frac{5}{8}\sigma^2.$$

$$\gamma_2 = E[(Y_t - 5)(Y_{t-2} - 5)]$$

$$= E[(W_t - \frac{1}{2}W_{t-1} + \frac{1}{4}W_{t-2})(W_{t-2} - \frac{1}{2}W_{t-3} + \frac{1}{4}W_{t-4})]$$

$$= \frac{1}{4}E[W_{t-2}^2]$$

$$= \frac{1}{4}\sigma^2.$$

$$\gamma_k = 0, \qquad k \geq 2.$$

Thus,

$$\rho_0 = 1$$

$$\rho_1 = \frac{\gamma_1}{\gamma_0} = \frac{-\frac{5}{8}\sigma^2}{\frac{21}{16}\sigma^2} = -0.4761905$$

$$\rho_2 = \frac{\gamma_2}{\gamma_0} = \frac{\frac{1}{4}\sigma^2}{\frac{21}{16}\sigma^2} = 0.1904762$$

$$\rho_k = 0, \qquad k \geq 2.$$

□

5.3.10 Observed Time Series from an MA(2) Process

An observed time series generated by an MA(2) process will suffer the same curse as the ones generated from the MA(1) and those generated from white noise: the sample ACF will have extraneous autocorrelations due to sampling, therefore they will not be exactly like the mathematical ACF.

The bottom images in Figure 5.10 show the sample ACF and sample PACF of a simulated MA(2). As we can see, the sample ACF will show significant autocorrelations at lags 1 and 2, but we will observe that there are a lot of artifacts that appear as significant autocorrelations when they are merely a result of finite sampling, and therefore are not real. Figure 5.10 helps the reader compare the sample ACF and sample PACF of white noise, an MA(1) and an MA(2). The differences when using sampled time series are better observed with very long time series, like those we generated for Figure 5.10 with program *ch5simcode.R*. If the sample size is not very large, then the

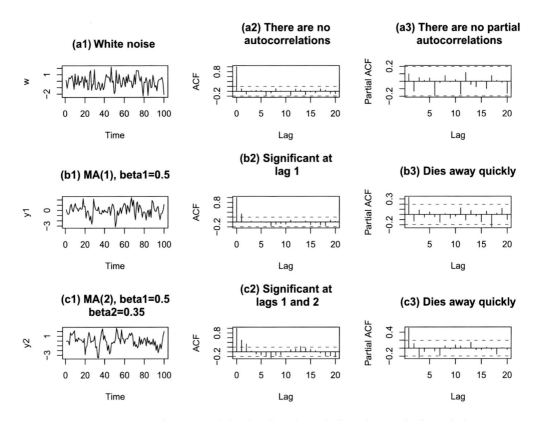

Figure 5.10 Sample autocorrelation functions do not look as clean as the theoretical ones. Sampling introduces a lot of chance variability that should be ignored. (a1) A randomly chosen member of the ensemble of white noise processes with $\sigma^2 = 1$ has a sample ACF (a2) and sample PACF (a3) slightly different from the true one seen in Figure 5.6. For the MA(1) process (b1) the sample ACF and the sample PACF (b2, b3) also look different than the theoretical ones seen in Figure 5.8. Similar conclusions can be reached for the time series generated from an MA(2) process (c1, c2, c3).

same stochastic process could result in very different sample ACF. Large samples in time series are as important to have as large samples in other statistics studies areas, because large samples will resemble more closely the aspects of the stochastic process generating the data.

5.3.11   Exercises

Exercise 5.14   Do an additive decomposition of domestic air passengers in data file *ch1passengers.csv* for January 2012 to December 2017 inclusive and extract the random component. Then find the sample ACF and sample PACF of the random part. Determine whether the sample ACF resembles the sample ACF of a time series generated by an MA(1) process. Explain how you find that out.

Exercise 5.15 The following information is given to you about a stochastic process: $\rho_0 = 1$, $\rho_1 = -0.6$, $\rho_k = 0$, $k > 1$; $\mu = 4$, $\sigma^2 = 1$. Which of the following models could represent this process, if any?

(a) $Y_t = W_t + 0.4683749 W_{t-1}$
(b) $Y_t = W_t - 0.4683749 W_{t-1}$

$E(W_t) = 0$, $\text{Var}(W_t) = \sigma^2$, $\text{Cov}(W_t, W_{t-k}) = 0$, $k = 1, 2, 3, \ldots$

Exercise 5.16 Let a stochastic process be represented by

$$X_t = 0.7 W_{t-1} + 0.5 W_{t-2} + 0.2 W_{t-3} + W_t,$$

where $E(W_t) = 0$, $\text{Var}(W_t) = \sigma^2$, $\text{Cov}(W_t, W_{t-k}) = 0$, $k = 1, 2, 3, \ldots$
Find the expected value, variance and autocorrelation function of this process.

After you derive them theoretically, simulate several realizations of the process and plot them and obtain three sample correlograms. Compare the sample correlograms with the mathematical one. Is the correlogram what you expected based on your mathematical results? Explain.

Exercise 5.17 Is the MA(2) process stationary? Show work to support your answer, up to lag 4 at least. An MA(2) process is

$$Y_t = W_t + \beta_1 W_{t-1} + \beta_2 W_{t-2},$$

where $E(W_t) = 0$, $\text{Var}(W_t) = \sigma^2$, $\text{Cov}(W_t, W_{t-k}) = 0$, $k = 1, 2, 3, \ldots$.
After your mathematical proof, simulate four times an invertible MA(2) process, and plot the ensemble of the four simulations (one plot with the four time series in the same plot. Use different colors). Also produce the mean, variance and acf of each member of the ensemble. Plot the four elements of the ensemble and their correlograms in one page, using `par(mfrow=c(4,2))`, that is, the four time series on the left, the corresponding correlograms on the right.

Exercise 5.18 (a) Is the MA(3) process stationary? Show work to support your answer, up to lag 4 at least. An MA(3) process is

$$Y_t = W_t + \beta_1 W_{t-1} + \beta_2 W_{t-2} + \beta_3 W_{t-3},$$

where $E(W_t) = 0$, $\text{Var}(W_t) = \sigma^2$, $\text{Cov}(W_t, W_{t-k}) = 0$, $k = 1, 2, 3, \ldots$
(b) After your mathematical proof, simulate with R, four times, an invertible MA(3) process of your choice, and plot the ensemble of the four simulations (one plot with the four time series in the same plot; use different colors). The code will go in your R script file, the plot in your pdf file, with comments. Also produce the mean, variance and ACF of each member of the ensemble.
(c) Plot the four elements of the ensemble and their correlograms in one page, using `par(mfrow=c(4,2))`, that is, the four time series on the left, the corresponding correlograms on the right. The code will go in your R script file and your plot in your pdf file. Comment. Also discuss why the plots of the samples are not exactly like what you found in the proof.

(d) Is what you see in the samples' correlograms what you expect based on your math derivation in part (a)? Explain.

5.4   Autoregressive Stochastic Processes

Autoregressive means regression of a variable on itself. Autoregressive models of order p or AR(p) relate the current value of a series, Y_t, to values of the series exhibited in the past p periods, namely

$$Y_t = \alpha_1 Y_{t-1} + \alpha_2 Y_{t-2} + \cdots + \alpha_p Y_{t-p} + W_t,$$

which may be expressed in polynomial form as follows:

$$(1 - \alpha_1 B - \alpha_2 B^2 - \cdots - \alpha_p B^p)Y_t = W_t.$$

Autoregressive processes are invertible, but they might be or might not be stationary.

Invertibility means that there is a linear process version of the AR process, which is more convenient sometimes, and that consists of a weighted linear combination of white noise terms. That version is obtained by substitution. We will demonstrate it for the AR(1) process in Section 5.4.1.

AR processes may be nonstationary, depending on the values of their parameters. They will be stationary if the mean absolute values of the roots of the backshift polynomial are larger than 1.

Example 5.9 Consider the AR(3) model

$$Y_t = \alpha_1 Y_{t-1} + \alpha_2 Y_{t-2} + \alpha_3 Y_{t-3} + W_t,$$

where $\alpha_1 = 0.01, \alpha_2 = 0.02, \alpha_3 = 0.005$.
We rewrite the model as

$$Y_t - \alpha_1 Y_{t-1} - \alpha_2 Y_{t-2} - \alpha_3 Y_{t-3} = W_t,$$

simplifying to get the backshift polynomial:

$$(1 - \alpha_1 B - \alpha_2 B^2 - \alpha_3 B^3)Y_t = W_t.$$

We then look at the roots of this polynomial in B by entering in R

```
Mod(polyroot(c(1, 0.01, 0.02, 0.005)))
```

which gives the following modulus of the roots of B: 5.201988, 7.390796, 5.201988, all of them larger than 1. Since all the roots are outside the unit circle, we say that the specific AR(3) process considered is stationary. □

To understand AR(p) processes, it is important that we first understand the most basic one, the AR(1). There are some regularities about AR models that follow nicely from the AR(1).

In practice, for AR(p) processes, the conditions for stationarity are obtained by writing in backshift polynomial notation and finding the roots of the polynomial in the backshift operator. The modulus of the roots of the polynomials should exceed 1 in absolute value. This will usually be true if some conditions on the model parameters are satisfied.

5.4.1 An Autoregressive Process of Order 1, AR(1)

The simplest case of an AR(p) process is the first-order autoregressive process, abbreviated as AR(1), which is defined by

$$Y_t = \alpha_1 Y_{t-1} + W_t.$$

A process like this implies that

$$Y_{t-1} = \alpha_1 Y_{t-2} + W_{t-1},$$
$$Y_{t-2} = \alpha_1 Y_{t-3} + W_{t-2},$$
$$\cdots,$$
$$Y_{t-k} = \alpha_1 Y_{t-k-1} + W_{t-k},$$

and so on, where W_t is white noise (everything in the series at time t that is not explained by the past value of the series). The W_t is independent of past values of all the values of Y appearing as regressors. This is the recursive version of the AR(1) process. An alternative expression is

$$(1 - \alpha_1 B)Y_t = W_t,$$

or

$$Y_t = \frac{1}{1 - \alpha_1 B} W_t.$$

If Y_t does not have mean 0, we assume that the mean has been subtracted and Y_t denotes the mean adjusted random process. That way, the mean of Y_t becomes 0 and studying the properties of the AR(1) process will be easier, for example:

$$E(Y_t) = \alpha_1 E(Y_{t-1}) + E(W_t) = 0.$$

We now find the variance of Y_t or γ_0.

$$\gamma_0 = \alpha_1^2 \gamma_0 + \sigma^2.$$

Solving for γ_0, we get

$$\gamma_0 = \frac{\sigma^2}{1 - \alpha_1^2} = \text{Var}(Y_t),$$

which requires that $\alpha_1^2 < 1$ or $|\alpha_1| < 1$ to have a finite variance.

To find the autocovariance at lag k, given that the expected value of Y_t is 0 for all t, multiply the Y_t by Y_{t-k}.

$$(Y_t - 0)(Y_{t-k} - 0) = \alpha_1(Y_{t-1}Y_{t-k}) + W_tY_{t-k}.$$

Then

$$E(Y_tY_{t-k}) = \alpha_1E(Y_{t-1}Y_{t-k}) + 0,$$

or

$$\gamma_k = \alpha_1\gamma_{k-1}, \qquad k = 1,2,3\ldots.$$

Setting $k = 1,2,3,\ldots$, we obtain the autocovariance function at lags 1, 2, 3,....

Setting $k = 1$, we get $(Y_t - 0)(Y_{t-1} - 0) = \alpha_1(Y_{t-1}Y_{t-1}) + W_tY_{t-1}$, which implies that

$$\gamma_1 = \alpha_1\gamma_0 = \alpha_1\frac{\sigma^2}{1-\alpha_1^2}.$$

Setting $k = 2$, we get $(Y_t - 0)(Y_{t-2} - 0) = \alpha_1(Y_{t-1}Y_{t-2}) + W_tY_{t-2}$, which implies that

$$\gamma_2 = \alpha_1\gamma_1 = \alpha_1^2\frac{\sigma^2}{1-\alpha_1^2},$$

and we can see that

$$\gamma_k = \alpha_1^k\frac{\sigma^2}{1-\alpha_1^2}.$$

Thus, the autocorrelation at lag k for an AR(1) process is given by

$$\rho_k = \frac{\gamma_k}{\gamma_0} = \alpha_1^k,$$

implying that the autocorrelation function of an AR(1) process at each lag k is just the value of the coefficient of the model, α_1 to the power k.

Since $|\alpha_1| < 1$, the magnitude of the autocorrelation function decreases exponentially as the number of lags, k, increases. If $0 < \alpha_1 < 1$, all autocorrelations are positive. If $-1 < \alpha_1 < 0$, the lag 1 autocorrelation is negative ($\rho_1 = \alpha_1$) and the signs of successive autocorrelations alternate from positive to negative, with their magnitudes decreasing exponentially.

Note that for α_1 near ±1, the exponential decay is quite slow (for example, $0.9^6 = 0.53$), but for smaller α_1, the decay is quite rapid (for example, $(0.4)^6 = 0.00410$). With α_1 near ±1, the strong correlation will extend over many lags and produce a relatively smooth ACF if α_1 is positive and a very jagged series if α_1 is negative. This will give the impression that the time series is nonstationary, because it is indeed close to being nonstationary.

The invertibility property of the AR(1) process can be seen by realizing that

$$Y_t = \frac{1}{1-\alpha_1B}W_t,$$

> The fact that the autocorrelation of an AR(1) at lag k equals the coefficient of the model, α_1 to powers of k, namely
>
> $$\rho_k = \alpha_1^k,$$
>
> makes the mathematical ACF of an AR(1) appear as the ACF of a time series with trend when the α is less than but very close to 1, even though the process is stationary.

but

$$\frac{1}{1 - \alpha_1 B} = 1 + \alpha_1 B + \alpha_1^2 B^2 + \alpha_1^3 B^3 + \ldots.$$

Thus

$$Y_t = W_t + \alpha_1 W_{t-1} + \alpha_1^2 W_{t-2} + \alpha_1^3 W_{t-3} + \ldots.$$

The stationarity condition for the AR(1) process in terms of model coefficients is therefore that $\mid \alpha_1 \mid < 1$.

An alternative way to see the invertibility is with an infinite series representation by using the method of substitution:

$$Y_t = \alpha_1 Y_{t-1} + W_t = \alpha_1 (\alpha_1 Y_{t-2} + W_{t-1}) + W_t = \alpha_1^2 Y_{t-2} + \alpha_1 W_{t-1} + W_t.$$

If we repeat the substitution many times, we end up with

$$Y_t = W_t + \alpha_1 W_{t-1} + \alpha_1^2 W_{t-2} + \cdots + \alpha_1^k W_{t-k}.$$

Assuming $\mid \alpha_1 \mid < 1$ and letting k increase without bound, we could obtain an infinite series representation. This representation emphasizes the need for the restriction $\mid \alpha_1 \mid < 1$.

In practice, we will check for stationarity of an AR(1) process by finding the root of the backshift polynomial. The process will be stationary if the absolute value of the root is above 1.

Example 5.10 Consider the model

$$Y_t = 0.8 Y_{t-1} + W_t.$$

We saw several realizations from this process in Figure 5.2. According to the theory, the theoretical ACF will consist of the values given by the following function:

$$\rho_k = 0.8^k,$$

which means that all the autocorrelations are different from 0 but will decay toward 0 as lag k increases.

Consider also the model

$$Y_t = -0.8 Y_{t-1} + W_t.$$

According to the theory, the theoretical ACF will consists of the values given by the same function as that for positive $\alpha_1 = 0.8$. However, the autocorrelations will alternate in sign, with positive autocorrelations when the lag k is an even number and negative autocorrelations when the lag k is an odd number.

Figure 5.11 shows the theoretical ACF of both models.   □

Example 5.11   The random walk is an AR(1) stochastic process that is nonstationary. When trends are not well modeled by polynomials, it is said that the trend is stochastic, changing piecewise along the path of the time series. The model is

$$Y_t = Y_{t-1} + W_t,$$

which by back substitution, that is, replacing Y_{t-1} by $Y_{t-2} + W_{t-1}$ and then replacing Y_{t-2} by $Y_{t-3} + W_t$ and so on, can be represented as a finite sum of white noise terms.

Using the polynomial version of the model,

$$(1 - B)Y_t = W_t,$$

we can see that the root of the polynomial is $B = 1$, so the model is not stationary.

An alternative way to prove this is by using the invertibility of AR processes. We rewrite the polynomial version as

$$Y_t = \frac{1}{1 - B}W_t$$
$$= (1 + B + B^2 + B^3 + \cdots + \ldots)W_t$$
$$= W_t + W_{t-1} + W_{t-2} + \cdots$$

Thus,

$$E(Y_t) = 0,$$

since W_t is white noise.

$$\gamma_0 = \text{Var}(Y_t) = t\sigma^2,$$
$$\text{Var}(Y_{t-k}) = (t + k)\sigma^2,$$
$$\gamma_1 = E[(Y_t - 0)(Y_{t-1} - 0)] = E[W_t W_{t-1} + W_t W_{t-1} + \cdots] = t\sigma^2,$$
$$\rho_1 = \frac{t\sigma^2}{(\sqrt{t\sigma^2})((t - 1)\sigma^2)},$$

and, in general,

$$\rho_k = \frac{t\sigma^2}{(\sqrt{t\sigma^2})((t + k)\sigma^2)} = \frac{1}{\sqrt{1 + \frac{k}{t}}},$$

which shows that all autocovariances and autocorrelations, including the variance, depend on time. A random walk process then is nonstationary.   □

5.4.2   Observed Data from an AR(1)

We use program *ch5simcode.R* to generate data from an AR(1) process. Assume the observed data follows an AR(1),

$$y_t = \alpha_1 y_{t-1} + w_t,$$

where w_t is white noise.

Contrast the ACFs obtained from the AR(1) processes in Figure 5.11 with the sample ACF s of the data generated from those processes in Figure 5.12. The figures in the latter corresponding to (a), (b), (c) and (d) are different due to sampling. Notice the extraneous autocorrelations in Figure 5.12.

5.4.2.1   Special Case: Observed Random Walk

The autocorrelations decay very slowly when the time series is a random walk process. In fact, when time series are not stationary, the autocorrelations displayed in the ACF, although not as extreme as for the random walk, decay very slowly.

A unique feature of a random walk process is that the first regular difference is white noise.

A way to determine whether an observed stochastic process is a random walk is to compare the correlogram of the series (which will have very slowly decaying autocorrelations) and the correlogram of the first regular difference (which would be like the correlogram of white noise, except for perhaps the 5 percent significant autocorrelations that happen just by chance and are scattered in the higher valued r_k).

5.4.3   The Autocorrelation Function of an AR(p) Process: Yule–Walker Equations

To find the autocorrelation at lag k for an AR(p) process of order $p > 1$, we multiply both sides by Y_{t-k} and take expectations. Assuming stationarity, zero mean and independence of W_t from any Y_{t-k}, the covariances are

$$\gamma_k = \alpha_1 \gamma_{k-1} + \alpha_2 \gamma_{k-2} + \cdots + \alpha_k \gamma_{k-p},$$

which makes the autocorrelation function, after dividing all terms by γ_0,

$$\rho_k = \alpha_1 \rho_{k-1} + \alpha_2 \rho_{k-2} + \cdots + \alpha_k \rho_{k-p}.$$

Setting values for $k = 1, \ldots, p$, we obtain the famous Yule–Walker equations for an AR(p) stochastic process. Remember that $\rho_0 = 1$. With these equations, it is possible to obtain the autocorrelations ρ_k as a function of the model coefficients.

For an AR(p) process, we only need to get p autocorrelations computed in order to have the system of p equations in p unknown model coefficients needed to obtain the model coefficients from the autocorrelations.

If we apply the Yule–walker recursion when $p = 1$, that is, to AR(1), we will get the results already obtained in Example 5.10.

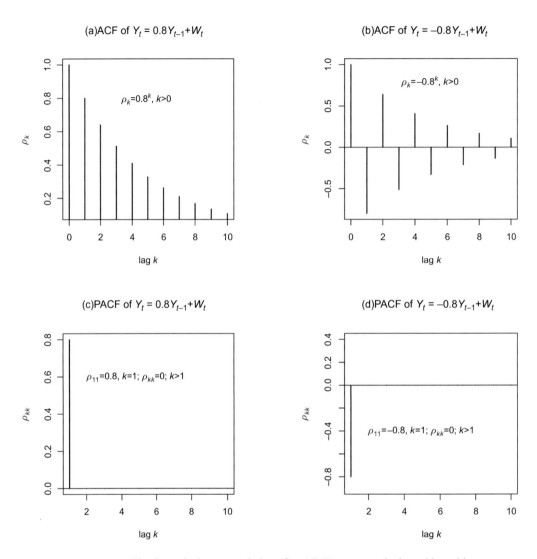

Figure 5.11 The theoretical autocorrelation of an AR(1) process, whether with positive coefficient or negative coefficient, has all autocorrelations different from 0. It will be hard to discern the order of the model by just looking at the ACF for these cases. But the PACF helps determine the order. As we can see in panels (c) and (d), only lag 1 has partial autocorrelation different from 0, indicating that the AR process is AR(1).

Consider next the AR(2). Note that in taking the variance of the AR(2) model, we must notice that the covariance between Y_{t-1} and Y_{t-2} must be taken into account, since these two variables are correlated.

Example 5.12   We will show now that the Yule–Walker recursion indeed gives the solution for the autocorrelation function of an AR(3).

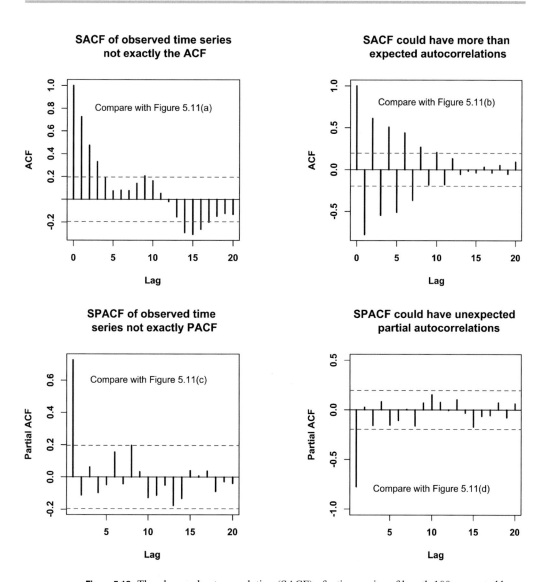

Figure 5.12 The observed autocorrelation (SACF) of a time series of length 100 generated by the AR(1) process of Figure 5.11 has artifacts not present in the theoretical autocorrelation. The sample PACF also is not as clearcut as the theoretical one for the same process in Figure 5.11.

Let Y_t be a centered time series (mean subtracted),

$$Y_t = \alpha_1 Y_{t-1} + \alpha_2 Y_{t-2} + \alpha_3 Y_{t-3} + W_t.$$

Then

$$E(Y_t) = 0,$$

$$\mathrm{Var}(Y_t) = \gamma_0 = \alpha_1^2 \gamma_0 + \alpha_2^2 \gamma_0 + \alpha_3^2 \gamma_0 + \sigma^2,$$

which implies

$$\gamma_0(1 - \alpha_1^2 - \alpha_2^2 - \alpha_3^2) = \sigma^2$$

or

$$\gamma_0 = \frac{\sigma^2}{(1 - \alpha_1^2 - \alpha_2^2 - \alpha_3^2)}$$

and

$$\rho_0 = \frac{\gamma_0}{\gamma_0} = 1.$$

$$\gamma_1 = E[Y_t Y_{t-1}] = \alpha_1 E[Y_{t-1}^2] + \alpha_2 E[Y_{t-2}Y_{t-1}] + \alpha_3 E[Y_{t-3}Y_{t-1}] + E[W_t Y_{t-1}]$$
$$= \alpha_1 \gamma_0 + \alpha_2 \gamma_1 + \alpha_3 \gamma_2,$$

which implies

$$\rho_1 = \alpha_1 + \alpha_2 \rho_1 + \alpha_3 \rho_2.$$

Similarly,

$$\rho_2 = \frac{\gamma_2}{\gamma_0} = \alpha_1 \rho_1 + \alpha_2 + \alpha_3 \rho_1,$$

$$\rho_3 = \frac{\gamma_3}{\gamma_0} = \alpha_1 \rho_2 + \alpha_2 \rho_1 + \alpha_3.$$

We can see the Yule–Walker recursion pattern. The reader must realize that $\rho_{-1} = \rho_1$ and that if the order of the model is p, all we need to know is $\rho_1, \rho_2, \ldots, \rho_p$ in order to find the coefficients of the model $(\alpha_1, \alpha_2, \ldots, \alpha_p)$. Similarly, if we know the coefficients of the model, we may find the autocorrelation coefficients up to lag p. In the example with $p = 3$,

$$\begin{pmatrix} \rho_1 \\ \rho_2 \\ \rho_3 \end{pmatrix} = \begin{pmatrix} 1 & \rho_1 & \rho_2 \\ \rho_1 & 1 & \rho_1 \\ \rho_2 & \rho_1 & 1 \end{pmatrix} \begin{pmatrix} \alpha_1 \\ \alpha_2 \\ \alpha_3 \end{pmatrix},$$

and we solve for ρ_1, ρ_2, ρ_3 as a function of the parameters $\alpha_1, \alpha_2, \alpha_3$.   □

Example 5.13

$$Y_t = 0.2Y_{t-1} + 0.1Y_{t-2} + 0.05Y_{t-3} + W_t$$

where $E(Y_t) = 0$.

$$\rho_1 = \frac{0.2 + 0.05(0.17689864)}{1 - 0.1} = 0.2563,$$

$$\rho_2 = \frac{(0.2 + 0.05)0.2}{1 - 0.1 - 0.2(0.05) - 0.05^2} = 0.17689,$$

$$\rho_3 = 0.12666.$$

$$\begin{pmatrix} 0.2563 \\ 0.17689 \\ 0.12666 \end{pmatrix} = \begin{pmatrix} 1 & 0.2563 & 0.17689 \\ 0.2563 & 1 & 0.2563 \\ 0.17689 & 0.2563 & 1 \end{pmatrix} \begin{pmatrix} 0.2 \\ 0.1 \\ 0.05 \end{pmatrix}.$$

□

5.4.4   Exercises

Exercise 5.19   Access the `AirPassengers` data set that comes with R. Do multiplicative decomposition and study the time plot, the sample ACF and the sample PACF of the random term. What can you say? Do you think they are like the ACF and PACF of an AR(1), or an MA(1)? Why? (Add the code needed to your R script file.)

Exercise 5.20   Generate time series from AR(1) stochastic processes, first from a process with $\alpha_1 = 0.1$, then from a process with $\alpha_1 = 0.5$, then from a model with $\alpha_1 = 0.9$. Generate at least 200 observations for each time series.

(a) Plot the time series, after making them R's time series objects.
(b) Plot the sample ACF and the sample PACF of each time series and describe the autocorrelation structure that matters.
(c) Compare the time plots, the sample ACF and the sample PACF of all three time series. Describe the similarities and the differences.

5.5   ARMA(p,q) Processes

In the previous two sections we studied MA(q) and AR(p) processes. In this section, we consider the autoregressive-moving average (ARMA) processes, which, in a variety of ways, depict a combination of the AR(p) and MA(q) processes. The model for an ARMA(p,q) process is

$$Y_t = \alpha_1 Y_{t-1} + \cdots + \alpha_p Y_{t-p} + \beta_1 W_{t-1} + \cdots + \beta_q W_{t-q} + W_t,$$

which, in backshift polynomial form, is

$$(1 - \alpha_1 B - \cdots - \alpha_p B^p)Y_t = (1 + \beta_1 B + \cdots + \beta_q B^q)W_t.$$

The model is stationary if the roots of $(1 - \alpha_1 B - \cdots - \alpha_p B^p)$ are outside the unit circle, and invertible if the roots of $(1 + \beta_1 B + \cdots + \beta_q B^q)$ are outside the unit circle.

Many authors have done the math to derive the covariance and autocorrelation function for an ARMA process. We refer the reader to those.

Interpreting the sample ACF and the sample PACF of an ARMA(p,q) process is not as straightforward as the interpretation for the separate AR(p) and MA(q) processes. But we will give some taxonomy that comes from years of experience.

5.6   Practical Guidelines

In view of all the models seen in this chapter and their ACF, it is interesting to provide some summary of where to focus our attention in the sample ACF and sample PACF of real time series, which, as we discussed earlier, are going to have a lot of distracting features hiding the signal in the data.

5.6.1   Lessons Derived from Theory

To understand the behavior of an observed stationary time series or of a particular transformation of a stationary time series, we examine the statistically significant auto-correlations in the sample ACF at the nonseasonal small lags k and at the seasonal lags.

We define the behavior of the sample ACF at the lower lags to be the behavior of this function at lags $k = 1$ through $k = 3$. Recall from Chapter 3 that an autocorrelation is statistically significant if the sample autocorrelation r_k is outside the two-standard-errors band. Since it makes more sense to use recent time periods in forecasting the future, we might want to look very carefully at lags 1, 2, and possibly 3 in the sample ACF very closely.

Next we define the behavior of the sample ACF at the seasonal level to be the behavior of this function at lags equal to (or nearly equal to) s, 2s, 3s, and 4s, where $s = 12$ for monthly data and $s = 4$ for quarterly data. We would just pay attention to the lag s and its very near lags, as the rest are just harmonics of the first. Those are the seasonal lags. Thus the exact seasonal lag for monthly data ($s = 12$) would be 12, 24, 36 and 48 and the near seasonal lags would be 10, 11, 13, 14, 22, 23, 25, 26, and so on. Looking at r_s at $s = 12$ and then lags $k = 10, 13$ would suffice to identify an annual seasonal component, for example. For any exact s or near seasonal lag k, we say a regular seasonal component exists if the r_s in the sample ACF is statistically significant. Since using the same month and quarter one year to forecast the next year seems reasonable, we might want to look at exact seasonal lags more closely.

We say that the sample ACF cuts off after lag k at the seasonal level if there are no spikes at exact seasonal lags or near seasonal lags greater than lag k in this function. Furthermore, we say that the sample ACF dies down at the seasonal level if this function does not cut off but rather decreases in a steady fashion at the seasonal level.

In general, to determine whether an observed time series is stationary, we need to look at the sample ACF. It can be shown that if the sample ACF of the time series values does both of the following:

- cuts off fairly quickly or dies down fairly quickly at the lower lags close to 0 level, and
- cuts off fairly quickly or dies down fairly quickly at the seasonal level,

then these values should be considered stationary. Otherwise, these values should, initially, be suspected as being nonstationary.

It is worth noticing that if the sample ACF or sample PACF cuts off fairly quickly at the nonseasonal level, it will often do so after a lag k that is less than or equal to 2. Moreover, if the sample ACF or sample PACF cuts off fairly quickly at the seasonal level, it will often do so after a lag that is less than or equal to s, the first seasonal lag, unless the time series is AR(p).

Once we decide that stationarity is present, we will then use both the sample ACF and the sample PACF to decide on the type of model to use for the regular part (low lags) and the seasonal part (seasonal lags).

Summarizing our findings in earlier sessions of this chapter, we would tentatively identify the following types of models:

- If the sample ACF cuts off, the process could be MA(q).
- If the sample ACF dies down, then we look at the sample PACF to see where it cuts off, to identify the order of an AR(p).
- If both the sample ACF and the sample PACF die down, then we determine which dies down faster and apply the preceding two rules to identify an ARMA(p,q) model.
- If both the sample ACF and the sample PACF cut off, then we determine which cuts off faster and identify an ARMA(p,q) model.

5.7   Transitioning to Statistical Inference

In Chapter 6, we will assume that the time series, or the random term of the time series if differencing was needed, that we will analyze have been generated by stationary stochastic processes like those seen in this chapter. It is worth, then, convincing ourselves that the information we will obtain from the estimation is indeed that of some stationary stochastic process. For that reason, in this section, we simulate an MA(1) process and an AR(1) process and then apply to the generated data the same estimation method that we will use in Chapter 6. The coefficients estimated by the procedure should be the same or very close to the coefficients of the simulated model. Indeed, we can see in Sections 5.7.1 and 5.7.2 that just with a finite sampled time series of 100 observations in either case, the estimates obtained for the parameters of the stochastic process generating the data are very close to the actual values of the parameters.

This approach to checking estimation is used in [30], and we believe it is very effective in convincing ourselves that if indeed the model assumption is correct, the statistical estimation procedure that we will use works and gives us some information about the stochastic process generating the data.

5.7.1   Simulating and Estimating an MA(1)

Figure 5.13 shows a realization of the MA(1) process,

$$Y_t = 0.4W_{t-1} + W_t,$$

$$y_t = 0.4w_{t-1} + w_t$$

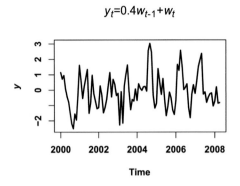

SACF of y

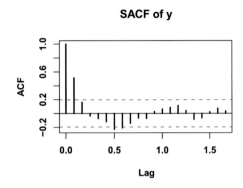

SPACF of y

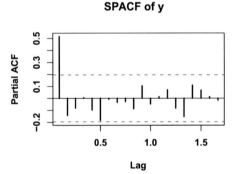

Estimated coefficient:
ma1
0.4797
s.e. 0.0757

Figure 5.13 A sample of 100 observations from an MA(1) process. The estimate of the MA model parameter, $\widehat{\beta}_1 = 0.4797$, is very close to the true value 0.4. The sample ACF (SACF) is what is expected, cuts off at lag $k = 1$, but with the added artifacts that sampling brings. The sample PACF dies away.

and its sample ACF and sample PACF. The latter are what we would expect of such a process, except for the extra detail that is believed to be an artifact of sampling. The sample ACF cuts off after lag $k = 1$. The coefficient estimate for β_1 (0.4797) is close to that of the data-generating model, but not quite. A sample much larger than 100 would produce an estimate closer to 0.4.

5.7.2   Simulating and Estimating an AR(1)

Figure 5.14 shows a realization of an AR(1) process,

$$Y_t = 0.8Y_{t-1} + W_t,$$

and its sample ACF and sample PACF . The latter are what we would expect of such a process, except for the extra detail that is believed to be an artifact of sampling. Thus, we tentatively tell the software to estimate an AR(1) model. The coefficient estimate (0.7730) is close to that of the model, but not quite. A sample much larger than 100 would produce an estimate closer to 0.8.

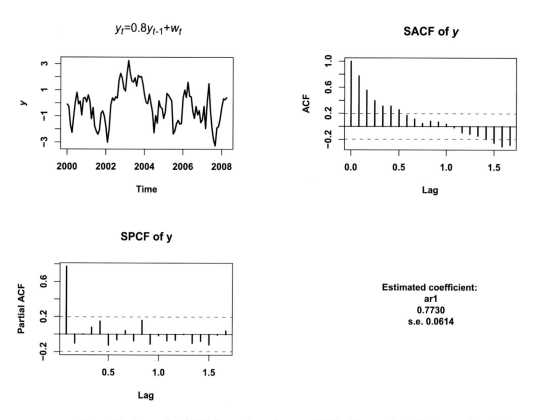

Figure 5.14 A sample of 100 observations from an AR(1). The sample ACF is as predicted, dies away, the sample PACF cuts off after lag $k = 1$ and the time process shows signs of cyclical behavior. The estimated coefficient, $\widehat{\alpha}_1 = 0.7730$, is very close to the true coefficient, $\alpha_1 = 0.8$.

5.8   Limitations of Stationary Stochastic Processes

AR(p), MA(q) and ARMA(p,q) models are static, time-invariant models in the sense that the model coefficients remain the same across time. In some cases, this is a perfectly reasonable assumption, and it is quite valid if we can assume that the process will not change over time. In other situations, it is not. In time series it is a very dangerous assumption. The passage of time alone always brings changed circumstances, new situations, fresh considerations [156]. Chapter 7 is about models that capture those changes over time, also known as dynamic linear models or state space models. We do not discuss them here in further detail because their theory, their simulation and applications differ from those of the stationary stochastic models studied thus far.

We will advance that, in general, the state space model is characterized by two principles. First, there is a hidden or latent process called the state process. The state process is assumed to be a Markov process, that is, a process where the future and past are independent, conditional on the present. The second condition is that the observations are independent, given the states [179].

Hidden Markov models are state space models that allow for a finite number of states that follows a discrete-valued Markov chain over time. We dedicate a separate chapter to them because their theory is different from that of the stationary models that we have discussed here.

5.9   The Spectrum of Stochastic Processes

The power spectrum is the principal function of interest in spectral estimation. It is a harmonic decomposition of variance. As such, it provides a clear interpretation for the decomposition of time series processes into trend, cyclical, seasonal and irregular components.

The power spectrum is a function of a stochastic stationary process. Each and every stationary process mentioned in this chapter can also be characterized by its power spectrum. The use of the spectrum, instead of the sample ACF and the sample PACF, is prevalent in signal analysis.

For a good discussion on the spectrum of a stationary time series, see [91] [98] [113] [158].

5.10   Problems

Problem 5.1   Consider the following stochastic model:

$$Y_t = W_t + 0.5 W_{t-1}.$$

We simulated 200 observations of this time series:

```
par(mfrow=c(2,1))
y=arima.sim(n=200,list(ma=0.5),innov=rnorm(200))
plot.ts(y,ylab="MA(1) process")
acf(y,main="autocorrelations",ylab="",ylim=c(-1,1))
```

Is the observed correlogram consistent with an MA(1) process? According to the result obtained for the r_k from this sample, what would be the estimate of the model coefficient β_1?

Problem 5.2   Suppose a time series is believed to follow the moving average model of order one (MA(1)).

$$Y_t = W_t + \beta W_{t-1}.$$

Show that the autocovariance at lag $k = 2$ is 0 and therefore the autocorrelation at lag 2 is 0.

Problem 5.3   Given a time series data set, there is a very tedious way to find out whether the time series is stationary. You split the series into equally long intervals and calculate the sample mean and the sample variance at each interval.

Also, compute the sample correlogram. For example, consider a time series X_t that consists of the annual prices of minis (cars) from 1959. Observed values are 496.95, 487.21, 468.82, 482.35, 401.48, 405, 386.35, 378.76, 395.3, 412.88, 415.85, 415.15.

For checking whether the mean is constant over time, we could divide the data into four intervals of four numbers each. Then you would find out that $\bar{x}_1 = 484.326$, $\bar{x}_2 = 429.61$, $\bar{x}_3 = 386.80$, $\bar{x}_4 = 414.626$.

The sample standard deviations at each of those chunks of data are $s_1 = 14.28$, $s_2 = 45.7$, $s_3 = 8.27$, $s_4 = 1.55$.

What conclusions do you draw from these results? Explain.

Problem 5.4 Consider an MA(2). Find the expected value, variance and covariance of this process, showing work.

Problem 5.5 An MA(1) stochastic process has autocorrelation coefficient $\rho_1 = 0.4$. Write the formula for the MA(1) model.

Problem 5.6 (i) Generate 100 observations from an MA(1) with parameter -0.8 and plot the sample ACF and the sample PACF . Interpret what you get (i.e., which r_k is statistically significant). (ii) Generate 100 observations from an MA(1) with parameter $+0.8$ and plot the correlogram and the partial correlogram. Interpret what you get (i.e., which r_k is statistically significant).

Problem 5.7 After studying a time series, we typed in R:

```
acf (y) $acf
```

and we got

lag k	
[1,]	1.000
[2,]	0.5075
[3,]	0.1059
[4,]	−0.1313
[5,]	−0.0353
[6,]	0.00503
[7,]	−0.019
[8,]	−0.233
[9,]	−0.371

Sketch the sample ACF and then scatterplots of y_t, y_{t-1} and y_t, y_{t-2}, using the information in the sample ACF. Given that this is from the ACF of observed data, what else would you need to identify a stochastic model?

Problem 5.8 Consider the following MA model of order 1, namely, MA(1). It is a stationary process:

$$y_t = W_t + \beta_1 W_{t-1},$$

where W_t is a white noise process.

(a) Type in R:

```
par(mfrow=c(3,3))   # to put the next graphs
```

(b) Write code to generate 100 observations from the following three processes. For each process, create the time plot, the sample ACF and PACF plots requested. The plots should go into your graph in the following order. You must create three sets of code, one set for each row.

   First row: time plot of MA(1) with $\beta = -0.3$, sample ACF of this process (with title indicating coefficient and process, MA(1)), and sample PACF of the process with title indicating process and coefficient.

   Second row: time plot of MA(1) with $\beta = -0.7$, sample ACF of this process (with title indicating coefficient and process, MA(1)), and sample PACF of the process with title indicating process and coefficient.

   Third row: time plot of MA(1) with $\beta = -0.9$ (notice minus), sample ACF of this process (with title indicating coefficient and process, MA(1)), and sample PACF of the process with title indicating process and coefficient.

(c) Save your plot. The plot will take a full page. The spikes of the sample ACF and sample PACF must be noticeable. If the r_k' are significant, we should be able to see that in the graph.

(d) Comment about your plots using technical time series language. What happens to the time plot as the negative coefficient gets larger in absolute value? What happens to the sample ACF? What happens to the sample PACF?

(e) Say whether there is something common in the three time plots or in the three sample ACFs or in the three sample PACFs.

(f) What is the difference, if any, between the general view of the sample ACF and the sample PACF at each coefficient value? That is, do they follow a different visual pattern? What is that difference and pattern?

(g) The sample ACFs of stationary (nonseasonal) processes are such that only the r_k from 1 to no more than 10 should be looked at. If a process is stationary, there should be nothing else, except perhaps a seasonal r_k if there is a stationary seasonal. Which of these lower lag r_k' are statistically significant in your plots, if any? (Do not say lag 0; we already know that.) Provide the numerical value of the autocorrelation coefficient of the statistically significant r_k.

(h) Provide the numerical value of the r_k that are significant in your three models. Double-check that the theoretical relation between the model formula and the autocorrelations holds. That is, if you use the theoretical formula for the ρ_k, what would be the value of the coefficient β_1?

Table 5.1 Table for Problem 5.10

Lag k	Covariance	Sample autocorrelation	Standard error
0	0.081742	1.00000	0
1	0.048996	0.79356	0.077152
2	0.036500	0.59118	0.116971
3	0.023432	0.37952	0.132702
4	0.016210	0.26254	0.139013
5	0.0081180	0.19148	0.142657
6	0.0076632	0.12412	0.143298
7	0.0077999	0.12633	0.143959
8	0.014969	0.24235	0.146968
9	0.021188	0.34318	0.151081
10	0.032241	0.52215	0.161467
11	0.042517	0.68862	0.178092
12	0.052832	0.85570	0.201081
13	0.041438	0.67114	0.214000
14	0.029927	0.48472	0.220438
15	0.017760	0.28765	0.222661
16	0.011111	0.17956	0.223523

Problem 5.9   Suppose the autocorrelation at lag $k = 1$ of an MA(1) process equals $\rho_1 = 0.7$. Find an estimate of the slope parameter and write the equation of the corresponding MA(1) model. Explain your answer.

Problem 5.10   A stationary time series x_t is observed. There are $n = 168$ monthly observations. The sample mean $\bar{x} = 5.166$ and the sample standard deviation is 0.2484. The sample autocovariance and autocorrelation coefficients r_k, up to lag $k = 16$ and the standard error of the $r_k, k = 0, 1, 2, ..., 24$, are given in Table 5.1. This output was produced by a SAS program, which uses the formula for standard error for any sample size, seen in Section 3.3.3, not the asymptotic constant standard error that R uses. Notice how, using that formula, the standard error increases as lag increases.

(a) Is ρ_{12} significantly different from 0? Write the null and alternative hypotheses, provide the test statistic, and justify your conclusion. Repeat for ρ_{10} and ρ_3.
(b) After studying the material in this chapter, what would you say is the stochastic process generating the data that resulted in the sample ACF of Table 5.1?
(c) Do a Ljung–Box test to determine whether the time series is white noise up to lag 6. Provide the null and alternative hypotheses, compute the test statistic, and justify your conclusion.

Problem 5.11   For the following stochastic processes, determine whether they are stationary and also derive the expected value, variance, and autocorrelation function.

(a)

$$Y_t = \frac{1}{2}Y_{t-1} + W_t$$

(b)

$$Y_t = Y_{t-1} - \frac{1}{4}Y_{t-2} + W_t$$

(c)

$$Y_t = 1/2Y_{t-1} + 1/2Y_{t-2} + W_t$$

(d)

$$Y_t = -1/4Y_{t-2} + W_t$$

Problem 5.12 Consider the following stochastic models:

$$Y_t = -0.9Y_{t-1} + W_t$$
$$Y_t = 0.4W_{t-1} + W_t.$$

(a) Write the backshift polynomial form of the model.
(b) Give the roots of the polynomial and conclude what they imply.
(c) Derive the autocorrelation function.
(d) Generate 100 observations from the process, and create the sample ACF. Describe what relevant r_k' values are practically and statistically significant.
(e) Do a Ljung–Box test of the generated data and interpret the results.

5.11 Quiz

Question 5.1 In computing the covariance function for a time series process, the following intermediate step came up:

$$\text{Cov}(Y_t, Y_{t-1}) = \text{Cov}\left[((-1/2)W_{t-1} + (1/4)W_{t-2}), \ (W_{t-1} - (1/2)W_{t-2})\right],$$

where W_t is white noise, that is, $E(W_t) = 0$ and $\text{Var}(W_t) = \sigma^2$ and $\text{Cov}(W_t, W_{t-k}) = 0$, $k = 1, 2, \ldots$.
That expression equals which of the following?

(a) $-\frac{5}{8}\sigma^2$
(b) 0
(c) $\frac{0.25\sigma^2}{\rho}$
(d) σ^2

Question 5.2 An MA process of order $q = 0$ has the same statistical properties as

(a) an AR process of infinite order.
(b) white noise.
(c) an MA process of order $q = 0$.
(d) an ARMA(1,1).

Question 5.3 The process

$$Y_t = 5 + W_t - \left(\frac{1}{2}\right)W_{t-1} + \left(\frac{1}{4}\right)W_{t-2}$$

is a stationary one. The $E(W_t) = 0$ and $\mathrm{Var}(W_t) = \sigma^2$, $t = 1, 2, \ldots$ What is the $\mathrm{Cov}(Y_t, Y_{t-3})$ equal to?

(a) $\frac{1}{4}\sigma^2$
(b) 0
(c) $1/3\sigma^2$
(d) 1

Question 5.4   Let

$$X_t = (1 + \theta_1 B + \theta_2 B^2)W_t,$$

where W_t is white noise. Which of the following is true about this stochastic model? Choose all that apply.

(a) It is stationary.
(b) It is not stationary.
(c) It is always invertible.
(d) It is an MA(2).

Question 5.5   Consider the following time series model:

$$X_t = W_t + \theta_1 W_{t-1},$$

where W_t is white noise.
   It is known that $\rho_1 = -0.4$.
   Thus, the value of θ_1 must be which of the following?

(a) -0.5
(b) 2
(c) 1
(d) 0.7

Question 5.6   Consider the AR(2) process

$$Y_t = \frac{1}{2}Y_{t-1} - \frac{1}{4}Y_{t-2} + W_t,$$

where W_t is white noise. Rewrite the model in terms of polynomial notation and find the roots of the backshift polynomial. The roots are

(a) outside the unit circle so the process is stationary.
(b) inside the unit circle so the process is not stationary.
(c) on the unit circle; it is a random walk.
(d) incalculable because roots are complex.

Question 5.7   Consider the process

$$Y_t = 3 + 2t + W_t$$

where W_t is white noise.

Which of the following is true?

(a) The process is mean stationary, but covariance nonstationary.
(b) The process is mean and covariance nonstationary.
(c) The process is mean and covariance stationary.
(d) The process is mean nonstationary but covariance stationary.

Question 5.8   Is the following AR model stationary?

$$Y_t = \frac{1}{2}Y_{t-1} + \frac{1}{2}Y_{t-2} + W_t$$

What are the roots ?

(a) Yes; roots larger than 1 in absolute value.
(b) No; one root is 1, and the other is 2 in absolute value.
(c) Yes; one root is 1, but the other is larger than 1 in absolute value.
(d) Yes; both roots are one.

Question 5.9   Consider the following MA(2) process:

$$Y_t = W_t - 0.2W_{t-1} + 0.4W_{t-2}.$$

Is this process invertible?

(a) It is not invertible because the root is inside the unit circle.
(b) It is not invertible and it is not stationary.
(c) It is invertible; it has a root outside the unit circle.
(d) Has a unit root; it is not invertible.

Question 5.10   Consider an observed time series, x_t, which is a realization of a random walk stochastic process. Which of the following is true? Circle all that apply.

(a) x_t is stationary.
(b) ∇x_t is stationary.
(c) x_t is just additive white noise.
(d) ∇x_t is white noise.

5.12   Case Study: COVID-19

Simulation has always been essential to any engineering process. That is, instead of static descriptions, engineering decision makers dynamically observe sequences of outcomes from the underlying process as represented by a simulation and make their decisions based on what the simulations lead them to expect about the future outcomes. Nowadays, advances in computing technology allow us to build simulations for virtually any scenario. For simple probabilistic situations, a description or illustration may be a wiser choice. But as complexity grows and uncertainties arise, simulations can help managers better understand statistical information and thus enable them to make better decisions [63, 64, 65].

When it comes to forecasting, simulation is also very important for decision making because, according to evidence from cognitive psychology research, different decision makers may make very different assumptions about a probabilistic forecast and interpret the same forecast very differently without being exposed to the realistic expectation of many different scenarios [65]. The method used for the presentation of the forecast (description, plot, simulation) affects how the forecast is interpreted [63].

Naive statisticians, individuals not trained in statistics, will interpret the data they see as representative of the processes that generate them, instead of one of the many possible members of the ensemble. (See Figure 5.1 and subsequent discussion throughout this chapter.) Simulation resembles what people do best: learn from many past experiences. By simulating many times the sequence of outcomes, the decision maker realizes that each time the sequence is different, and therefore understands the uncertainty embedded in a forecast of the future without needing too much statistical sophistication, except perhaps a small training in the law of large numbers to understand the reason why the simulation should be repeated many times.

The reader will often hear in the media about "the models" when referring to the weather forecast or seismic forecasts. That is because atmospheric and oceanic sciences rely on many models to get an average of all the forecasts, as much as Economics and financial institutions do. In fact, the practice of using ensembles and averages as an indication of what is going on is the most prevalent practice in Machine Learning, Deep Learning, Neural Networks and all modern approaches to prediction.

5.12.1 Simulating COVID-19

Simulation has played a very important role in decision making during the COVID-19 pandemic. Given the lack of past information about this disease, it is natural to use simulation of well-known epidemic disease models, dynamic models such as Susceptible-Exposed-Infectious-Removed (SEIR) models, to foresee the possible scenarios [16]. Simulations based on these models, however, do not incorporate chance variation (the equivalent of our white noise term), so their predictions about trends are purely deterministic. Simulation is different from estimating a model and forecasting with that model.

5.12.2 Forecasting Approaches to COVID

With the small amount of data available a few months into the pandemic, some authors used ARIMA models, accounting for uncertainty, to predict, instead of simulation [58].

A problem with using the approach of predicting the future based on the pattern of the time series over a small period of time is that the model includes the initial trend, thus making the model assume that the short-term trend will persist and hence predominate in the predictions. The reader will understand this better after seeing the forecasting done with ARIMA models in Chapter 6. A more realistic approach to take, if we were to use stochastic models to simulate, would be to simulate several possible

AR scenarios, with different model parameters. These models, as we have seen in this chapter, can result in very different temporary trend patterns that are more realistic, given that random interventions of different kinds take place, and given that many factors are at play such as, for example, festivities that make people forget they have to wear masks, weather, air quality and many other unknown factors.

Papastefanopoulos and colleagues [145] compare time series methods in regard to predicting the percentage of active COVID-19 cases with respect to the total population. The ten countries with the greatest number of total confirmed cases were selected to experiment with and draw comparisons from, as they accounted for more than 70 percent of the confirmed cases globally. They showed that ARIMA and exponential smoothing have outperformed more sophisticated methods when predicting the number of active cases at any given point in time.

It is not unusual to use time series methods to track other epidemics, but the authors estimated the percentage of active COVID-19 cases with respect to the total population for the ten countries with the most active cases at the time of their writing. The authors consider that a way to be aware of the scale of the spread of COVID-19 at the time of their writing is to able to accurately estimate the number of active cases at any point in time. Although they used different time series approaches, we will focus on only one. Time series forecasting with the stochastic models learned in this chapter is particularly useful in two cases: when there is little or no knowledge available on the underlying data-generating distribution/process or when there is no explanatory model that is able to adequately relate the prediction variable. The authors use ARIMA, the Holt–Winters additive model (HWAAS), TBAT, Facebook's Prophet, DeepAR and N-Beats.

The authors do not provide the ARIMA model that they fit, and judging by their Python code at their github, an ARIMA model is fitted using automatic procedures.

5.12.3   Simulating a COVID ARIMA model

Since an ARIMA model does very well for Italy in [145] and the authors of that paper choose that model automatically, we will fit an automatic ARIMA model and then simulate from this model to present an ensemble of scenarios. We will use the same windows of the data used by the authors of [145] to trace active cases, and create an ensemble of potential sequences that a model like this could give rise to. Then we will compare with the trends that have been observed after the period covered by the data.

The authors use a data set accessible on Kaggle.com for model development and analysis purposes, the "Novel Corona Virus 2019 Dataset" [89], and population by country data set [157]. The former contains (i) the number of confirmed COVID-19 cases; (ii) the number of recovered COVID-19 patients; (iii) the death toll due to COVID-19.

For each country, 104 instances were created in [145], each representing the percentage of active coronavirus cases as a fraction of the total population for a single day in the corresponding country. The number of active cases in each country for each day was calculated by subtracting both the recovered patients and the number of

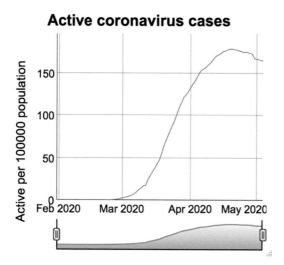

Figure 5.15 Daily active COVID-19 cases in Italy between February 21 and May 5, 2020 [89, 157].

deaths from the confirmed cases, while the number of cases per 100,000 population that these active cases correspond to was calculated by dividing each day's active-cases value by the population of the respective country. For training and validation purposes, 97 of these instances were used by [145] (72 for training and 25 for validation), while a window of seven days was kept aside to evaluate the performance of the predictive models. In terms of the evaluation metric, the root mean squared error (RMSE) was used to assess the predictive power of each of the approaches [145].

We use the same data set for Italy to fit an automatic ARIMA model and then simulate several scenarios generated by that model, an ensemble of scenarios for a longer period of time. The data set and the change in the values of cases (the new cases) are plotted in Figures 5.15 and 5.16.

The automatic model selected for the number of cases by the forecast package [77] is ARIMA(0,2,1). Chapter 6 will explain to the reader what that expression means. For now, suffice it to say that the package is saying that we should regularly difference the data twice to make it stationary. Differencing was covered in Chapter 4. Then the package says that an MA(1) model should be fitted to the differenced data. The model is

$$y_t = -0.5049 y_{t-1} + w_t,$$

where y_t denotes the differenced cases. This is the model that we will simulate to obtain many possible scenarios. The reader may visit the book's website to play with different simulation scenarios.

ARIMA.sim will not simulate seasonal models. The R package forecast has a function that allows one to do that. But in this case study, that function will also allow

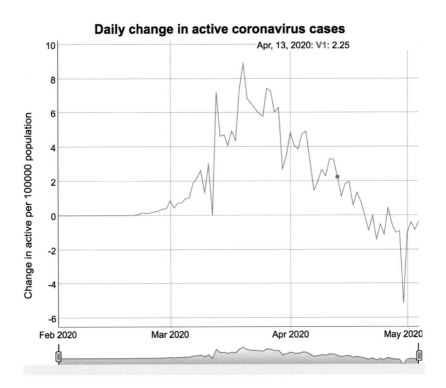

Figure 5.16 New daily COVID-19 cases in Italy between February 21 and May 5, 2020 [89, 157].

us to simulate future values based on an ARIMA model of any kind, seasonal or not, given some initial values. We chose as initial values the last two values of the fitted data. The simulated future paths of active cases after May 2020 can be seen in Figure 5.17. As we can see in that figure, if the model fitted to the data was the true model, the paths that COVID cases could have taken are very varied.

In an image like Figure 5.17, we can see the advantage of simulation. It allows us to see the uncertainty about the future that is generated by the fact that COVID-19 is a stochastic process. The forecast that we can make with the model fitted is just the case in "the middle" of the ensemble, an average of all potential forecasts. The curious reader can try to generate more paths and plot them, and also check the actual COVID-19 data for Italy after May to see which of the paths were followed by the natural process.

5.12.4   Exercises

Exercise 5.21   Select another country from the data set where we obtained the COVID-19 data and repeat the analysis that we have done, but for that country. Compare the model fitted automatically by the forecast package [77], and the paths simulated 40 steps ahead with the forecast, also done 40 steps ahead.

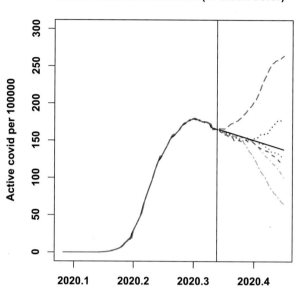

Several simulated future paths if model is true, and shorter-term forecast (in black color)

Figure 5.17 Simulated future paths of COVID-19 in Italy after May 5 suggest that many paths are possible. The forecast using the model estimated seems to be the average of those paths.

5.13 Appendix A: Review of Introductory Probability

In introductory probability, we represent a random variable with uppercase letters, and a realization of the random variable, an observed value, by a lowercase letter. Some authors of time series books do not make that distinction and use lowercase letters for both [179], but we have been making it throughout this book and will continue to do so.

A function of a random variable is itself a random variable. In particular, linear functions are commonly used in applications of probability theory.

The following review of the properties of expectations will help in understanding the derivations we do in this chapter.

Linearity of expectations: If $Y = a+bX$, where a, b are constants and X is a random variable with $E(X) = \mu$ and $\text{Var}(X) = \sigma^2$, then Y is also a random variable. $E(Y) = a + bE(X)$ and the $\text{Var}(Y) = b^2\text{Var}(X)$, because the expectation of a constant is the constant, and the variance of a constant is 0.

Using variance to compute expectations: If $Y = a + bX^2$, where a, b are constants and X is a random variable with $E(X) = \mu$ and $\text{Var}(X) = \sigma^2$, then Y is a random variable and $E(Y) = a + b(\text{Var}(X) + [E(X)]^2)$.

Expectation of a linear function of random variables: If $X_1, X, \ldots, X_n$ is a set of random variables, and $a_a, a_2, \ldots, a_n$ is a set of constants, then $E(a_1X_1 + a_2X_2 + \cdots, +a_nX_n) = a_1E(X_1) + a_2E(X_2) + \cdots, +a_nE(X_n)$.

Covariance and correlation between two random variables: If we have a two-dimensional vector random variable $\begin{pmatrix} X \\ Y \end{pmatrix}$, the covariance $\text{Cov}(X, Y) = E[(X - \mu_x)(Y - \mu_y)]$ and the correlation $\rho = \dfrac{\text{Cov}(X, Y)}{\sigma_x \sigma_y}$.

Independence: If two random variables X and Y are independent, $E(XY) = E(X)E(Y)$. Their covariance and correlation are 0.

These concepts generalize to a vector random variable,

$$\mathbf{Y} = \begin{pmatrix} Y_1 \\ Y_2 \\ \cdots \\ Y_t \\ \cdots \end{pmatrix}.$$

The expected value of the random vector Y representing the stochastic process is merely the vector of expectations of its elements:

$$E(\mathbf{Y}) = \begin{bmatrix} E(Y_1) \\ E(Y_2) \\ \cdots \\ E(Y_t) \\ \cdots \end{bmatrix} = \begin{bmatrix} \mu_1 \\ \mu_2 \\ \cdots \\ \mu_t \\ \cdots \end{bmatrix}.$$

The variance and covariance can be represented in the variance-covariance matrix of vector Y:

$$\Sigma_Y = \text{VarCov}(\mathbf{Y}) = E[(\mathbf{Y} - E(\mathbf{Y}))(\mathbf{Y} - E(\mathbf{Y}))^T]$$
$$= \begin{bmatrix} \sigma_1^2 & \gamma_{1,2} & \gamma_{1,3} & \gamma_{1,4} & \cdots & ..\gamma_{1,n} \\ \gamma_{2,1} & \sigma_2^2 & \gamma_{2,3} & \gamma_{2,4} & \cdots & \gamma_{2,n} \\ \gamma_{3,1} & \gamma_{3,2} & \sigma_3^2 & \gamma_{3,4} & \cdots & \gamma_{3,n} \\ \gamma_{k,1} & \gamma_{k,2} & \cdots & \sigma_4^2 & \cdots & \gamma_{i,n} \\ \cdots & \cdots & \cdots & \cdots & \sigma_t^2 \cdots & \cdots \\ \cdots & \cdots & \cdots & \cdots & \cdots & \cdots \end{bmatrix}.$$

We usually concentrate on the correlation between two random variables rather than their covariance, if only because the correlation does not depend on the units of measurement and the covariance does. The autocorrelation matrix is

$$\rho(Y) = \frac{1}{\sigma^2} E[(\mathbf{Y} - E(\mathbf{Y}))(\mathbf{Y} - E(\mathbf{Y}))^T]$$

$$
= \begin{bmatrix}
1 & \rho_{1,2} & \rho_{1,3} & \rho_{1,4} & \cdots & \cdot\cdot\rho_{1,n} \\
\rho_{2,1} & 1 & \rho_{2,3} & \rho_{2,4} & \cdots & \rho_{2,n} \\
\rho_{3,1} & \rho_{3,2} & 1 & \rho_{3,4} & \cdots & \rho_{3,n} \\
\rho_{i,1} & \rho_{i,2} & \cdots & 1 & \cdots & \rho_{i,n} \\
\cdots & \cdots & \cdots & \cdots & 1 \cdots & \cdots \\
\cdots & \cdots & \cdots & \cdots & \cdots & \cdots
\end{bmatrix},
$$

where $\rho_k = \mathrm{Corr}(\mathrm{Cov}(Y_t, Y_{t-k})) = \frac{\mathrm{Cov}(Y_t,Y_{t-k})}{\sqrt{\sigma_t^2,\sigma_{t-k}^2}} = \frac{\mathrm{Cov}(Y_t,Y_{t-k})}{\sqrt{\sigma^2,\sigma^2}} = \frac{\mathrm{Cov}(Y_t,Y_{t-k})}{\sigma^2} = \frac{\gamma_k}{\gamma_0}.$

Obviously, the diagonals are $\rho_1 = \frac{\mathrm{Cov}(\mathbf{Y}_t,\mathbf{Y}_t)}{\sigma_t,\sigma_{t-k}} = \frac{\sigma_t^2}{\sigma_t^2} = 1.$ The autocorrelation matrix is also symmetric and positive definite.

5.14 Appendix B: A Time Series as a Vector Random Variable

We have said that, in theory, we assume that a time series is an indexed collection $\{Y_t, t \in T\}$ of random variables having finite second moments; that is, $E(Y_t^2) < \infty$ for each element of the index set T, where T is considered the set of all integers; that is, we assume that the phenomenon being observed has been going on for a long time and will continue indefinitely [131]. Our observed time series is just a finite window of this indefinite process.

If we are interested in the joint behavior of the random variables that make a time series, then we will need the concepts of vector random variables, their expectation, their variance-covariance matrix, and their correlation matrix, which are also probability concepts. A finite discrete time series is really a vector random variable. The expectation and the variance-covariance matrix will suffice to characterize the joint behavior of the random variables that a time series is if the time series process is a stationary stochastic process of the Gaussian family.

A time series $\{Y_t, t \in T\}$ is said to be a Gaussian time series if the joint distribution of any finite number of Y_t is multivariate normal.

The multivariate normal plays a crucial role in the theory and analysis of time series.

5.14.1 Expectation and Covariance Matrix of a Time Series

It may be more convenient to represent what we just said in vector notation, if only because it allows us to see all the parameters of the models that must be estimated when we work with time series in a compact form. Consider again the time series, a sequence of random variables $Y_1, Y_2, \ldots, Y_t, \ldots$ as a vector random variable (where we assume the time index starts at $t = 1$).

The Autocorrelation Function (ACF) of a stationary stochastic process is just a listing of the autocorrelations depicted in the autocorrelation matrix, which is a symmetric matrix. When we saw in Chapter 3 the correlogram, the $r'_k s$ were estimates of the off-diagonal upper elements of the autocorrelation matrix.

> The expectation vector and the variance-covariance and autocorrelation matrices are the most important aspects of a Gaussian time series. For time series for which that is not the case, we may need other moments in order to characterize them.

> The reason for studying autocorrelations is that in the stationary models studied in this chapter, the autocorrelation coefficients in the correlation matrix are related to the coefficients of the stationary stochastic process models and vice versa.
> If we believe our observed random components of a time series are generated by Gaussian stochastic stationary models, then the sample ACF and the sample PACF will tell us a lot about the models. That is why we studied them in Chapter 3.

6   ARIMA(p, d, q)(P, D, Q)$_F$ Modeling and Forecasting

6.1   Introduction to ARIMA(p, d, q)(P, D, Q) Methodology

Now that we have become acquainted with time series data, autocorrelation, partial autocorrelation, the backshift and difference operators, the concept of stationarity, and the components of time series and stationary stochastic models, we are ready to start statistical modeling with the Autoregressive Integrated Moving Average (ARIMA) methodology, also known as Box–Jenkins methodology for univariate time series. ARIMA is a generic name for a family of forecasting models that are based on the AR, MA and ARMA stationary processes studied in Chapter 5. These models have been widely used in time series analysis since the 1970s.

A stochastic process X_t is ARIMA$(p, d, q)(P, D, Q)_F$ if $\nabla^D(\nabla^d)X_t$ is a stationary time series with autocorrelations at small lags ($k = 1, 2, 3, ..$) captured by an ARMA(p, q) process and autocorrelations at seasonal lags ($F - 1, F, F + 1, ...$) captured by an ARMA$(P, Q)_F$ process, where F denotes the frequency of the time series. Everything we have studied in earlier chapters is needed to do modeling with the ARIMA modeling approach.

- The main goal in ARIMA modeling is to fit to the historical data such a good model that there is no more room for improvement, that is, a model that gives white noise residuals and has the smallest number of coefficients.
- A secondary goal is to use the good model to forecast what will happen in the unknown future.

The notation, ARIMA$(p, d, q)(P, D, Q)_F$, is a concise description of the differencing done and the autocorrelation and model identification analysis of the random (stationary) term to capture the signal hidden in the historical time series. The notation conveys information about whether differencing was done to make the time series stationary or not (d, D), and the order of the AR, MA or ARMA model identified and fitted to the stationary series. The notation does not say anything about white noise residuals, forecast performance or variance-stabilizing transformations done to the time series prior to autocorrelation analysis. The meaning of the letters in the ARIMA expression is as follows:

- d = an integer, denoting the order of regular differencing needed to remove the contribution of the long-term trend to the nonstationarity of a time series. If differencing is not needed, $d = 0$.
- D = an integer, denoting the order of seasonal differencing needed to remove the contribution of the regular seasonality to the nonstationarity of a time series. If seasonal differencing is not needed, $D = 0$.
- p = an integer, denoting the order of the AR(p) model identified and confirmed to capture short-term correlations (k lags close to 0, e.g., $k = 1, 2, 3$) of the stationary time series obtained after differencing, if there is differencing at all. If no AR term is found, $p = 0$.
- P = an integer, denoting the order of AR(P) seasonal model identified and confirmed to capture seasonal correlations (k = 11, 12, 13, if monthly time series) remaining after seasonal differencing, if differencing is applied. If no AR term is found, $P = 0$.
- q = an integer, denoting the order of the MA(q) model identified and confirmed to capture short-term correlations (lags $k = 1, 2, 3$), after differencing, if differencing is done. If no MA term is found, $q = 0$.
- Q = An integer, denoting the order of the MA(q) model identified and confirmed to capture the seasonal correlation left after differencing, if differencing is done. If no MA term is found, $Q = 0$.
- F = the frequency of the time series (monthly, quarterly, weekly, etc.). See Section 1.9 for more details about frequency of a time series and its handling with R.

The ARIMA modeling approach was introduced by Box and Jenkins in the 1970s in their seminal work, now available in its third edition with coauthor Gregory C. Reinsel [7]. The method became, and still is, the most popular, advanced and robust time series modeling and forecasting approach for univariate time series. It is so much so that it is a benchmark against which all new modeling approaches for time series are compared. ARIMA can be found explained in most books on time series. In some of those books, [30] [32] [7], authors make a distinction between the following:

- stationary models or ARMA(p, q) models,
- nonstationary models or ARIMA(p, d, q),
- seasonal ARIMA(p, d, q)(P, D, Q)_F_ models.

But clearly, the most general notation ARIMA(p, d, q)(P, D, Q)_F_ encompasses all of those cases, and which of those cases we are talking about will be clear by the values assigned to the p, d, q, P, D, Q.

ARIMA modeling and forecasting is based on the idea that time series can be made stationary by operations such as differencing. Once this has been done, the model-fitting techniques designed for data from stationary stochastic processes like those of Chapter 5 can be applied to the differenced data. The differencing operations are then inverted (aggregated or integrated, hence the I) and used for automatic back-transformation of the forecast and the fitted values [60].

The ARIMA notation is so well settled in the practice of time series analysis that in Base R, for example, a single programming statement with just the arima () function will do the d and D that we order it to do, will fit the identified regular and seasonal AR and MA to the differenced data and then will invert the d and D differencing, if any was done, to calculate the fitted values and to forecast.

A single Base R function applies the differencing of the data that we have established with our exploratory analysis and fits the model that we have identified with that same exploratory analysis. This function gives an object containing the parameter estimates, standard errors, and predicted values, among other things. The function is the arima () function:

```
model1= arima("time series", order = c(p,d,q),
                seas    =list(order = c(P,D,Q),12),
                          include.mean=FALSE)
              summary(model1)
```

Example 6.1 Suppose x_t is an observed nonstationary monthly time series with seasonality and variance proportional to the mean. We log it, to stabilize the variance, and obtain $x_t^* = \log(x_t)$, which we make a ts () object. After exploratory analysis with the sample ACF and sample PACF, we decide the order of differencing and identify a tentative model. Hence we ask R to regularly difference x_t^* once and then seasonally difference it once, and then to fit to the differenced series an AR(1) for the low lags k, and an AR(1) for the seasonal lags as follows:

```
x.t.star=log(x)
acf(x.t.star)
    #reveals nonstationarity
acf(diff(x.t.star,lag=1,differences=1));
              pacf(diff(x.t.star,lag=1,differences=1))
# with acf and pacf identified ARIMA(1,1,0)(1,1,0)
model1= arima(x.t.star, order = c(1,1,0),
seas    =list(order = c(1,1,0),12),
                include.mean=FALSE)
```

where x.t.star is x_t^*, a ts () object, and model1 is the object where we store all the objects that the arima () function creates such as the estimates of the model coefficients, standard errors, the predicted values of x_t^*, the *AIC*, and, if we make include.mean=TRUE, also the mean of the series that it calls (inappropriately) "constant" but should not be interpreted as the constant of the model [75]. The next diagram indicates what we did:

$$(1 - \alpha_{12}B^{12})(1 - \alpha_1 B) \ \underbrace{(1 - B^{12}) \ \underbrace{(1 - B)x_t^*}_{\substack{\text{Regular} \\ \text{differencing of } x_t^*}}}_{\text{Seasonal differencing of } (1-B)x_t^*} = w_t.$$

$$\underbrace{\qquad\qquad\qquad\qquad\qquad\qquad}_{\substack{\text{Multiplicative seasonal model} \\ \text{applied to } (1-B^{12})(1-B)x_t^*}}$$

Notice that we do not feed `arima()` differenced data. Rather, we just feed it the x_t^* and inform the software to do the d and D differencing that we think is necessary based on what we discovered in the exploratory analysis.

The reader is invited to practice using the `arima()` command by running the programs corresponding to the complete studies presented in Section 6.2.   □

Many beginners make the mistake of applying the `arima()` function to already differenced data and still keep the d and D larger than 0 in the R statement. That is a big mistake to be avoided because then the differencing is done twice and will result in ridiculous-looking predictions and forecasts.

When forecasting, it is a common mistake for beginners to forget the variance-stabilizing transformation done to some time series at the beginning of a data analysis. Therefore, they produce forecasts for the variance-stabilized time series and compare them with the original time series, reaching therefore the wrong conclusion and causing loss of faith in ARIMA. A logged quantity will be at the bottom of the graph, the unlogged one at the top. We must reverse the variance-stabilizing transformation to obtain a forecast of the level of the series. Figure 6.3, for example, shows intervals and forecast leveled with the data.

ARIMA modeling started in the 1970s with a distinctly engineering language, but as its applicability grew, it has lent itself to all kinds of extensions. Over time, ARIMA models were generalized in order to encompass special time series data. Examples are:

- ARFIMA (FI meaning fractionally integrated, which means that d, D are nonintegers), also known as FARIMA. These models were developed in order to fit long-range dependence or nonexponential decays in the autocorrelations that occur in geophysical and other time series. Cowpertwait and Metcalfe's chapter 8 [30] is a very approachable introduction to these models and their application using R.
- GARCH models are ARIMA models of the stochastic variance of a time series. The relative emphasis placed on GARCH in time series books varies depending on how much the book leans on applications to financial time series, which tend to need this kind of model. Cryer and Chan's [32] chapter 12 is a gentle introduction to these models and their application with R.

FARIMA and GARCH cannot be implemented in R with the function `arima()` but there are specialized functions in R packages that the reader can find in the R Archive [159] after consulting those sources.

More recently, ARIMA modeling has been at the forefront of the debate on how to do machine learning, neural networks and deep learning for time series data. Something that the reader will notice in learning about all the variations is that Box–Jenkins methodology is at their core.

Despite its wide applicability, ARIMA models have received criticism for being linear, assuming a constant coefficient structure over time and not being flexible enough for single-handed forecasting. However, it is becoming common knowledge that predicting the future is very hard to do using only very summarized features of the time series like some modern machine learning methods do, and that has led to introducing a lot of ARIMA modeling into new unsupervised and supervised machine learning methods. The reader should expect a lot of progress in this arena in the next few years.

Several examples will clarify the process of modeling with the Box–Jenkins methodology. Section 6.2 contains two complete analyses.

6.2   The Practice of ARIMA Modeling and Forecasting

Nothing makes beginners more proficient in modeling time series with ARIMA than going over several examples of data analysis that lay out all the steps of a modeling and forecasting effort. Complete introductory examples are hard to find in the literature on time series modeling and forecasting, but they prove very useful in learning time series analysis. To guide the reader in this endeavor, we present in Section 6.2.1 a complete example that includes most of the steps involved in fitting a model to annual historical unemployment rate and forecasting with it. The narrative will be more transparent if the reader opens program *firstarimaexample.R* and executes it while reading this discussion. Then, in Section 6.2.2, we go one step further and guide the reader in the fitting of an ARIMA model to nonstationary seasonal data, and use the model to forecast. Program *secondarimamodel.R* will allow the reader to replicate the analysis. We recommend that the reader run each of those R programs while following our narrative.

6.2.1   Modeling and Forecasting Annual Data

Annual time series display patterns on a very aggregate time scale. We do not expect to see patterns at the weekly, monthly, or daily scale when we are using annual time series. In other words, no seasonality of any frequency will be expected.

The patterns that we would expect in annual time series are long-term trends in mean and the cyclical scale patterns that take a few years to complete, such as, for example, business cycles or meteorological cycles. Some shocks or unexpected interventions or events could also be present. Thus, knowing that the time series is annual already tells us that we should only look for those patterns in the middle of the noise.

Both cycles and trends take place over several years and, for that reason, sometimes they manifest themselves in the same way in the sample ACF.

Example 6.2   In this example, an ARIMA model for a training set of historical annual unemployment rate in the United States will be identified and fitted and the model will be used to forecast the level of unemployment a few years ahead. To validate the model for forecasting, forecasts will be compared with the actual values of the series for the forecasted years (the so-called test set part of the unemployment data).[1] In the discussion that follows, we will indicate what we base our decisions on (time plot or sample ACF or both sample ACF and sample PACF, or Ljung box tests, or coefficients' p-values, or other). The reader should review the earlier chapters of the book where those concepts were introduced. The R program *firstarimaexample.R* contains the code producing the supporting material for our discussion.

The Data

The unemployment rate time series that we will study here has been featured one way or another in several time series books [151]. The time series is included with the R package urca [151], which provides logged annual unemployment rate in a time series called npext, dating from 1860 to 1988. But the unemployment values for 1860 to 1889 are missing (NA in R). Thus, we subset the data set from 1890 to 1988. This leaves us with an annual time series, log.unemploy, of 99 observations.

> **Recommended practice:** It is very important to identify and let the audience know the time index of the time series at the outset of a report or presentation. It is not the same to talk about unemployment rate in the period 1890–1988 as to talk about unemployment rate in the period 1990–2021. Can the reader think of at least one reason why?

> **Recommended practice:** It is very important to read the metadata (the information about the data). For example, data nepxt in the R package urca contains logged unemployment rate, not unemployment rate.

The first values of the logged data are:

```
1.3862944,  1.6863990,  1.0986123,  2.4595888,  2.9123507,....
```

If we take the exponential of those values, we obtain the actual unemployment rates, which for the first values of the data are

---

[1] In practice, when we forecast, we do not know the actual future values of the time series. We forecast the next period and wait until the next period happens to assess how well we forecasted. But, when we are learning how models work, it is standard practice to divide the available data into a training set, used to identify and fit a model, and a test set. That way, we can evaluate the forecasting performance of the model before we use it. In machine learning jargon, fitting a model to the training set is called "learning the model."

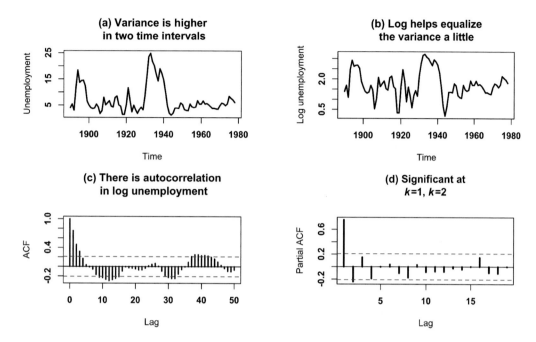

Figure 6.1 Time plot of unemployment rate and log unemployment rate (panels (a) and (b)) (1890–1977). Correlogram and partial correlogram of log unemployment data (panels (c) and (d)). The sample ACF and sample PACF combination suggests an AR(p) model.

```
4.0,   5.4,   3.0, 11.7, 18.4, 13.7,....
```

But we will not see the latter when reading nepxt into R without first doing some *data management* using the software.

If the reader wonders why the author [151] gave the logged data instead of the raw unemployment rate, Figure 6.1 explains why. In that figure, panel (a) shows the raw data, without log, and we can see that the variability is higher in some intervals. There is no constant variance. As we saw in Chapter 5, we need to have constant variance (variance stationarity) in the data to apply stationary stochastic process models. Panel (b) shows the logged data, and we can see that the variability is more homogeneous in the logged data. More on this topic will be discussed in a few paragraphs.

Split the Data into Training and Test Sets

We then create the training set, an object called y in the R program that contains logged unemployment rate for the years 1890 to 1978, a total of 89 observations, and make it an R ts() object of frequency equal to 1 since y_t is annual. That is the data that we use to fit the in-sample (training, estimation or learning set) model. In machine learning jargon, we say that we learn the model with observations in the training set. The leftover data, 10 observations, will be used to compare with the out-of-sample (test set) forecasts produced by the in-sample model.

> **Recommended practice:** Before using a model for high-stakes tasks, we validate the model first by splitting the data into a training set and a test set. The model will be fitted to the training set and validated with the test set by comparing the forecast with the test data. It is important to do the split at the beginning, before the exploratory analysis starts. And we must document for others the time period of the training and the test sets.

Figure 6.3 gives an idea of what we mean when we refer to the training and test data. The training data set is the historical data indicated by the long two-sided arrow, up to where the vertical dotted line is in Figure 6.3. The test set is the historical data to the right of the vertical dotted line. We reserve that data to compare it with the forecast we make using the model fitted to the training data. Figure 6.3 is the image where we will show the forecast when we get to the forecasting stage later on in this section.

Notice how, in the R program, y is the logged unemployment rate in the training set, read and cleaned, that is, omitting the NA values that came with the raw data obtained from the urca library. Remember how important it is to be making it easy to trace your steps.

Predifferencing Transformations

At the risk of sounding redundant, we will discuss a little more the log transformation. Why did the author [151] log the data and choose to publish the logged data? The time plot of the training set of unemployment rate data (the exponentiated logged unemployment) seen in panel (a) of Figure 6.1 reveals a lack of trend, but, upon inspection, a variance-stabilizing pre-transformation is needed before we continue working with these data because the variability is not constant. The variance is much higher during the late 1900s and during the 1930s. We notice that the Great Depression years in the 1930s is an exceptional period in US unemployment, and perhaps an analysis without those years would make more sense, as they were extremely unusual years in the economic history of the United States, and not part of the US general unemployment patterns. But we keep them in the training set.

Before the sample ACF is looked at, the variance needs to be stabilized. For that reason, we believe, Pfaff chose to report the logged time series, which can be seen in Figure 6.1, the logged data in panel (b). The variability in log unemployment rate is more homogeneous over the whole period than the variability of raw unemployment rate.

We tried other transformations of the data in panel (a) of Figure 6.1, such as square root or quartic root transform to see if that stabilizes the variance more satisfactorily than log; but our transformations did not give different results, so we continue the analysis with the log transform and agree with Pfaff that logged data is the right pre-differencing transformation to do to raw unemployment.[2]

---

[2] Alternatively, we could try to identify and fit a different type of model like the state space models or time series regression models, or the additive models used nowadays in automatic forecasting.

Determine d and D

> **Recommended practice:** After the variance of the unemployment rate in the training data is stabilized with log transform, we must look at the correlogram of the variance stabilized training set. If it reveals nonstationarity, we difference until the sample ACF reveals stationarity, after which we decide the values of d, D.

Once the variance is stabilized, and only then, it makes sense to look at the correlogram of the logged unemployment rate in the training set, y_t, to see whether the training set is mean stationary or whether we need to difference. As we can see in Figure 6.1 (c), the correlogram of the logged unemployment in the training set dies away, not as fast as one would like so as to call the process stationary, but not as slow as to call it nonstationary. It could be the correlogram of an AR process. Recall, from Chapter 5, that an AR(1) process with negative model coefficient could have a sample ACF like that in Figure 6.1 (c). So, tentatively, we conclude that differencing is not necessary for y_t. Since we did not difference, we keep calling the time series in R y. Because this is annual data, we do not need to worry about seasonality. But it is possible that the time series has cycles. We are still inclined to think that this is an AR process with lots of memory and do not difference.

Identify a Model for the Training Set

As a consequence of our analysis of the sample ACF, we conclude that $d = 0$ and $D = 0$, since we decided not to regularly or seasonally difference the training set. Thus, we have by now figured out that logged unemployment is ARIMA$(p, 0, q)(0, 0, 0)$. We still need to find the values of p and q.

> **Recommended practice:** Model identification is done after the need for pre-transforming to stabilize variance has been done and d and D have been determined and applied, the latter if differencing is needed.

Model identification of the logged unemployment rate entails looking at the sample ACF and the sample PACF (the correlogram and the partial correlogram) of the logged unemployment rate in the training set. Panels (c) and (d) in Figure 6.1 indicate that the sample PACF of y_t cuts off at lag 4, but the partial autocorrelations at lags 3 and 4 are borderline significant. Given the dying away of the correlogram and cutoff behavior of the partial correlogram, and given what we learned about model identification in Chapter 5, an AR model for y_t, to start with, makes sense. We contemplate different neighboring AR models: AR(1), AR(2), AR(3) and AR(4).

Fit the Identified Model(s) to the Training Set

> **Recommended practice:** Once the model has been identified, we must be careful about what is fed to R's `arima()` function. We retrace the steps taken. We keep the order of the model fitted small.

The next step is to fit the models contemplated to y_t. We use the `arima()` function in R to fit the models.

R's `arima()` function is applied in program *firstarimaexample.R* to the logged but not-differenced unemployment, y_t:

```
arima(y, order = c(3,0,0),
seas  =list(order = c(0,0,0),1),
              include.mean=FALSE)
```

`order=c(3,0,0)` means that the 0 in the middle orders R to not do regular differencing. The 3 and the last 0 tell R that the undifferenced log unemployment should be fitted a regular AR(3) model to capture short-term autocorrelations.

`seas =list(order = c(0,0,0),1)` means that the 0 in the middle tells R that it should not do seasonal differencing of logged unemployment. The other two 0's are telling R that the undifferenced logged unemployment should be neither a seasonal AR nor a seasonal MA. The 1 after the comma means that the data is annual.

Check the Residuals of the Fitted Model

Usually, we will try a few models of no more than order 1, 2, 3, or 4, and keep for further analysis only those that give white noise residuals, because that indicates that the model has captured all the autocorrelations present in the time series and there is no more room for improvement. Of those that yield white noise residuals, select the one with the smallest number of parameters to avoid overfitting.

> **Recommended practice:** We check first the residuals of the fitted model. If they are not white noise, the modeling is not finished. All autocorrelations must have been captured by the model. The sample ACF of the residuals should be the sample ACF of white noise. We double-check with the Ljung–Box test.

The AR(1) and AR(2) models give sample ACF of residuals that have significant autocorrelations at lag $= 1$. So their model residuals are not white noise. Only AR(3) and AR(4) have residuals with sample ACF of white noise. Nothing is gained from increasing the order after 3, because the AR(4) model has white noise residuals but more coefficients, and the fourth coefficient, α_4, is not significantly different from 0.

> **Recommended practice:** Overfitting is a big temptation in time series analysis. It creates major problems:

- It will cause an inability to accurately make predictions on new data. Overfitting results in the forecasting error being much larger than the model fitting standard error.
- It will render the standard errors useless for determining whether a coefficient is significantly different from 0.

We avoid overfitting. Models of low order giving white noise residuals should be preferred.

To prevent overfitting, a common practice is, for example, if we selected AR(3), to fit AR(3) and AR(2). If AR(2) has the same corresponding coefficients as AR(3) and the third coefficient is not significant, then that is a sign of overfitting with an AR(3).

Recommended practice: If more than one low-order model gives white noise residuals, we use the AIC (Akaike's Information Criterion) or other model selection criteria to select the smallest model.

The AIC is a method of selecting p, P, q, Q based on information theory. The formula is

$$AIC = -2\log(L) + 2(p + q + P + Q + k),$$

where $k = 1$ if there is a constant term in the model, and 0 otherwise, and L is the maximized likelihood of the model fitted to the differenced data, if any differencing was done. The objective in model selection is to minimize the AIC.

So we select AR(3) based on the residuals being white noise and having AIC smaller than the AIC of the AR(4) model.[3]

Figure 6.2 shows the sample ACF of the residuals, $y_t - \hat{y}_t$. The sample ACF looks like the sample ACF of data generated by a white noise process (see Chapter 5). This indicates that the model is good. The histogram of the residuals (not shown) reveals that they are close to normally distributed. To confirm the white noise conclusion, we should do the Ljung–Box white noise test.

The Ljung–Box test up to lags 6, 12, 18 and 24 have p-values larger than 0.05, suggesting that the residual series is indeed white noise and all ρ_k are significantly not different from 0.

If the residuals of a model fitted to a time series are not white noise, we would try to use the sample ACF and sample PACF of the residuals to try to identify a better model for the time series. But that is not the case in the current example.

---

[3] There are packages in R, some tied up to books authored by the authors of the packages, that do similar analysis to the one done in this book, but using different code. For example, the `arma` function in the `tseries` library gives the p-values directly.

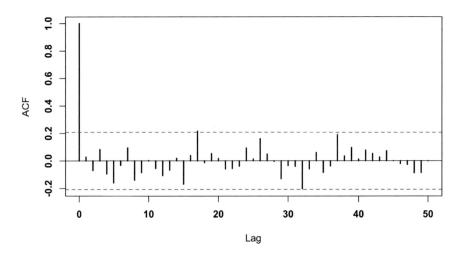

Figure 6.2 ACF of the residuals of the AR(3) model fitted to the logged mean subtracted unemployment data and used for forecasting.

Presenting the Results

The expression for the model assumed based on our analysis, in extended or forecasting form, is

$$y_t^* = \alpha_1 y_{t-1}^* + \alpha_2 y_{t-2}^* + \alpha_3 y_{t-3}^* + w_t,$$

where y^* is logged unemployment. In polynomial form,

$$(1 - \alpha_1 B - \alpha_2 B^2 - \alpha_3 B^3) y_t^* = w_t$$

and we can confidently say, for now, that our model is ARIMA$(3,0,0)(0,0,0)$.

We now populate the model with the values of the coefficients given by R, and their standard errors. We calculate the p-value of the coefficients by determining the number of standard errors away from 0. If the coefficient divided by the standard error is larger than 2 or smaller than -2, then we say that the coefficient is statistically significant. Notice that this means that we are assuming that the residuals are normally distributed.

The selected fitted model of y_t^* with standard errors and p-values is

$$\hat{y}_t^* = 1.1075 y_{t-1}^* - 0.4296 y_{t-2}^* + 0.2952 y_{t-3}^*$$
$$s.e(\hat{\alpha}) = 0.1009 \qquad 0.1439 \qquad 0.1015$$
$$p\text{-}value = \; < 0.001 \qquad\quad < 0.05 \qquad\quad < 0.05.$$

The estimate of the variance of the residuals is $\widehat{\sigma^2} = 0.1925$, and AIC $= 116.94$.

The modulus of the roots of the back-shift polynomial

$$(1 - 1.1075B + 0.4296B^2 - 0.2952B^3)$$

are 1.0234, 1.819275, and 1.819275, all larger than 1, so the AR(3) model fitted is stationary.

We would not have selected the model if it had not had white noise residuals.

Forecasting

Satisfied with the model fitted, we decide to forecast y_t 10 years ahead from 1978 to 1988 and obtain prediction intervals for the forecasts. Plotting the data, forecasts and prediction intervals for forecasts is standard and allows us to evaluate the forecast informally.

To forecast, we refit the model with the `arima()` function in R, call it `model3` and then use the `predict` function in R to forecast logged unemployment rate, which is the time series fed to `model3`:

```
forecast=predict(model3,n.ahead=10,se.fit=TRUE)
```

We are assuming that the reader is looking at the program *firstarimaexample.R* to see other lines of R commands needed to put together the final picture of the forecasts seen in Figure 6.3. As we said, R will forecast the logged unemployment, but it does not give forecasts for the actual unemployment levels as predictions. We should always check what R is giving.

Recommended practice: Once the forecast is obtained, steps must be retraced in order to remind ourselves whether the time series was transformed to stabilize the variance. If that is the case, the reverse operation must be applied to the forecast in order to obtain the forecast in the same units as the time series.

So we need to create an object that is the exponentiated R predictions to return the forecast of unemployment levels:

```
my.predict=ts(exp(forecast$pred),start=1979)
```

This code is calculating the forecast of the unlogged unemployment rate, namely

$$\hat{y}_t^{**} = e^{\widehat{y_t^*}} = e^{1.1075y_{t-1}^* - 0.4296y_{t-2}^* + 0.2952y_{t-3}^*}.$$

The values of the unlogged forecasted unemployment levels and the test set level of unemployment are as follows:

Unemployment (1979–88)	5.8	7.1	7.6	9.7	9.6	7.5	7.2	7.0	6.2	5.5
forecast		5.82	5.84	5.81	5.73	5.67	5.64	5.62	5.60	5.59	5.58

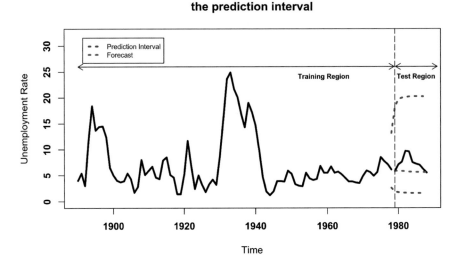

Figure 6.3 Forecast (in red) and confidence interval for y^{**}, the actual value of the time series in the future (in blue). The test set data is in black.

Assessing the Accuracy of the Forecasts

It is hard to assess the accuracy of the forecast with a list of numbers. A plot will not allow us to do that closely either. But let's look first at the plot and then we will calculate the Root Mean Square Error (RMSE) of the forecast.

Figure 6.3 shows the training time series and the forecasted values for the test set of 10 years, and the prediction interval for the true future value of the series. We observe in Figure 6.3 that the observed test set is inside the interval, which is good news. The interval gives us a measure of how uncertain we are about our forecasts.

After obtaining the forecasted values, and comparing them with the data for 1979– 88 that we had set aside as test data, we calculate the mean absolute error and the root mean square error of the forecast. Calculating the latter gives us a measure to compare white noise residual models in order to choose the model that gives the best forecast error according to this criterion.[4]

The RMSE of the forecast is

$$RMSE = \sqrt{\frac{\sum_{t=79}^{88}(y_t^{**} - \hat{y}_t^{**})^2}{10}} = 2.1015.$$

$\square$

---

[4]Some forecasters use automatic forecasting, disregarding the good statistical qualities of a given model, and focusing only on the forecasting performance. Larger models always fit better, in sample, but may be unable to make predictions on new data. Automatic forecasting should not be done blindly. The case study in Section 6.11 is precisely about this topic.

The variance is not constant

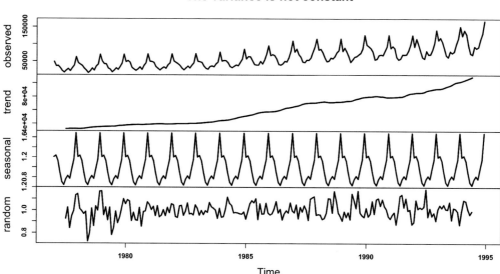

Figure 6.4 Time plots of the training set and the components of the multiplicative decomposition of overseas visitors.

6.2.2   Modeling and Forecasting Nonstationary Seasonal Time Series

Nonstationary seasonal time series usually require a little more work than annual time series before we settle on a model. Besides the predifferencing transformations to stabilize the variance, we must decide what differencing must be done to account for the long-term trend and the regular seasonal short-term trend. With seasonal data, we can expect trends but also seasonalities at different levels depending on the granularity of the data. If the data is daily, we could expect weekly and monthly seasonality; if the data is weekly, we could expect monthly seasonality. With monthly data, as we saw earlier, we can expect an annual seasonal component. That does not mean they are there in the data, but we should check. Thus, we should look for all those signals in the noise to account for them before we proceed with the modeling and forecasting task.

We will now look at a complete example of ARIMA modeling and forecasting a monthly time series.

Example 6.3   In this example, we will be concerned with the number of overseas visitors to New Zealand. The reader will find the example more useful if R program *secondarimaexample.R* is executed while reading this analysis. We will proceed as we did in Example 6.2, splitting the data into a training and a test set, and explaining all the steps of the analysis.

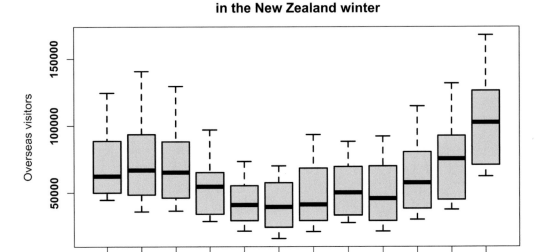

Figure 6.5 Seasonality in the training set of overseas visitors is evident by the dip in visits in the winter. (Winter in the Southern Hemisphere occurs during the summer months of the Northern Hemisphere.)

The Data

The number of overseas visitors to New Zealand is a monthly time series that comes with the book by Cowpertwait and Metcalfe [30] and is used here with the authors' permission. The time series runs from January 1977 to December 1995 and has 228 observations. We will denote the observed values by x_t.

Split the Data into Training and Test Sets

We choose the training set to be from January 1977 to December 1994. The test set is from January 1995 to December 1995. We leave out the last 12 observations as test data to compare with the forecasts and validate the model. So the training time series has 216 observations and the test set 12 observations.

We can see a view of the components of the multiplicative decomposition of the visits in the training set in Figure 6.4, and we can see a seasonal box plot of the training set in Figure 6.5. We observe in the time plot of the observed training set in Figure 6.4 that the number of overseas visitors has been increasing over time, and decreasing every year during the winter months, which, in New Zealand, because it is in the Southern Hemisphere, are the months of June to September. The variance of the time series seems to be increasing proportionally with the trend. A variance-stabilizing transformation is needed.

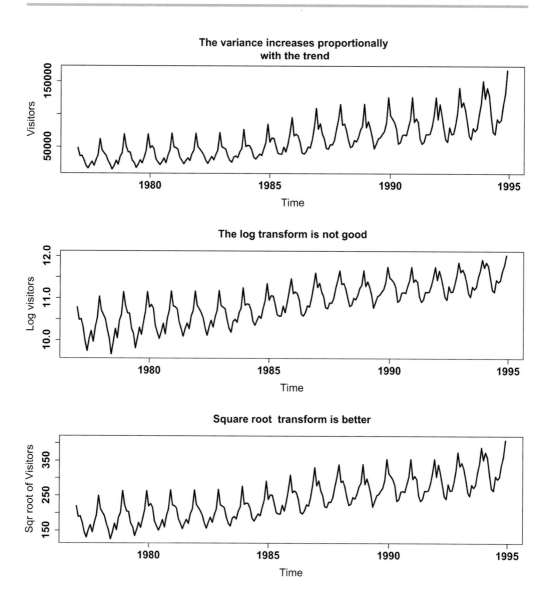

Figure 6.6 View of the time plots of the training set of overseas visitors as recorded, and after variance-stabilizing transformations.

Variance-Stabilizing Transformation

Figure 6.6 shows the time plot of the levels of overseas visitors in the training set. We can see some nonlinearity in the long-term trend and a seasonality that seems to be increasing proportionally to the trend. The variance is increasing proportionally to the trend, so the time series is heteroscedastic. We must do a pre-differencing transformation to stabilize the variance.

We tried log and square root of the training set data and realized that the square root does a better job at stabilizing the variance. As seen in Figure 6.6, the seasonal

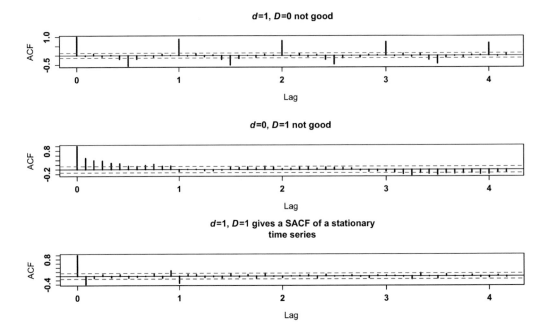

Figure 6.7 The sample ACF is the plot to observe in order to determine whether the differencing has achieved stationarity. Notice that R's sample ACF marks the cycle 1 (lag 12), cycle 2 (lag 24), and so on.

swings are more uniform with the square root transform. So, we continue the analysis with the square root transform. We denote the square root transform by $x_t^* = \sqrt{x_t}$.

Recommended practice: We keep track of transformations done to the original time series by denoting the resulting time series with a different letter or symbol than the untransformed data, as we did. If x_t is the number of overseas visitors in thousands, then the square root of the number of overseas visitors is $x_t^* = \sqrt{x_t}$. When explaining our model to someone, it is crucial to make it clear whether the raw or a transformed version of the training set is the subject of our analysis.

Determine d and D

We first tried first-order regular differencing of the square root transform, $y_t = x_t^* - x_{t-1}^*$. We also, separately, tried seasonal differencing ($x_t^* - x_{t-12}^*$). And then we tried seasonal differencing of the regular difference, which would be seasonally differenced y_t, to obtain $z_t = y_t - y_{t-12}$. In order to determine the suitability of each differencing to achieve stationarity, we looked at the correlogram of each. Figure 6.7 shows that seasonal differencing of the first-order regular differenced training set does a better job at giving stationarity. There are a few statistically significant lags near lag 0, and the seasonal frequency is significant only where 1 is in the horizontal axis. (Recall that in R's sample ACF plot, if `frequency` is larger than 1, the full cycle appears as the lag number 1; in our case here, 1 represents lag $k = 12$, since the training set is monthly data.)

The findings lead us to conclude that $d = 1$ and $D = 1$. That is, seasonal differencing of the regular difference of the square root of the training data is the data to which we will fit a stationary model. In ARIMA notation, the work done so far by us can be summarized as

$$z_t = (1 - B^{12})(1 - B)x_t^* = (1 - B - B^{12} + B^{13})x_t^* = x_t^* - x_{t-1}^* - x_{t-12}^* + x_{t-13}^*.$$

We then can say that the square root of overseas visitors is regularly integrated of order 1 and seasonally integrated of order 1, that is, it is stationary after regular and seasonal differencing of order 1.

By now, therefore, we know that we have the following model: ARIMA$(p, 1, q)$ $(P, 1, Q)_{12}$. We still have to figure out p, q, P, Q.

Identify a Model for the Training Set

After being satisfied with the differencing done, we proceed to do model identification by looking at the sample ACF and the sample PACF combined. Identification requires both, particularly if the time series is a realization of an AR process. According to Figure 6.8, the sample ACF cuts off at lag 1 and the sample PACF dies away at low lags k. Therefore, these are the features of an MA(1) for the short-term correlations. On the other hand, at the seasonal frequency, which appears in the plot at lag $k = 1$, both the sample ACF and the sample PACF cut off, but the sample ACF cuts off faster. According to the reference list in Chapter 5, we would tentatively recommend an MA(1) model for both the regular and the seasonal correlations.

Thus, we tell the software to fit ARIMA$(0, 1, 1)(0, 1, 1)_{12}$ to the square root of the level of overseas visitors in the training set, that is, to $x_t^* = \sqrt{x_t}$.

This means that the model for z_t is

$$z_t = (1 + \beta_{12}B^{12})(1 + \beta_1 B)w_t.$$

The reader is encouraged to try to set up small variations of this model because, as we have said elsewhere in this book, the same sample ACF and sample PACF could be representing more than one model.

Fit the Identified Model(s) to the Training Set

Based on all the operations done to the square root of the training set data, the instructions to give to R are

```
model1=arima(sqrt.osvisit.ts,order=c(0,1,1),
              seas=list(order=c(0,1,1),12))
```

This R command is applied to the square root of the training data and summarizes the operations thus far done:

(1)   First-order regular differencing of the square root of the training set x_t^* (that gave us y_t earlier).

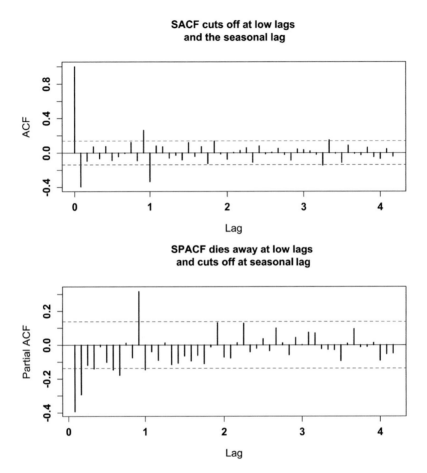

Figure 6.8 Identification of a model for the differenced square root of the number of overseas visitors requires looking to both the sample ACF (SACF) and the sample PACF. Notice that R's sample ACF marks the cycle 1 (lag 12), cycle 2 (lag 24), and so on.

(2) Seasonal differencing of the regular difference y_t once (that gave us z_t).

(3) Fitting an MA(1) model to capture the short-term correlations remaining in the seasonal difference of the regular difference of the square root of the training data and an MA(1) to capture the remaining seasonal correlation in the same data.

Notice that R will do the differencing for us in the background and then it will undo it (integrate, sum back) to forecast when we order it to do so using model1 as object.

Check Residuals and Restart If Needed

Upon checking the residuals of the model fitted, we discover that they do not have the sample ACF of white noise. That means that we need to go back and reidentify the model. But, for that, we use the sample ACF and sample PACF of the residuals to see what else we should add to the initial model. We proceed to try to identify a better model for the seasonal difference of the regular difference of the square root of the training data.

The sample ACF of the residuals of `model1` dies away and sample PACF cuts off at the seasonal lags. So we decide to add an AR(1) term to the seasonal part of the model. The smaller lags $k = 1, 2$, have nonsignificant correlation, indicating that the MA(1) for the small k autocorrelations, or regular part, was well fitted. So we try model 2, ARIMA$(0, 1, 1)(1, 1, 1)_{12}$; but this, although an improvement, does not give the sample ACF of residuals being white noise either. Neither does ARIMA$(1, 1, 1)(0, 1, 1)_{12}$, although the latter has lower AIC. So we try to fit ARIMA$(1, 1, 1)(1, 1, 1)_{12}$ and we are satisfied with the results obtained with this last model, because the residuals of this model are the residuals of a white noise process, and AIC is slightly lower than for the last model mentioned. The only borderline significant autocorrelations occur at randomly occurring lags and are the expected five percent significant autocorrelations due to sampling. Consequently, we fit ARIMA$(1, 1, 1)(1, 1, 1)_{12}$ and keep this model for forecasting out-of-sample.

Using ARIMA modeling notation, these two expressions combine into the following polynomial expression, since ARIMA does multiplicative seasonal modeling:

$$(1 - \alpha_{12}B^{12})(1 - \alpha_1 B)z_t = (1 + \beta_{12}B^{12})(1 + \beta_1 B)w_t.$$

To ask R, using the `arima()` function, to fit an ARMA(1,1) to the regular part (the short-term autocorrelations) of the z_t and an ARMA(1,1) to the seasonal part of z_t we use the following execution statement:

```
model3=arima(sqrt.osvisit.ts,order=c(1,1,1),
seas=list(order=c(1,1,1),12))
```

where `order=c(1,1,1)` says that we regularly differenced the x_t^* (the `sqrt.osvisit.ts`) once ($d = 1$) and then fitted an ARMA(1,1) to capture the autocorrelation at lower lags ($p = 1, q = 1$). The argument `seas=list(order=c(1,1,1),12)` says that we seasonally differenced ($D = 1$) the first regular difference of the square root of the level of overseas visitors and then identified an ARMA(1,1) for capturing the autocorrelations in the resulting seasonal part ($P = 1, Q = 1$).

R will respond with the parameter estimates. We show the output produced by executing the last statement in R in order to indicate what each element is.[5]

```
           ar1          ma1        sar1       sma1
        0.2840      -0.8338     0.2078    -0.6175
s.e.    0.0972       0.0582     0.1510     0.1278
sigma^2 estimated as 58.38:
log likelihood = -702.75,   aic = 1415.5
```

---

[5]The parameter estimates that R gives for the seasonal part usually have the s in front. For example, `sar1` is the AR coefficient for the seasonal part, $\hat{\alpha}_{12}$. The parameters of the regular part will have no s in front, for example, `ma1` is the MA parameter of the regular part, $\hat{\beta}_1$.

where `sigma`^{2} is a measure of how good the fit of this model is to the training data. The smaller `sigma`^{2} is, the better the fit. That is also the estimate of the variance of the W_t, also known as the sample variance.

`ar1, ma1` are $\hat{\alpha}_1, \hat{\beta}_1$, respectively; they are the values of the coefficients of the ARMA model fitted to capture the lower lag autocorrelations. Then `sar1, sma1` are $\hat{\alpha}_{12}, \hat{\beta}_{12}$, the values of the coefficients of the ARMA model fitted to capture the seasonal lag autocorrelations.

All coefficients are statistically significant. We know because dividing each coefficient by its standard error gives us a number larger than 2.

In addition to the above, model 3 has the lowest AIC of all three models attempted.

Presenting the Results

In polynomial notation, the model fitted that gives satisfactory white noise residuals is

$$(1 - 0.2078B^{12})(1 - 0.2840B)z_t = (1 + (-0.6175)B^{12})(1 + (-0.8338)B)w_t.$$

We check that the process is invertible and stationary by checking with R's `polyroots()` function that all the roots are outside the unit circle in absolute value. Therefore, the model is stationary in the AR part and invertible in the MA part. The final model is obtained after algebraically disentangling all the polynomials. We must multiply the polynomials until we end up with just an expression for x_t^*, the square root of visits as a function of all the other terms. We illustrate how that happens. We first multiply the polynomials:

$$(1 - \alpha_1 B - \alpha_{12}B^{12} + \alpha_{12}\alpha_1 B^{13})z_t = (1 + \beta_1 B + \beta_{12}B^{12} + \beta_1\beta_{12}B^{13})w_t.$$

Then, we apply the backshift operator to the variables and leave only z_t on the left-hand side:

$$z_t = \alpha_1 z_{t-1} + \alpha_{12} z_{t-12} + \alpha_{12}\alpha_1 z_{t-13} + w_t + \beta_1 w_{t-1} + \beta_{12} w_{t-12} + \beta_{12}\beta_1 w_{t-13},$$

$$\hat{z}_t = 0.2840z_{t-1} + 0.2078z_{t-12} + 0.059z_{t-13} - 0.6175w_{t-12} - 0.8338w_{t-1}$$
$$+ 0.5149w_{t-13}.$$

But R's next step will consist of undoing all the differencing to obtain a forecasting model for the square root of the number of visitors, x_t^*.

$$x_t^* - x_{t-1}^* - x_{t-12}^* + x_{t-13}^* = 0.2840(x_{t-1}^* - x_{t-2}^* - x_{t-13}^* + x_{t-14}^*) + 0.2078(x_{t-12}^* - x_{t-13}^*$$
$$- x_{t-24}^* + x_{t-25}^*) + 0.059(x_{t-13}^* - x_{t-14}^* - x_{t-25}^* + x_{t-26}^*) - 0.6175w_{t-12} - 0.8338w_{t-1}$$
$$+ 0.5149w_{t-13},$$

or

$$\hat{x}_t^* = x_{t-1}^* + x_{t-12}^* - x_{t-13}^* + 0.2840(x_{t-1}^* - x_{t-2}^* - x_{t-13}^* + x_{t-14}^*) + 0.2078(x_{t-12}^* - x_{t-13}^*$$
$$- x_{t-24}^* + x_{t-25}^*) + 0.059(x_{t-13}^* - x_{t-14}^* - x_{t-25}^* + x_{t-26}^*) - 0.6175\hat{w}_{t-12} -$$
$$0.8338\hat{w}_{t-1} + 0.5149\hat{w}_{t-13}.$$

This can be simplified by putting terms with the same lag together:

$$\hat{x}_t^* = 1.2840x_{t-1}^* - 0.2840x_{t-2}^* + 1.2078x_{t-12}^* - 1.4328x_{t-13}^* + 0.225x_{t-14}^* -$$
$$0.2078x_{t-24}^* + 0.1488x_{t-25}^* + 0.059x_{t-26}^* - 0.8338w_{t-1} - 0.6175w_{t-12} + 0.5149w_{t-13}.$$

Forecasting

The training set ends in December 1994 and that is $t = 216$. Our first forecast will be for January 1995 or $t = 217$. To forecast the first value out of sample, the value of January 1995 or $t = 217$, the formula would be

$$\widehat{x}^*_{217} = 1.2840x^*_{216} - 0.2840x^*_{215} + 1.2078x^*_{205} - 1.4328x^*_{204} + 0.225x^*_{203} - 0.2078x^*_{193} + 0.1488x^*_{192} + 0.059x^*_{191} - 0.8338w_{216} - 0.6175w_{205} + 0.5149w_{204}.$$

Upon substituting the past values of the square root of the training set where x^* are and the value of the residuals of the fitted model where w_t appear, we obtain

$$\widehat{x}^*_{217} = 1.2840 * (409.53) - 0.2840 * (362.81) + 1.2078 * (352.77) - 1.4328 * (388.81) + 0.225 * (341.84) - 0.2078 * (334.82) + 0.1488 * (375.31) + 0.059 * (321.4903) - 0.8338 * (1.01) - 0.6175 * (10.7) + 0.5149 * (0.47) = 366.7322.$$

Any small difference of this value from the one obtained from the predicted value produced by R using `model3` is due to rounding errors.

Finally, to obtain the fitted values in the same units as the original time series, we must undo the square root and square the fitted values and the forecast:

$$\widehat{x}_t = (\widehat{x}^*_t)^2.$$

We square because the reverse operation of square root is to square.

The forecasts obtained with R are for the square root of the number of visitors. But we want to forecast the number of visitors. So we compute:

$$\widehat{x}_{217} = (\widehat{x}^*_{217})^2 = 366.7322^2 = 134492.5$$

and

$$\widehat{x}_{218} = (\widehat{x}^*_{218})^2,$$

and so on, using the program *secondarimaexample.R*.

To measure how good the forecasts are, we must compare the forecasts of number of visitors with the actual values of the number of visitors series.

Table 6.1 contains the lowest value of the prediction interval for the out-of-sample values, the observed test data, the forecasts from `model3` and the highest value of the prediction intervals. As we can see, the forecast overpredicts the test data throughout, and there is only once instance in which the interval does not contain the test data.

We may compute the Root Mean Square Error of the forecast as follows:

$$RMSE = \sqrt{\frac{\sum_{t=Jan1995}^{Dec1995}(x_t - \widehat{x}_t)^2}{12}} = 6175.554.$$

Table 6.1 Forecast of 1995:1 to 1995:12 levels of overseas visitors using an ARIMA(1, 1, 1)(1, 1, 1)$_{12}$ model fitted to the square root of visitors.

Year	Month	CIlow	test set	forecast	CIhigh
1995	1	124307.57	137792	135091.76	146324.48
1995	2	136744.05	144540	149160.23	162115.88
1995	3	124607.26	132570	136900.98	149772.96
1995	4	96920.68	110097	108064.15	119813.85
1995	5	74828.29	80179	84865.23	95533.64
1995	6	71853.60	73931	81890.41	92583.15
1995	7	92486.64	102908	104043.52	116280.59
1995	8	87567.52	92582	99026.57	111190.01
1995	9	90700.60	95115	102561.07	115150.11
1995	10	110481.89	116850	123755.08	137781.02
1995	11	127725.81	135015	142201.98	157455.06
1995	12	163766.31	187216	180365.07	197764.91

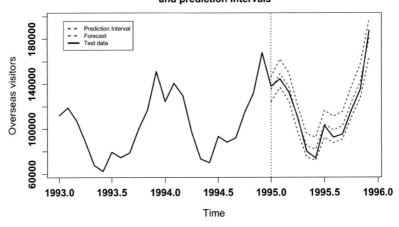

Figure 6.9 The forecast of the overseas visitors is shown together with the test data set, the training data set, and the confidence intervals for future values.

Figure 6.9 shows the forecast results.

6.2.3   Exercises

Exercise 6.1   Consider time series x_t and consider the following two scenarios.

(a) We tell R that we are ready (after identification) to fit the following model to x_t^*, the log of x_t: ARIMA(0, 1, 1)(0, 1, 1)$_{12}$. Describe the steps followed by R to fit and obtain the predicted values of x_t^*. Write the final forecasting equation for x_t.

(b) We tell R that we are ready (after identification) to fit the following model to x_t^*, the log of x_t: ARIMA(0, 1, 2)(0, 1, 2)$_{12}$. Describe the steps followed by R to fit and obtain the predicted values of x_t^*. Write the final forecasting equation for x_t.

Exercise 6.2 Write the ARIMA$(p, d, q)(P, D, Q)_F$ notation form, the polynomial form and the forecasting form of an appropriate model fitted to a stationary monthly time series x_t in which only the value of the month of the previous year influences the current monthly value.

Exercise 6.3 Write the ARIMA$(p, d, q)(P, D, Q)_F$ notation form, the polynomial form and the forecasting form of an appropriate model fitted to a stationary monthly time series x_t in which only the value of the month of the previous year influences the current monthly value after regularly differencing the time series once to remove the long-term trend.

Exercise 6.4 Suppose we were told that the following model is the model applied to an observed time series x_t that has mean 0:

$$\widehat{x_t} = 0.5x_{t-1} + x_{t-1} - 0.5x_{t-2} + 0.3w_{t-1}.$$

Describe this model in polynomial form and in the ARIMA$(p, d, q)(P, D, Q)_F$ notation form.

Exercise 6.5 Suppose we have a quarterly time series, x_t. We had to square root it to achieve constant variance. We denote the square-rooted time series x_t^*. Because the time series had trend and seasonal, we differenced and concluded that regular differencing made the time series stationary. We denoted the differenced-stationary time series y_t. Looking at both the sample ACF and the sample PACF, we identified an ARMA(3,1) for the lower-lag autocorrelations. No model was needed for the seasonal part, as the autocorrelations were close to zero around the seasonal lag.

Write the compact notation for everything we did to x_t^* and present the backshift polynomial form and the forecasting form of the model.

Exercise 6.6 In the program *global.R*, the reader will find an analysis of a monthly time series.

Figure 6.10 shows a set of plots, each depicting different information. The following questions refer to the plots and what you would do based on your interpretation of the plots.

(a) Is the time series seasonally integrated of any order $D > 0$? Explain.
(b) Is the time series integrated of any order $d > 0$? Explain.
(c) Is any of the plots referring to a stationary time series? Which? Explain.
(d) Is any of the plots referring to a nonstationary time series? Which? Explain.
(e) What, in the shortcut expression ARIMA$(p, d, q)(P, D, Q)$, do you know, at least tentatively, and could fill in? Explain. Justify your answer by referring to what you see in the plots.

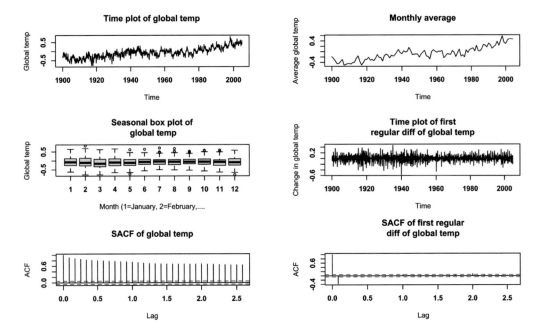

Figure 6.10 Several views of the global temperature time series from different perspectives. Source of the data: Cowpertwait and Metcalfe [30] (used with permission of the authors).

Exercise 6.7 A monthly time series that we will denote by x_t was pre-transformed, differenced and applied a proper model after identification. The summary of what was done and estimated is given in the following expression:

$$(1 - 0.2191B)(1 - B^{12})(1 - B)z_t = (1 - 0.8245B)(1 - 0.3745B^{12})w_t,$$

where z_t is the $log(x_t)$.

(a) Substitute for each of the letters in the expression ARIMA$(p, d, q)(P, D, Q)$.
(b) Write the model in forecasting form for z_t and for z_{t+1}.
(c) Write the model in forecasting form for x_{t+1}.

Exercise 6.8 A monthly time series that we will denote by x_t for the number of overseas visitors to New Zealand over the period 1977 to 1995 was pre-transformed, differenced and fitted a model after identification. The summary of what was done and estimated is given in the following expression:

$$z_t = z_{t-1} + 0.0250z_{t-1} - 0.0250z_{t-2} + w_t,$$

where z_t is the $log(x_t)$.

(a) Substitute for each of the letters in the expression ARIMA$(p, d, q)(P, D, Q)_{12}$ to summarize what was done.
(b) Write the model in backshift polynomial form.

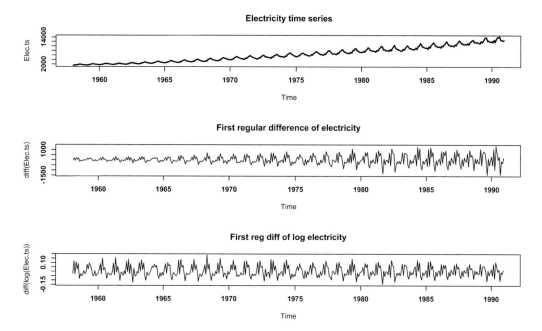

Figure 6.11 China's monthly total electricity production in gigawatts per hour (not seasonally adjusted). Source: FRED, CHNPRENELO1MLM, 1999-01-01 to 2020-10-01.

Exercise 6.9 Look at Figure 6.11. It contains China's electricity production. You may look at the R script file in program *china.R* to see how we got those plots. What intentions might we be having in obtaining those plots? What will they help us decide?

Exercise 6.10 Consider now the electricity data in the same program *china.R*. After taking the log, we first regularly differenced the data, and someone identified an MA(1) model for the differenced logged data. The operations are summarized in ARIMA$(0, 1, 1)(0, 0, 0)_{12}$. Do all of the preceding to the data and determine whether the resulting residuals are white noise. Explain your results and your reasoning.

Exercise 6.11 Consider the following ARIMA$(1, 0, 0)(0, 0, 0)_F$ model

$$x_t = \alpha_1 x_{t-1} + w_t,$$

which was fitted to an observed time series after proper exploratory analysis and identification. The estimate of the parameter α_1 is $\hat{\alpha}_1 = 0.3$ and the estimated standard error is $s.e._{\hat{\alpha}_1} = 0.08765$. The sample mean of the observed time series is $\bar{x} = 200$ and the sample variance is $s^2 = 16$. There are 100 observations in this time series.

(a) Construct a 95 percent confidence interval for α_1 and interpret it.
(b) Is α_1 significantly different from 0?
(c) Is the fitted model stationary?
(d) Write the forecasting model to forecast x_{101}, knowing that $x_{100} = 180$.

Exercise 6.12   Using the `osvisit` time series analyzed in Section 6.2.2, predict how many visitors will visit New Zealand in 1996.

Exercise 6.13   Consider the electricity data seen in Figure 6.11. An investigator provided the following information in a report.

(a) The electricity data is dominated by a trend of increasing electricity production over the period, so we must difference it; but we see increasing variance, so we first took the log and then differenced.

(b) We fitted an ARIMA(0,1,1)(0,0,0) model to log electricity as a first tentative model after viewing the correlogram and partial correlogram.

That model was chosen because the correlogram cuts off at lag 2 and the partial correlogram dies away after a few lags.

(c) The resulting prediction model is $\hat{z}_t = z_{t-1} + w_t - 0.2934w_{t-1}$, where $z_t = log(electricity)$.

The final prediction is calculated using e^{z_t}.

(d) The model residuals indicate that we did a good job at modeling.

The reader will evaluate critically everything done by this researcher and determine whether and how the analysis should be modified.

Exercise 6.14   Write the forecasting equation for $\hat{x}^*_{218}$ in Section 6.2.2 and calculate the value of $\hat{x}_{218}$, the number of overseas visitors predicted at time $t = 218$. What year and month is $t = 218$?

Exercise 6.15   A time series called `sunspot.month` comes with R. It contains 2,988 monthly numbers of sunspots, from January 1749 to September 2013. It is recorded from the World Data Center, also known as SIDC. According to R's documentation, this is the version of the data that will occasionally be updated when new counts become available. It is one of R's `ts()` objects, and it is univariate.

(a) Select the last 678 observations and use 666 of those for the training set and the last 12 for the test set.

(b) Check the time plot, seasonal box plot and the correlogram of the training data set. Should you pre-transform due to heteroscedasticity? Is there seasonality? Should you difference? If so, do as needed.

(c) With the stationary time series that results, check the sample ACF and the sample PACF and identify a stochastic model for the time series.

(d) Fit the model and check the correlogram of the residuals. If it is not the correlogram of white noise, check both the correlogram and partial correlogram and identify what you missed in the modeling you did. Refit a modified model if needed.

(e) With the model fitted, forecast sunspots 12 months ahead. Plot the forecast, the actual value of the test data, the confidence intervals, and compute the root mean square error of the forecast.

6.3   Estimation

When it comes to explaining how parameters of ARIMA models are estimated, the time series literature is very diverse in its approach, depth and difficulty for beginners. What to recommend depends on the background of the reader and the depth and breadth of the recommended material. A common question asked by all resources talking about estimation is, "Once we have specified all the arguments in the ARIMA$(p, d, q)(P, D, Q)_F$, how do we estimate the p autoregressive parameters and/or the q moving average parameters of the I(d) and/or I(D) series and the noise variance?" We certainly estimate the parameters of the differenced time series, in all cases, if $d > 0$ and/or $D > 0$. Let Y_t represent the stationary process.

Cryer and Chan [32]'s chapter 7 surveys several methods of estimation and provides examples. Depending on the reader's background in mathematical statistics, some may be more approachable than others. For example, they describe the method of moments, which will be easy to follow if the reader has felt at ease with our discussion of the relation between autocorrelation coefficients and model parameters in Chapter 5. This method is good for estimating the AR(p) model parameters but not the MA(q) ones [32] [179]. For that reason, least squares estimation is considered a better method. From the least squares point of view, time series estimation is not too different from estimation in statistical analyses where least squares is used. The parameters that minimize an objective function, the sum of squared errors, in-sample is minimized.

Let us illustrate with a simple example. Suppose that, after differencing, the mean adjusted time series y_t is identified to follow the model

$$y_t = \alpha_1 y_{t-1} + \beta_1 w_{t-1} + w_t.$$

Then we can write

$$w_t = [y_t - (\alpha_1 y_{t-1} + \beta_1 w_{t-1})].$$

To estimate the parameters, we would minimize

$$w_t^2 = [y_t - (\alpha_1 y_{t-1} + \beta_1 w_{t-1})]^2.$$

This is basically the principle of least squares, widely used in regression. Given the estimates $\widehat{\alpha}_1, \widehat{\beta}_1$, we could estimate the variance of w_t, $\hat{\sigma}^2$, the standard error of the parameters, and other summary statistics. The estimates, however, must be obtained numerically, with approximating algorithms because when putting the MA term as a function of y_t, the coefficients are exponentiated, that is, we get a nonlinear equation in the parameters.

If the w_t are normally distributed, which is usually assumed, a more comprehensive approach is maximum likelihood estimation. Given a model, we compute the likelihood and then obtain the maximum likelihood estimates. To obtain maximum likelihood estimates, programs often use iterative approximation algorithms common in optimization. Two common numerical optimization routines for accomplishing maximum likelihood estimation are Newton–Raphson and the method of scoring [179] or Kalman filter.

An overview of the maximum likelihood method for the simple example given earlier follows.

The likelihood function is

$$f(\alpha_1, \beta_1, \sigma^2 \mid w_1, \ldots, w_n) = \frac{1}{(2\pi\sigma^2)^{n/2}} \exp\left[-\frac{1}{2\sigma^2}\sum_{i=1}^{n} w_i^2\right],$$

which gives a log likelihood

$$\log f(\alpha_1, \beta_1, \sigma^2 \mid w_1, \ldots, w_n) = -\frac{n}{2}\log(2\pi\sigma^2) - \frac{1}{2\sigma^2}\sum_{i=1}^{n} w_t^2.$$

Maximizing the log likelihood with respect to the three parameters gives us the estimate of each parameter. There is, however, a problem at the start of the series. We have to make some assumptions about initial values. The simplest is to assume values up to lag p are 0:

$$\log f(\alpha_1, \beta_1, \sigma^2 \mid w_1, \ldots, w_n) = -\frac{n}{2}\log(2\pi\sigma^2) - \frac{1}{2\sigma^2}\sum_{t=p+1}^{n} w_t^2.$$

But this approach, although giving reasonable parameter estimates, makes the parameter estimates conditional on the initial values being 0 and forces us to lose some observations. Box and Jenkins proposed using backcasting, which consists of back forecasting the initial values. This technique gives us an improvement in estimation [85].[6]

There are other approaches to finding initial values for the approximation algorithms. The following example presents a simple one.

Example 6.4 A time series of sales was differenced to make it stationary. After differencing it, an analysis of the sample ACF and the sample PACF resulted in the identification of an ARIMA(0,1,1)(0,0,0) model:

$$x_t = w_t + \beta_1 w_{t-1}.$$

To find some initial estimate of the MA parameters (β_1) the values of the autocorrelations were obtained. It was found that $r_1 = 0.3066$.

Using that correlation value, and results seen in Chapter 5, we know that in theory

$$\rho_1 = \frac{\beta_1}{1 + \beta_1^2}.$$

We can solve for β_1 if we use r_1 as an estimate of ρ_1. The two solutions are $\beta_1 = 2.926$ or $\beta_1 = 0.3417$. We choose the one that satisfies the invertibility condition $\mid \beta_1 \mid < 1$. We then use that β_1 as the initial value for the estimation algorithms. ☐

---

[6]Backcasting is a widely used technique nowadays that consists of the opposite of forecasting. In backcasting, the past values that would have given rise to future values are predicted.

6.4   More on Model Diagnostics

Model diagnostics refers to statistics that tell us how well a model fits the historical time series in the training data.

> Goodness of fit, in-sample error and accuracy of the model are interchangeable terms for the assessment of how well a model fits the training data. They should not be confused with forecast error, which measures how well the fitted model forecasts the future.

Most goodness-of-fit measures are based on the residuals that arise from the model fitted to the historical training data. We have discussed often that those residuals must be white noise as indicated by their sample ACF. A common assumption made about the residuals is that they also are normally distributed. This assumption makes the summary statistics that are often used for goodness-of-fit easier to interpret.

Several models with white noise and normally distributed residuals could differ in the root mean square error of the fit. That measures the estimate of the variance of the random term in the model, as in any regression. Smaller root mean square error in-sample means better fit, on average. The estimate of the standard deviation of the $\hat{w}_t$,

$$\hat{\sigma} = \sqrt{\frac{\sum_t (z_t - \hat{z}_t)^2}{n - k}},$$

where z_t is the value of the stationary historical time series in the training set, is the right value to look at. The $\hat{z}_t$ is the fitted value at time t, and k is the number of coefficients. We want $\hat{\sigma}$ to be small for a good model fit. But sometimes it is hard to tell what small means unless you are comparing it with the root mean square error of some other model. So we often just use it to compare models.

As we have seen earlier, the autocorrelation of the residuals is checked by means of the correlogram or sample ACF of the residuals or by the white noise test. The residuals are the difference between the fitted values (also usually called the predictions) in sample and the training time series.

If the residuals of the model fit pass the diagnostic check, and the model coefficients are significantly different from 0, then we can proceed to use the model to forecast future values of the series for which we do not know the value.

6.4.1   Model Selection

It is a fact that several models fitted to the training data could pass the diagnostic scrutiny, produce white noise residuals and fit a time series well. However, the true model may not be among them. In order to reduce arbitrariness in the selection of an ARIMA model and to prevent overfitting, it is important to have criteria to determine which model to choose among those that fit the training data well.

> We want to choose the smallest possible model that gives white noise residuals and is a good fit to the data. Low orders for AR and MA are preferred. This is called the principle of parsimony, also known as the KISS principle, which stands for "Keep It Sophisticatedly Simple."

> The reader should be aware that selecting the model that provides the best fit to historical data generally does not result in a forecasting method that produces the best forecasts [125]. Concentrating too much on the model that produces the best historical fit often results in overfitting. That is why we need model selection criteria.

Akaike's Information Criterion

Akaike's Information Criterion (AIC) is a statistic that gives us a measure of the trade-off between goodness of fit and model size. The criterion penalizes the sum of squared residuals for including additional parameters in the model. This is how this criterion works:

- Counts the number of parameters in the model. For example, if it is ARIMA(1, 1, 1)(1, 1, 1)$_{12}$, there are four parameters (two for the regular part and two for the seasonal part). Note that *d* and *D* values are not parameters. Let the number of parameters be denoted by *m*.
- Calculates the maximum likelihood estimates of those parameters. We will denote that $\widehat{L}$. R also calculates the log likelihood.
- R calculates AIC for each model as follows:

$$AIC = 2m - 2log(\widehat{L}).$$

- We then compare the AIC for different models.
  We will choose the model with white noise residuals that minimizes the AIC. The smaller the AIC, the better the model. The criterion penalizes large models while rewarding goodness-of-fit as determined by the likelihood function. AIC should be applied only to models that give white noise residuals.

> We should not compare the AIC of models that do not give white noise residuals.

If the AIC is negative, we follow the rules used for numbers (do not look at the absolute value). Larger negative is smaller.
That is, AIC = −300 is smaller than AIC = −10.
- If the AICs of the best two models are very closed, consider the two models for forecasting, then choose the one that gives the best forecast.

6.4.2 Comparing Models Selection Criteria

Some model-selection criteria are better than others, and, of course, there is a very active area of research in time series focused on designing ever better measures. A lot of interesting points are brought up in the discussion on measures. From a practical point of view, it is very common to use AIC, SIC (Schwarz Information Criterion) and BIC (Bayesian Information Criterion). In practice, it is good to look at all these measures, knowing their limitations because. AIC, for example, is said not to penalize enough for large models.

6.4.3 Exercises

Exercise 6.16   Fit the following models to the square root of the overseas visitors training time series that we analyzed in 6.2.2. Then, for each model, do the following:

(a) Check the roots of the model polynomials. The model must show that there is invertibility and stationarity in the fitted part.

(b) Check the sample ACF of the residuals and determine whether it is the sample ACF of white noise. If not, discard the model.

(c) If white noise residuals, record the AIC and calculate the RMSE of the forecast for the months of 1995.

(d) Record the estimate of σ^2 in each case.

(e) Determine the model that has the lowest σ^2, that is, the model that best fits the training data.

(f) Determine the model with the lowest RMSE.

(g) Determine the model with the lowest AIC.

    Here are the models to consider:

    Model 1: $\text{ARIMA}(1, 1, 0)(0, 1, 0)_{12}$
    Model 2: $\text{ARIMA}(0, 1, 0)(1, 1, 0)_{12}$
    Model 3: $\text{ARIMA}(1, 1, 0)(1, 1, 0)_{12}$
    Model 4: $\text{ARIMA}(0, 1, 1)(0, 1, 1)_{12}$
    Model 5: $\text{ARIMA}(1, 1, 0)(0, 1, 1)_{12}$
    Model 6: $\text{ARIMA}(0, 1, 1)(1, 1, 0)_{12}$
    Model 7: $\text{ARIMA}(1, 1, 1)(1, 1, 1)_{12}$
    Model 8: $\text{ARIMA}(1, 1, 1)(0, 1, 1)_{12}$

Exercise 6.17   The `AirPasssengers` time series comes with R. Use as training set all but the last 12 observations.

(a) Determine pre-transformation and differencing needed to make the time series stationary.

(b) Fit an appropriate ARIMA model that gives you white noise residuals, the latter justified by the sample ACF. Provide a measure of the goodness-of-fit of the model fit to the training data. Report the AIC.

(c) Write in notation form $\text{ARIMA}(p, d, q)(P, d, Q)_F$ what you did.

(d) Write in polynomial notation form what you did.
(e) Validate the model by comparing the forecasted values for the test set period and the test set values. Calculate the root mean square error of the forecast.
(f) Compare the measure of goodness of fit of your model to the training data and the root mean square error of the forecast. Are they very different? Is there an indication of overfitting?

6.5   Forecasting

Forecasting is a common data science task that is central to many activities within an institution. Comparing and evaluating forecasting procedures is a growing field of research. But forecasting is not easy, as we have seen. Analysis that can produce high-quality forecasts is quite rare because forecasting is a specialized skill requiring substantial experience. Nevertheless, the demand for forecasts is on the rise, as the case study in Section 6.11 will illustrate. Producing forecasts at scale is a specialized industry gaining importance. That does not mean that we will rely on completely automatic forecasting techniques, which are often too inflexible to incorporate useful assumptions and heuristics.

Forecasting and the analysis of time series are two distinct activities. A forecast is a view of an uncertain future. Time series analysis is a description of what has already happened.

There are two types of forecasts: point forecasts and interval forecasts. Within each type, we find forecasting at scale and forecasting one time series at a time when faced with a demand to forecast many time series. And within each of those subcategories, we find short-term, medium-term and long-term forecasting. That a model fits a time series well is no guarantee that the fitted model will predict future observations accurately in any of those categories. Hence the challenge. Those using the forecasts are very concerned about the accuracy of the forecast and not so concerned about the accuracy of the fit of the model to the historical data.

Competing models may be judged on the basis of their forecasting performance.

6.5.1   Forecasting Rules for ARIMA Models

(a) If we have past values of the x_{t-k} that appear on the right-hand side of the forecasting equation, we use them for the x_{t-k}'s.
(b) If we do not have past values of the x_{t-k} that appear on the right-hand side of the forecasting equation, use the forecasted values.
(c) For forecasting, w_{t-k}'s are equal to $x_{t-k} - \hat{x}_{t-k}$ if x_{t-x} is available. If the latter is not available, then the best value to substitute for w_{t-k}'s is 0.

We remind the reader of the process of point forecasting with an example.

Example 6.5   Suppose n is the total number of observations of the time series and suppose the model fitted to the time series is

$$\hat{z}_t = 1.5537z_{t-1} - 0.6515z_{t-2}$$

and z_t has mean 0.

We know $z_{n-1} = 6.63$ and $z_n = 6.20$. Then the point forecasts for three times ahead are

$$\hat{z}_{n+1} = 1.5537(6.2) - 0.6515(6.63) = 5.313495$$
$$\hat{z}_{n+2} = 1.5537(5.313495) - 0.6515(6.2) = 4.216277$$
$$\hat{z}_{n+3} = 1.5537(4.216277) - 0.6515(5.313495) = 3.089088.$$

Notice that for predicting one period ahead, we use actual values on the right-hand side. After that, we must start using the predictions from the earlier periods. These forecasts are for the estimated expected value. If we wanted to obtain the point forecast of the actual value of the series, we would include the residual at time n in the first equation.   □

6.5.2   Assessing the Accuracy of the Forecasts

Point forecasts of values for which we don't have raw data are called out-of-sample forecasts. These are the most interesting. But recall that, to validate a model, we customarily leave the last values of the time series out of the analysis to use as a test set against which we can compare our forecasts. The process is called validation, although it involves forecasting.

We measure the accuracy of the forecasts by the root mean square error (RMSE) statistic, or the root mean absolute deviation (RMAD) statistic, to name a few criteria. If z_t is the time series to forecast, m is the number of periods to forecast, and n is the length of the training set, we define these statistics for the forecast as

$$RMSE = \sqrt{\frac{\sum_{t>n}(z_t - \hat{z}_t)^2}{m}}$$

$$RMAD = \sqrt{\frac{\sum_{t>n}|z_t - \hat{z}_t|}{m}}.$$

When comparing models, these two statistics could give us very different conclusions about which model to select.

The RMSE and the RMAD are both scale-dependent measures of forecast accuracy. Their values are expressed in terms of the original units of measurement or the square of the original units. Accuracy measures that are scale dependent do not facilitate comparison of a single forecasting technique across different time series, or comparisons across different periods [125]. The Mean Percent Error (MPE) and the mean absolute percentage error (MAPE) are alternatives that solve that problem:

$$MPE = \frac{1}{m}\frac{\sum_{t>n}(z_t - \hat{z}_t)^2}{z_t}$$

and

$$MAPE = \frac{1}{m}\frac{\sum_{t>n}|z_t - \hat{z}_t|}{z_t}.$$

When forecasting in practice, we do not have raw data for the time period being forecasted. So, the only measure of uncertainty we have is the standard errors of the forecasts, which usually show that the initial uncertainty propagates and grows over time. This manifests itself in fanning out confidence intervals of the forecast.

If we have forecasted the z_t for $t > n$, the following expression gives us the confidence interval for the forecasts:

$$\widehat{z}_{t+k} \pm t^* s.e._{\hat{z}_{t+k}}.$$

Besides the usual assumptions about the w_t of having mean 0, constant variance at all t and 0 correlation, in order to do statistical inference, that is, tests, confidence intervals and so on, the w_t must be normally distributed. So one additional last thing to do is to check that the residuals are normally distributed.

> When validating the model, namely, comparing the forecasting performance of the models fit to historical time series to the test set data, it is convenient to compare short-term forecasting accuracy across models. The same goes for medium-term and long-term accuracy. Some models may outperform others in one of those but not in the other measures. It is also convenient to compare the forecasts according to different accuracy measures. Some models may outperform others based on some measures while the competitors outperform according to other measures.

6.6   Volatile Time Series

Many time series, especially financial ones such as stock returns and exchange rates, exhibit (random) changes in variance over time. That is not news. We have seen plenty of time series with nonconstant variance already. But we are referring now to a different type of nonconstant variance. These changes tend to be serially correlated, with groups of highly volatile observations occurring together. This is highly plausible, since if a financial market is in a state of uncertainty, perhaps brought about by some international crisis, it will take some time for the price to settle down [60] and to revert to its regular variance.[7] Figure 6.12 is a typical image of a volatile time series.

The term "volatility" is thus associated with nonstationary variance, but what is the difference between a volatile time series and a time series that that has nonconstant variance but is not volatile?

We distinguish between two types of variance nonstationarity: heteroscedastic and conditionally heteroscedastic (volatile) time series.

- *Heteroscedastic time series* such as, for example, `AirPassengers` and `osvis-its` have variance that changes in a regular, predictable way, proportional to the trend. We handle this variance nonstationarity by preidentification transformations,

---

[7] The reader should be aware that in the financial world, the variance of a time series is already called volatility as well. So, from a financial point of view, a period of nonconstant variance is a period of nonconstant volatility.

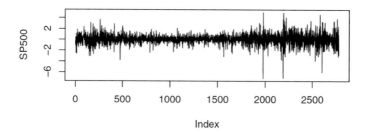

Figure 6.12 Standard and Poors time plot reveals volatility. Index denotes values of t.

such as taking logs, square root or some other suitable transformation that stabilizes the variance, as we have seen in this chapter.

- *Conditional heteroscedastic or volatile time series*, on the other hand, exhibit random periods of increased variance, as in Figure 6.12. Conditional means that next period's variance is conditional on the information this period. Log or similar transforms will not help stabilize the variance when there is conditional heteroscedasticity.

6.6.1 Detection of Volatility

How can we determine whether a time series is of the heteroscedastic or conditional heteroscedastic variety? The answer lies in the sample ACF. But the sample ACF of what?

In volatile time series, the mean adjusted data has the sample ACF of white noise, but the sample ACF of the square of the mean adjusted time series depicts autocorrelations. In contrast, the square of a white noise residual has the sample ACF of white noise.

> To detect volatility (conditional heteroscedasticity) we look at the sample ACF of the mean-adjusted time series and the sample ACF of the square of the mean-adjusted time series.
>
> The correlogram of a volatile time series looks like the correlogram of white noise, but the correlogram of the square of a volatile time series is the correlogram of a stationary time series with significant autocorrelations. We handle this by modeling the variance nonstationarity itself.

Example 6.6 The Standard and Poor's 500 (S&P 500) Index is calculated from the stock prices of 500 large corporations. The time series is defined as $100ln(SPIt/SPIt - 1)$, where SPIt is the value of the S&P 500 Index on trading day t. The series (from January 2, 1990 to December 31, 1999) is available in the MASS library within R [200]. The package MASS must be installed before accessing the data in that library. Consider the time plot and correlogram of the daily returns of the S&P 500 Index in Figure 6.12 and Figure 6.13 obtained with program *garchprograms.R*.

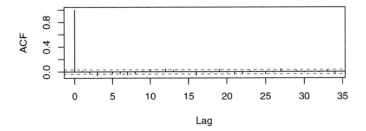

Figure 6.13 Correlogram of S&P 500 data.

Judging by the time plot in Figure 6.12, the series appears to be a realization of a stationary process. However, on closer inspection, it seems that the variance changes randomly. The correlogram of the S&P 500 in Figure 6.13 does not differ significantly from the correlogram of white noise. However, the returns are uncorrelated but not independent.

> We detect the volatility by looking at the correlogram of the squared mean-adjusted values of the S&P 500 data (Figure 6.14).

We can see in Figure 6.14 the correlogram of the squared time series, which resembles the correlogram of an AR process. The correlogram reveals that the squared returns for the S&P 500 are autocorrelated. The combination of the correlogram and the sample PACF suggest that an ARMA model would be tentatively appropriate. □

We usually would not do anything in practice to a white noise time series. But because of the volatility, a model is needed to capture the autocorrelation in the variance. The model fitted is called an ARCH(p) model. In fact there are a series of models, volatility models, that are specialized in volatile time series. ARCH(p) and GARCH(p, q) are one type of model. Other types are known as stochastic volatility (SV) models.

6.6.2   ARCH(*p*) Models for Volatility

Suppose we have a time series $Y_t = \mu_t + \epsilon_t$ where $\mu_t = 0$, a time series like the S&P 500 in Figure 6.12. Then the mean-adjusted Y_t equals ϵ_t. If the ACF of mean-adjusted Y_t is the ACF of white noise and the ACF of Y_t^2 is the ACF of a stationary autocorrelated process, then a model that gives faith to these features is an ARCH(p) model:

$$\epsilon_t = w_t \sigma_t = w_t \sqrt{\left(\alpha_0 + \sum_{i=1}^{p} \alpha_i \epsilon_{t-i}^2\right)},$$

where w_t is white noise with 0 mean and unit variance; the α parameters are positive, to ensure positive variance; and the slope α parameters are less than 1 to ensure stability.

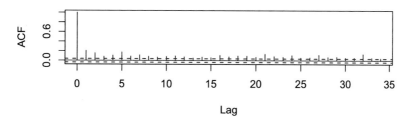

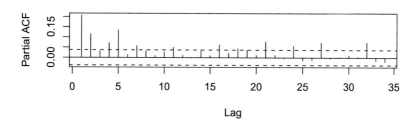

Figure 6.14 Correlogram of the squared mean-adjusted values of the S&P 500 is like the correlogram of an autoregressive process.

Then, the unconditional variance is

$$\sigma_t^2 = \alpha_0 + \sum_{i=1}^{p} \alpha_i \epsilon_{t-i}^2.$$

Thus, ARCH(p) assumes an autoregressive model for the variance process.

Example 6.7   A time series ϵ_t is first-order autoregressive conditional heteroscedastic, denoted $ARCH(1)$, if mean-adjusted

$$y_t = \epsilon_t = w_t \sqrt{\alpha_o + \alpha_1 \epsilon_{t-1}^2},$$

where $\{w_t\}$ is white noise with zero mean and unit variance, and α_o and α_1 are model parameters. $E(\epsilon_t) = 0$ since the data is mean adjusted. To see how this introduces volatility, square the equation to calculate the variance:

$$
\begin{aligned}
\mathrm{Var}(\epsilon_t) &= E(\epsilon_t^2) \\
&= E(w_t^2) E(\alpha_0 + \alpha_1 \epsilon_{t-1}^2) \\
&= E(\alpha_0 + \alpha_1 \epsilon_{t-1}^2) \\
&= \alpha_0 + \alpha_1 \mathrm{Var}(\epsilon_{t-1}),
\end{aligned}
$$

since w_t has unit variance and ϵ_t has zero mean. Notice how the variance ends up being modeled as an AR(1).   ☐

Decay in the autocorrelations of the mean-adjusted squared residuals should indicate whether an *ARCH* model is appropriate or not. The model should only be applied to a prewhitened (mean-adjusted) series $\{\epsilon_t\}$ that is uncorrelated and contains no trends or seasonal changes, such as might be obtained after fitting a satisfactory ARIMA model. Thus, ARCH(p) may be an appropriate model for residuals.

6.6.3 GARCH(p, q) Models

An extension of the ARCH(p) model, widely used in financial applications, is the generalised *ARCH* model, denoted $GARCH(q,p)$, which has the $ARCH(p)$ model as the special case $GARCH(0,p)$. A time series $\{\epsilon_t\}$ is $GARCH(q,p)$ if

$$\epsilon_t = w_t \sigma_t = w_t \sqrt{h_t},$$

where

$$h_t = \alpha_0 + \sum_{i=1}^{p} \alpha_i \epsilon_{t-i}^2 + \sum_{j=1}^{q} \beta_j h_{t-j}$$

and α_i and β_j ($i = 0, 1, \ldots, p; j = 1, \ldots, q$) are model parameters.

GARCH models can be fitted using the `garch()` function in the `tseries` library. R fits $GARCH(1, 1)$ by default, which often provides an adequate model, but higher-order models can be specified with the parameter `order=c(p,q)` for some choice of p and q. We can use `trace=F` to suppress output and a numerical estimate of gradient, `grad=numerical`, that is slightly more robust (in the sense of algorithmic convergence) than the default.

Example 6.8   For example, the $GARCH(1, 1)$ is represented as $\epsilon_t = w_t \sqrt{h_t}$, where $h_t = \alpha_0 + \alpha_1 \epsilon_{t-1} + \beta_1 h_{t-1}$ with $\alpha_1 + \beta_1 < 1$ to ensure stability. Thus, the generalized GARCH model has an AR and an MA component.

Using code in program *garchprograms.R*, we simulate this model and then find the sample ACF of the mean-adjusted generated data. It looks like the correlogram of white noise, but the sample ACF of the square of the mean adjusted generated time series looks like the correlogram of autocorrelated data.

Using also the code in program *garchprograms.R*, we fit a $GARCH(1, 1)$ model to the data resulting from the above simulation using the `garch` function, and we get the following 95 percent confidence intervals (CI) for the parameters (answers will be slightly different for each person): CI for ϵ_0: (0.0882, 0.109); CI for ϵ_1: (0.3308, 0.402); CI for β_1: (0.1928, 0.295). They each contain their corresponding true parameter, as they should, since we generated the data.   □

Example 6.9   This example is from Cowpertwait, chapter 7, page 155, exercise 6. This question uses the data in stockmarket.dat, which contains stock market data for seven cities for the period January 6, 1986 to December 31, 1997. We download the data into R. The first three rows are:

```
    Amsterdam    Frankfurt   London   HongKong
1   275.76       1425.56     1424.1   1796.59
2   275.43       1428.54     1415.2   1815.53
3   278.76       1474.24     1404.2   1826.84
    Japan        Singapore   NewYork
    13053.8      233.63      210.65
    12991.2      237.37      213.80
    13056.4      240.99      207.97
```

The Amsterdam time series has a trend in mean and volatility. We can see the trend in a time plot of the data in Figure 6.15. After first-order regular differencing, we notice that the trend in mean disappears but there is visible volatility. A comment on R is worthwhile at this point. Notice that when a time series object of daily data is created, putting `freq=12` or `freq=4` in the `ts()` function will result in strange data. We must always be careful and check what type of data we have before plugging in options in our R function. For this particular time series, the best is to not put in anything if we decide to use the `ts()` class in R.

Ignoring the volatility, as if we had not seen it, we fit the following models to the Amsterdam series and select the best fitting model: ARIMA(0, 1, 0), ARIMA(1, 1, 0), ARIMA(0, 1, 1), ARIMA(1, 1, 1). The best model according to the AIC criterion is just a random walk, ARIMA(0, 1, 0) with an AIC of 18416.58.

But although the correlogram of the residuals of the best fitting model shows that the residuals are white noise, the correlogram of the squared residuals has a long-term pattern of significant autocorrelations. This result, white noise residuals but autocorrelated squared residuals, suggests that there is volatility, as we indicated earlier. So we decided to model the change in the time series as a GARCH model. We fitted the following GARCH models to the first regular difference of the time series, which is the change, and select the best-fitting model among GARCH(0, 1), GARCH(1, 0), GARCH(1, 1), and GARCH(0, 2). The winner is GARCH(1,1). It has the lowest AIC (16142.38). The fitted model is

$$\hat{\epsilon}_t = w_t\sqrt{0.20807 + 0.07213\epsilon_{t-1}^2 + 0.91192h_{t-1}}.$$

Plotting the correlogram of the residuals from the best fitting GARCH model and the correlogram of the squared residuals from the best fitting GARCH model, we discover that they both look like the correlograms of white noise. The GARCH model fits the differenced stock market data for the Netherlands very well. □

6.6.4   Residuals of a GARCH(p, q) Model

We have already analyzed the S&P500 series. The residual series of the *GARCH* model $\hat{w}_t$ are calculated from

Differenced time series
looks volatile

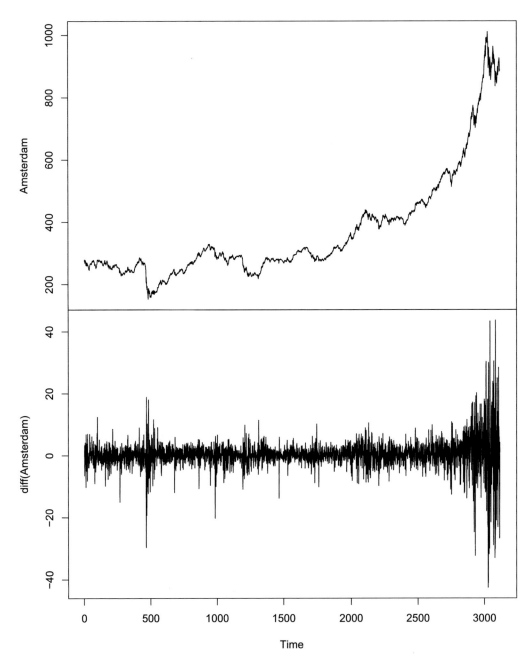

Figure 6.15 The first regular difference of Amsterdam time series reveals volatility.

$$\hat{w}_t = \frac{\epsilon_t}{\sqrt{\hat{h}_t}}.$$

If the GARCH model is suitable, the residual series should appear to be a realization of white noise with zero mean and unit variance. In the case of a $GARCH(1, 1)$ model,

$$\hat{h}_t = \hat{\alpha}_0 + \hat{\alpha}_1 \epsilon_{t-1}^2 + \hat{\beta}_1 \hat{h}_{t-1},$$

with $\hat{h}_1 = 0$ for $t = 2, \ldots, n$.[8]

The calculations are performed by the function `garch`. The first value in the residual series is not available (NA), so we remove the first value using -1 in the code, and the correlograms are then found for the resultant residual and squared residual series. The reader may see this in program *garchprograms.R*

There are many ARCH family models. ARCH models were first proposed by Engle [43], and GARCH models were proposed by Bollerslev [6]. Several textbooks cover these models including Tsay [193] and Pfaff [151]. Readers interested in financial time series will find additional references in those sources. In that literature, time series with volatility are often referred to as time series with *jumps*. For the concerns shared by financial analysts that need to analyze high volumes of financial and other economic data on a constant basis (nowcasting), see the papers presented at a recent NBER-NSF SBEIS Conference [130], in particular papers on variations of GARCH [162] [106]. These papers are rather advanced for someone just learning time series for the first time, but they will give the readers a flavor of how older methods are being adapted to deal with the massive amounts of data brought about by the IoT.

6.6.5 Exercises

Exercise 6.18 Answer the following questions about GARCH processes.

(a) What is the difference between a heteroscedastic and a conditionally heteroscedastic time series (assume both series have mean 0)?
(b) How does conditional heteroscedasticity get detected?
(c) What solution is there for conditional heteroscedasticity?
(d) What are GARCH(p) models? To what time series do we fit GARCH models? What should we expect to see after we have fitted a GARCH(p) model?
(e) What is volatility? Would log remove it?

6.7 Unit Root Tests

Sometimes, a nonstationary time series with trend is such that it suggests that some of the coefficients of the AR part of the model (some of the α's) are 1 if we were to fit an AR model directly to the nonstationary data. So, some authors have suggested to do the following test:

---

[8]Notice that a residual for time $t = 1$ cannot be calculated from this formula.

$$H_o : \alpha_1 = 1$$
$$H_a : \alpha_1 < 1$$

with test statistic

$$t = \frac{\hat{\alpha}_1 - 1}{s.e.(\hat{\alpha}_1)}.$$

The sampling distribution of this test statistic is very complicated, but we can obtain the p-values for the test using R. If the p-value is larger than 0.05, this means that there is statistically significant evidence that there is a unit root and therefore differencing is a good idea.

We will have more to say about unit root tests in Chapter 8.

6.7.1 Forecasting in the Age of Big Data

There is a large number of businesses specializing in forecasting in the big data context [50]. In some of them, like Facebook, the goal is to make it easy for persons that do not know time series to produce forecasts at scale, for several notions of scale. The idea is to let machines do the hard work of model evaluation and comparison when there are hundreds or even thousands of forecasts to be made, and for a large variety of forecasting problems. Those businesses approach forecasting as a curve fitting problem, with functions that have arguments that anyone can easily change. Paradoxically, in the world of automatic forecasting, the arguments are called parameters, which is a term that in Statistics means something totally different. In the case study of this chapter, Section 6.11, we talk a little more about these methods.

Outside the commercial world, academic researchers have contributed open source packages that allow automatic forecasting as well. Hyndman and colleagues [77, 78], in the `forecast` package, do automatic forecasting, using `auto.arima()`, which fits a range of ARIMA models and automatically selects the best one. Within that package, based on [80], the `ets` function fits a collection of exponential smoothing models and selects the best, the `snaive` function fits a naive model and `tbats` fits a model that allows for more than one type of seasonality.

Much progress is being made in the machine learning research world to use Artificial Intelligence and Neural Networks to not only obtain automatic forecasts one at a time but to forecast thousands of time series automatically, going beyond the curve fitting approach of Facebook, Amazon or Google, who each open some of their algorithms for the community to use. Until recently, however, due to the slow progress of Machine Learning in time series, unsupervised learning, and, in particular, features were the most common approach to massive amounts of time series. Although Machine Learning relies a lot on summary features of data, and we have been using summary features in earlier chapters to cluster time series, the summary features are good just for that, to cluster, unless we use the components of the time series and other outside variables as features, like Prophet does. Machine learning methods were not designed for time series, as we have mentioned elsewhere in the book, and machine learning methods were not made to forecast until recently.

In this section, we will use the forecasts of ARIMA modeling as summary features that will help us cluster several series.

6.7.2   Using Features of ARIMA Forecasts

Clustering methods for time series based on the similarity between fitted models are called model-based methods. Different approaches to model-based clustering can be found. Some have used the possibility of representing any fitted ARIMA model by its autoregressive expansion. The distance between the autoregressive expansions are then used to cluster the time series. Others use the residuals of the fitted models for clustering the time series. And others use the forecasts produced by the models to cluster the time series. The latter could be interesting if we are trying to design an app that predicts parking availability a week from now in downtown Birmingham, for example. The reader interested in a detailed bibliographical survey of the efforts made using those methods is referred to [112].

In this section, we are going to use a simplified variant of the forecast method to address the prediction of the future occupancy rates of car parks in a city.

Example 6.10   The data set Parking in Birmingham, introduced in Chapter 4, is sensor data published by Birmingham City Council in the United Kingdom, licensed under the Open Government License v3.0. It includes the car parks operated by NCP (National Car Parks) in that city, and is updated every 30 minutes from 8 AM to 5 PM. In their study, Stolfi and colleagues [182] worked with data from October 4, 2016, to December 19, 2016 (11 weeks). Those authors cleaned (filtered) the data to account for faulty sensors. After cleaning, the remaining data consisted of 32 car parks and 36,285 occupancy measures. That is the data available at the UCI Machine Learning Repository [111], to be used in this example.

The data are counts of the number of vehicles occupying 32 off-street car parks obtained via sensors that monitor such data. The capacity of each park, as well as the time at which the occupancy was last updated by the sensors are also variables in the data set. Chapter 4 explains what we did to the data and how we created a smaller data set representing just six parks.

Our analysis is for illustrative purposes, so we will omit parks with any missing value to avoid making arbitrary decisions as to how we are going to impute the missing values. However, alternative studies could be done where we use different fractions of the data for each park and predict the 11th week based on what we can use from each park. We will leave that as an exercise.

Analysis of a data set like the Birmingham data allows the prediction of car park availability, a subject that has been studied in a context of smart cities, especially now when most parking facilities have installed sensors as part of their infrastructure [182]. The availability analysis can be converted to an information system for customers to allow them to find suitable parking in the least amount of time and at the lowest fuel and stress costs. Intelligent Parking Reservation architectures have been proposed based on analysis of data like this. Architecture is just a fancy name to describe all the components, that is, the data sources obtainable through the Internet, the data analysis algorithms, and the user interface (which could be a mobile app or a web page).

The reader may imagine needing to park in downtown Birmingham at 10 AM a week from now and being able to look at the Birmingham app to look for predicted occupancy that week.

For the parks with complete data, we will use the first 10 weeks of data as training set and the last week as test set. That is, we will predict the whole week for the parks we study, after fitting to each park an appropriate ARIMA model. The features of the forecasts then will be subject to `kmeans` cluster analysis.

Program *birmingham.R* part II contains the R code for the analysis.   □

6.8   Average (Consensus) Forecasts

Forecast combinations often produce better forecasts than methods based on the "best" individual model [3, 189]. That is supported by the fact that in most forecasting competitions, the winner tends to be the average forecast. This is an average of the forecasts produced by different types of models used by the contestants. This practice of averaging forecasts is called the *ensemble method* of forecasting. The averaging could be done using the median, the mean or a trimmed mean, or some form of weighted mean.

For example, suppose that for a time series we produced forecasts for the four quarters of 2005, using several modeling approaches. That is, we obtained an ensemble of forecasts. And suppose that the following table shows the actual time and data during the test period in the first two columns, the forecasts obtained with each of those methods in columns 3 to 5, and the average of all the forecasts in the last column. Each number in the average forecast column is the average of the numbers corresponding to forecasts present in that row. In the last row of the table, we show the RMSE of the forecast. Notice how the average forecast has the lowest RMSE.

Timestamp Year:quarter	Test data	ARIMA forecast	Prophet forecast	Exponential forecast	Average forecast
2005:1	25	20	24	23	22.33
2005:2	21	19	23	19	20.33
2005:3	15	18	21	16	18.33
2005:4	23	16	22	21	19.67
RMSE		4.53	6.96	4.52	**2.727**

Average forecasts have been used routinely in economic and financial forecasting. Containing forecasts from 50-plus economists employed by some of America's largest and most respected manufacturers, banks, insurance companies, and brokerage firms, Blue Chip Economic Indicators provides forecasts of change in US GDP calculated by each panel member, plus an average or consensus of their forecasts. Tables with

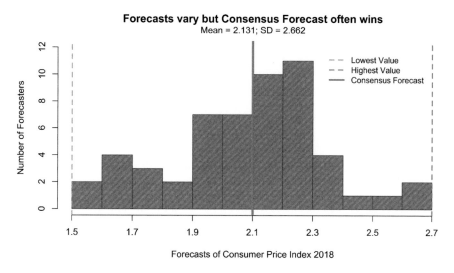

Figure 6.16 Data from the forecasts of the change in the CPI from a panel of forecasters published by the Blue Chip Economic Indicators [96].

the individual and consensus forecasts for real GDP and 15 other macro variables can be found in their reports [96] [97]. To give a simplified example of how those tables work, let's pick up a few forecasts from the tables for macro variable "Consumer Price Index, percent change 2018 from 2017." Action Economics, the table says, forecasted 2.3 percent change; Wells Fargo forecasted 2.3 percent; Amherst Pierpont Securities forecasted 2.7 percent; Conference Board forecasted 1.5 percent; US Chamber of Commerce forecasted 1.8 percent; and so on. Suppose those are the only companies forecasting Consumer Price Index change. Then the average of their forecast for Consumer Price Index change would be 2.12. That would be the consensus forecast for Consumer Price Index. Of course, the Tables published in Blue Chip Economic Indicators contain forecasts calculated by many more companies (54 to be exact), so the average or consensus forecast is the average of 54 forecasts. It is worth looking at the variability of all forecasts produced. We took all the forecasts produced for Consumer Price Index by all the companies. Figure 6.16 shows a histogram that indicates the variability in the forecasts produced by all Blue Chip companies listed in [96] [97]. As we can see, the forecast could be as low as 1.5 percent or as high as 2.7 percent change in CPI.

Consensus forecasts are also called *ensemble forecasts* and are not just used in financial and economic settings. Each week during the COVID-19 pandemic, the COVID-19 ForecastHub generated ensemble forecasts of cumulative and incident COVID-19 deaths and incident cases. That forecast is useful because it gives a sense of the general consensus forecast across all teams predicting COVID-related data. As indicated in the Hub's Ensemble Model section, *ensemble forecasts are often more accurate than any individual model that went into the ensemble* [29]. The COVID-19 ForecastHub is a central repository of forecasts and predictions from over 50 international research groups.

6.9   Problems

Problem 6.1   Sketch how the theoretical sample autocorrelation plot and the theoretical sample partial autocorrelation plot would look for an ARIMA(1,0,1)(0,0,0) model for a univariate monthly time series.

Problem 6.2   Consider the model

$$y_t = y_{t-1} + y_{t-12} - y_{t-13} + w_t - 0.5w_{t-1} - 0.5w_{t-12} + 0.25w_{t-13}.$$

Specify the model in backshift polynomial form and in the ARIMA$(p, d, q)(P, D, Q)_F$ notation.

Problem 6.3   A model is written in backshift polynomial form as follows:

$$(1 - 0.2191B)(1 - B^{12})(1 - B)y_t = (1 - 0.8245B)(1 - 0.3745B^{12})w_t.$$

Expand it to represent y_t as a function of all the other terms.

Problem 6.4   Which of the following can help us determine whether a univariate time series is white noise? Circle what applies.

(a) Ljung–Box test
(b) Run test
(c) Durbin–Watson test
(d) Sample ACF of the data
(e) Sample PACF of the data
(f) The sample ACF of the data and the sample ACF of the squared data
(g) The sample ACF of the squared data

Problem 6.5   A GARCH(0,1) process representation of the residuals of a model is

$$\epsilon_t = w_t\sqrt{\alpha_0 + \alpha_1\epsilon_{t-1}^2}.$$

Show that this implies that the residual variance behaves like an AR(1) process.

Problem 6.6   Show that an AR(1) process satisfies the Wold Decomposition theorem.

Problem 6.7   In Chapter 1, we looked at the unemployment rate for bachelors and for the population at large. The data used there was seasonally adjusted. It is possible to find the same data but not seasonally adjusted. That is a more desirable situation when we want to see how a modeling approach works. It is also the case that in many instances aiming for the most granular level possible improves forecasting performance. Too much granularity is not good, as it has too much noise (seconds, minutes, even hours). Thus, in this problem, the reader is asked to fit an appropriate ARIMA model to the data series LNU04027662 to be found in FRED [49]. If you did the case study in Chapter 1, Section 1.14, you may directly access the time series using the Quandl API and that way you do not have to download to your computer the data from FRED.

(a) What does the metadata in FRED say about the time series?

(b) To create the training set, remove the last 12 months. The last 12 months will be the test set.

(c) Indicate and demonstrate whether the training set needs any pre-transformation to stabilize the variance.

(d) Perform differencing, if any is needed. Demonstrate stationarity.

(e) Identify a model or a few models that could be good candidates. Justify your choice.

(f) Estimate the model(s) and make selections based on the residuals and model selection criteria. Justify and demonstrate that your choice is right.

(g) With chosen model(s) forecast the unemployment rate and calculate several measurements of forecast performance.

(h) Plot the training set, the test set, the forecast and the prediction intervals for the future data in the same plot.

Problem 6.8   In Examples 6.2.1 and 6.2.2 we obtained the out-of-sample predictions or forecasts. We used there the root mean square error of the forecast, RMSE, as an evaluation metric. Calculate the forecast error in each case using instead the MAD (Mean Absolute Deviation), the MAPE and the MASE evaluation metrics.

Problem 6.9   Consider the *viscosity.txt* data set, which you may read into R as follows:

```
data=scan("viscosity.txt")
```

Identify and fit an appropriate ARIMA$(p, d, q)(P, D, Q)$ model that gives you good residuals. Write your model in backshift polynomial form. Determine if the residuals of your model are volatile (but you do not need to fit a Garch model).

The model must be justified by what you see in a sample ACF and sample PACF, so make sure you describe what makes you choose that model. Automatic model selection is not allowed in this question. Explain briefly, but in detail, what you do and what you find. Justify your conclusion.

Problem 6.10   Traveler's rest Inc. operates four hotels in a Midwestern city. The analysts in the operating division of the corporation were asked to develop a model that could be used to obtain short-term forecasts, up to one year, of the number of occupied rooms in the hotels. These forecasts were needed by various personnel to assist in decision making with regard to hiring additional help during the summer months, ordering materials that have long delivery lead times, budgeting local advertising, expenditures and so on.

The available historical data consisted of the number of occupied rooms during each day for the previous 15 years, starting on the first day of January. Because monthly forecasts were desired, these data were reduced to monthly averages by dividing each monthly total by the number of days of the month. The first observation is for January 1977.

At the outset it was decided to perform all analyses with the data from the first 14 of the previous 15 years so that forecasts for the 15th year could be used as a check on the validity of the model for forecasting.

(a) Read the data into the software and split it into a training set (in-sample) containing the first 14 years of data and a test set (out-of-sample) containing the data for
the last year. Observe the training time series using appropriate plots and tests,
if needed. Explore and determine what course of action we should take. Take the
needed action prior to fitting any model. Explain. Use the following link to access
the data.

```
rooms=scan("rooms.txt")
```

(b) Fit a model properly identified. Justify your choice of model based on
both the sample ACF and the sample PACF. Write all the actions in the
ARIMA$(p, d, q)(P, D, Q)$ notation and in the backshift polynomial notation.

(c) Estimate the model that you identified in part (b). Substitute the coefficient values
in the backshift polynomial expression of the model that you wrote in part (b).
Comment on the t-test statistics, and the estimate of σ^2 and the AIC.

(d) Diagnose whether the model is good. Justify your answer with appropriate plots
and summary statistics.

(e) Forecast the last 12 months, compare with the test data and calculate the RMSE.

Problem 6.11   Let U_t be the unemployment rate in Canada data set that can be
accessed using the package `vars`. The sample ACF of the first regular difference
of U_t suggests that the first regular difference is stationary, and the combined sample
ACF and sample PACF suggest that the first regular difference of U_t follows an AR(1).
We decide that model identification is completed and that "the regularly differenced"
U_t is an ARIMA(1,1,0)(0,0,0). We used the R program called *ch6unemploycanada* to
do the job and to forecast and compute the RMSE with the model fitted.

There is something wrong with the conclusion we reached. Explain what that might
be. Use the code as needed and write down the alternative model you may suggest, if
any, using the ARIMA$(p, d, q)(P, D, Q)$ notation.

Problem 6.12   The data `JohnsonJohnson` is a quarterly data set that is hosted by
the `datasets` package of R and can be accessed from base R just by typing its
name.

(a) Write R code to access it and check what type of R object it is, where it starts,
where it ends, and split it into a training data set that excludes last year's four
quarters.

(b) Explore the training set to determine whether it is seasonal, whether the time plot
shows trend, and do an additive or multiplicative decomposition, as needed.

(c) Write a brief summary of what you find.

(d) After you are done with all of the above, conduct all the necessary work needed to
fit an ARIMA$(p, d, q)(P, D, Q)$ to the `JohnsonJohnson` data set, including pre-
differencing transformations, differencing, identification, and estimation steps, as
needed. If you pre-transform the time series, indicate whether you think the time
series is conditionally heteroscedastic or simply heteroscedastic.

(e) After you are happy with a fitted model, check volatility of the residuals. Write comments summarizing what you did and your results.

(f) Forecast four quarters ahead. Do a plot of the training data, the forecast and the test data.

Problem 6.13   We will be interested again in the time series that contains the amount of domestic passengers in the US from October 2002 to June 2019, which was analyzed in Chapter 1.

(a) Split the time series into a training set, from October 2002 to June 2018, and a test or holdout set, July 2018 to June 2019, and perform whatever is needed to be able to use the domestic air passengers series as a numeric `ts()` object. Call your numeric time series $x11$.

(b) Perform the whole ARIMA cycle with the numeric time series $x11$, training part: training set-time plot, variance-stabilizing transformation, if needed; differencing, if needed; identify a model for the resulting series, check that the model is good, forecast and produce a plot of the training, forecast, and prediction intervals and compute the standard error. Separately produce a plot of just the test data, the out-of-sample forecast and the confidence interval bands. Scale the plot so that we can see clearly whether the test data is within the intervals. Compute the room mean square error of the forecast.

(c) Explain what you did and why, making sure that you justify every step you take. When done explaining, indicate the model in backshift polynomial form.

(d) Do the actual test data fall inside the prediction intervals? Interpret the prediction intervals.

Problem 6.14   The UCI Machine Learning Repository contains many time series data sets [39]. Chapter 4 contains an introduction to this repository. In this exercise, the air quality data set is featured.

(a) After reading the documentation carefully and reading the data into R, determine which variable has the most missing values and which has the least? Write down an R program that will find out and convert the missing values to missing values that R will understand, if any.

(b) Make the data set a time series object of frequency 24, `start=c(1,1)`. If you use `ts()` applied to a data set with several numerical random variables, then it is a multiple time series object, to the eyes of R.

(c) Use the `aggregate()` function to convert the hourly observations to average daily observations.

(d) Do a time plot of an averaged variable of your choice and describe what you see.

Problem 6.15   The following model was fit to a time series of 120 weekly sales of absorbent paper towels:

$$\widehat{z}_t = 0.30688 z_{t-1}$$

$$s.e = 0.08765$$

$$p - value = 0.0007,$$

where $z_t = y_t - y_{t-1}$, $y_t = log(x_t)$, and x_t is actual sales over 100,000 rolls.

(a) Write down the forecasting equation for x_t.
(b) Is the slope coefficient significantly different from 0? Justify your answer.

Problem 6.16 The following model was fit to monthly seasonal data

$$z_t^{**} = \alpha_1 z_{t-1}^{**} + \alpha_3 z_{t-3}^{**} + \alpha_5 z_{t-5}^{**} + w_t - \beta_{12} w_{t-12},$$

where z_t^{**} is the seasonal difference of the quartic root of the original observed time series x_t.

(a) Expand the model to express it as a function of the quartic root of x.
(b) Write down the expression for the forecast of the quartic root of x at time period 169.
(c) Write down the final expression for the forecast of x_t at time period 169.
(d) Suppose the parameter estimates are $\hat{\alpha}_1 = 0.23242$, $\hat{\alpha}_3 = -0.22301$, $\hat{\alpha}_5 = -0.15263$ and $\hat{\beta}_{12} = 0.47634$. Suppose also that the last observed values of the original time series x_t are

t	156	157	158	159	160	161	162	163	164	165	166	167	168
x	813	811	732	745	844	833	935	1110	1124	868	860	762	877

What is the value of the forecast for x_t at time period $t = 169$? At time period $t = 170$?
(e) The standard error of the estimate $\hat{\alpha}_5$ is 0.07984. Should the corresponding variable be kept in the model? Why?

Problem 6.17 What are the differences among the Ljung–Box test for residuals, the t-test for the ARMA model parameters, and the t-test for the autocorrelations observed in the sample ACF, and what do these tests tell us about ARIMA modeling? How would you respond to a lack of statistical significance in each of these separately? How would you respond to lack of statistical significance in some of the three, but not in the others? Consider all possible cases. How is lack of significance determined in each case? Provide an example.

Problem 6.18 Let x_t be an observed time series that is heteroscedastic. The log of this series is taken. Let y_t be the log of x_t. A model for y_t is written in polynomial form as follows:

$$(1 - 0.2191B)(1 - B^{12})(1 - B)y_t = (1 - 0.8245B)(1 - 0.3745B^{12})w_t.$$

Expand the model and write the forecasting equation to forecast x_{t+1}.

6.10   Quiz

Some questions may have more than one answer.

Question 6.1   The covariance of a random walk at lag k, $\text{Cov}(x_t, x_{t+k})$, equals which of the following?

(a) $\sigma^2 t$
(b) σ^2
(c) kt
(d) 0

Question 6.2   In univariate ARIMA modeling, to determine stationarity by visual inspection before a model is identified, we need to look at

(a) the time plot of the residuals.
(b) the root mean square error.
(c) the sample ACF.
(d) the sample PACF.

Question 6.3   Is the following statement TRUE or FALSE?
    "In univariate ARIMA modeling, we difference the time series with the objective of making the time series white noise before identifying a model."

(a) TRUE
(b) FALSE

Question 6.4   In ARIMA modeling, to identify an AR, MA, or ARMA model for a stationary time series, we

(a) need to see both the sample ACF and the sample PACF of the observed univariate time series.
(b) need to make sure that we have first made the time series stationary.
(c) need only the sample ACF.
(d) need only the sample PACF.

Question 6.5   Let x_t be a time series that is heteroscedastic. The log of this series is taken. Let y_t be the log of x_t. A model fitted to y_t is written in polynomial form as follows:

$$(1 - 0.219B)(1 - B^{12})(1 - B)y_t = (1 - 0.8245B)(1 - 0.3745B^{12})w_t.$$

Expand and resolve the backshift polynomials to write the equation for y_t as a function of lagged values of y_t and lagged and present values of w_t. What is the value of the model coefficient for the term y_{t-13}?

(a) 0.219
(b) −0.375
(c) 0.3087
(d) 1.219

Question 6.6   An observed time series y_t had the following model, ARIMA$(2, 1, 1)(1, 1, 1)_{12}$, fitted to it, and the residuals were white noise. The original time series y_t was

(a) nonstationary.
(b) differenced regular and seasonally before maximum likelihood was used to estimate the model.
(c) stationary.
(d) a random walk.

Question 6.7   We simulated 1,000 observations from the following model:

$$y_t^* = 0.8y_{t-1}^* + 0.2y_{t-2}^* + 0.2w_{t-1} + 0.1w_{t-2} - 0.5w_{t-3},$$

where y_t^* is the seasonal difference of the regular difference of y_t. The process represented by this model for y_t^* is

(a) ARIMA$(1,1,1)(1,1,2)$ and y_t^* is not stationary.
(b) ARIMA$(2,1,3)(0,1,0)$ and y_t^* is stationary.
(c) ARIMA$(1,1,1)(0,1,1)$ and y_t^* is stationary.
(d) ARIMA$(2,1,3)(0,12,0)$.

Question 6.8   In the time series literature, that a time series y_t is integrated always means that

(a) the time series is white noise.
(b) some differencing will make the time series stationary.
(c) y_t is not stationary.
(d) the time series was the result of integrating the normal distribution.

Question 6.9   Consider the ARIMA$(2,0,2)(0,0,0)$ process for an annual time series y_t:

$$y_t = 0.6y_{t-1} + 0.2y_{t-2} + w_t + 0.3w_{t-1} - 0.4w_{t-2},$$

where w_t is white noise.
Suppose that $y_{100} = 4$; $y_{99} = 5$; $\widehat{w}_{100} = 1$; $\widehat{w}_{99} = 0.5$.
According to this model, the value of y_{101} should be which of the following? (Note: for things whose value we do not know, we may use their expectation.)

(a) 2.5
(b) 3.5
(c) 0
(d) 5.1

Question 6.10   An ARIMA$(1,1,1)(0,0,0)$ process fitted to an annual time series x_t is represented by

$$y_t = \alpha_1 y_{t-1} + \beta_1 w_{t-1} + w_t,$$

where

$$y_t = x_t - x_{t-1}$$

is the differenced data.

The model was fitted to a time series, and the following estimated model resulted:

$$\hat{y}_t = 0.750 y_{t-1} + 0.5324 w_{t-1}.$$

The standard errors are 0.1949 (for $\hat{\alpha}_1$) and 0.1740 (for $\hat{\beta}_1$), respectively. Which of the following statements is an appropriate conclusion?

(a) We are 95 percent confident that α is between 0.36799 and 1.132.
(b) We are 95 percent confident that $\hat{\alpha}$ is between 0.36799 and 1.132.
(c) We are 95 percent confident that 0 is in the 95 percent confidence interval for α.
(d) If the true α is 0.5, then we would reject the null hypothesis that $\alpha = 0.5$.

Question 6.11   A GARCH(0,1) process representation of a mean adjusted time series y_t is $y_t = w_t \sqrt{\alpha_0 + \alpha_1 y_{t-1}^2}$. This representation implies

(a) that the variance of the process behaves like an AR(1) process.
(b) that the variance of the process behaves like an ARMA(1,1) process.
(c) that the process is ARIMA(1,1,1)(0,0,0).
(d) that the process is nonstationary.
(e) that the process is MA(2).

Question 6.12   Which of the following statements is true? Circle all that apply.

(a) The partial autocorrelation function is all I need to look at to determine whether a time series is stationary or nonstationary.
(b) If a time series is nonstationary, then I will have to look at the PACF to identify the model I want to fit.
(c) A random walk and a volatile time series are both nonstationary.
(d) A model that is well fitted will have residuals that are stationary and not necessarily white noise.
(e) Stationarity may refer to variance or to mean stationarity. There are different solutions for each.

Question 6.13   The following quarterly univariate data, y_t, is the subject of this problem:

```
y=c(3.3602,-3.1769,0.3484,7.469,4.4963,-0.4621,
0.7218,6.9484,5.2374,2.9242,4.7006,11.2793,5.1637,
1.5441,12.121,9.6588,8.0922,3.9653,11.4177,13.2088)
```

A unit root test of the data was done with R; the results are as follows:

```
adf.test(y)

Augmented Dickey-Fuller Test

data:   y
Dickey-Fuller = -6.2209, Lag order = 2, p-value = 0.01
```

The conclusion extracted from this test and the right course of action are:

(a) The data is not mean stationary, and therefore we have to difference the data.
(b) The data is not mean stationary, and therefore we have to take the log of the data.
(c) There is statistically significant evidence that the data are mean and variance stationary. So, we have to difference it.
(d) There is statistically significant evidence that there is no mean nonstationarity. We do not have to difference. But we should take the log.
(e) We should just leave the data alone based on the result of this test only.

Question 6.14   If

$$(1 - B^{12} - 0.219B + 0.219B^{13} - B + B^{13} + 0.219B^2 - 0.219B^{14})y_t$$
$$= (1 - 0.8245 - 0.3745B^{12} + 0.8245 * 0.3745B^{13})w_t,$$

then the forecasting equation for y_{t+1} and the ARIMA$(p, d, q)(P, D, Q)$ expression for this model are:

(a)
$$y_{t+1} = 1.219y_t - 0.219y_{t-1} + y_{t-11} - 1.219y_{t-12} + 0.219y_{t-13}$$
$$- 0.8245w_t - 0.3745w_{t-11} + 0.3087753w_{t-12} + w_{t+1}$$

(b)
$$y_{t+1} = y_t - y_{t-1} + y_{t-11} - 0.219y_{t-12} + 0.219y_{t-13}$$
$$- 0.8245w_t - 0.3745w_{t-11} + 0.3087753w_{t-14} + w_{t+1}$$

(c)
$$y_{t+1} = y_t - y_{t-1} + y_{t-11} - 0.219y_{t-12} + 0.219y_{t-13}$$
$$- 0.8245w_t - 0.3745w_{t-11} + 0.3087753w_{t-14} + w_{t+1}$$

(d)
$$y_{t+1} = y_t - y_{t-11} - 0.219y_{t-12} + 0.219y_{t-13}$$
$$- 0.8245w_t - 0.1233w_{t-11} + 0.3087753w_{t-14} + w_{t+1}$$

6.11 Case Study: Automatic Forecasting at Scale

The demand for high-quality forecasts often far outstrips the pace at which they can be produced. According to [186] and many others in the modern forecasting industry, "the analysts responsible for data science tasks throughout an organization typically have deep domain expertise about the specific products or services that they support, but often do not have training in time series forecasting." Consider, for example, Facebook, which must forecast the number of Facebook events in the next days in thousands of locations, one time series from each location. That is many more forecasts than analysts with a variety of backgrounds can do manually. How does Facebook do that? How can Facebook forecast the number of events in thousands of different locations, each with its own time series and its own holidays? Facebook is not the only high-tech company confronting that dilemma. Other high-tech giants like Google, Uber, Netflix, Wikipedia, and retailers such as Walmart and Amazon, for example, depend on good forecasting of thousands of time series at different levels of granularity (daily, weekly, hourly,...) and at different scale (regionally, by county, etc.). There are not enough time series experts to attend to such high demand for forecasts. That is why, nowadays, with higher frequency than in the past, forecasting methods are judged not only by how well they predict the future but by how well they can be used by individuals who have not been trained in time series analysis. Automation of forecasts is on the rise. Machine learning approaches to modeling and forecasting a time series, supervised machine learning methods, are on the rise precisely because the high volume, velocity and variety of time series data need use of such methods.

R is more ready to do unsupervised machine learning in time series like the clustering examples that we have seen in past examples than supervised machine learning. To satisfy the massive demand for forecasts, many organizations have an in-house automated forecasting tool, a unified approach to forecasting that addresses common problems in data sets related to humans, such as seasonality, missing data, holidays and changes in trends, data like the ones we analyzed in the case study in Section 3.11, the case study in Section 4.7, and some exercises done throughout past chapters. The high-tech companies' automated tools automate the cleaning, smoothing, temporal and geographical aggregation of the data, and some do ensemble forecasting, combining forecasts from different models, which allows quantification of the uncertainty in the forecasts. A huge industry exists to help organizations build their in-house automatic forecasting tools. For example, representatives of this industry and academia meet annually at the International Symposium of Forecasters and other conferences such as the Annual Practitioners' Conference, publish academic research and practitioners' results in the *International Journal of Forecasting* and *Foresight*, organized and published, respectively, by the International Institute of Forecasters [83]. R came late into time series analysis, and it is also late for machine learning methodology for time series that can respond to the demands that we are now talking about. But it is trying. We will discuss three steps in that direction in this case study.[9]

6.11.1 The Automatic Forecasting Industry

As we just said, the bulk of progress made in supervised machine learning for time series, albeit slow, is happening mostly outside but also inside R. Businesses and the high-tech giants need to forecast sales of over a thousand product lines, but businesses are not alone in the need for forecasting massive amounts of time series. Air quality monitoring and ocean temperature, for example, are followed over time at multiple locations, giving rise to thousands of climatic time series. Astronomical data sets representing many different aspects of astronomical interest are continually analyzed. Smart cities keep track of parking availability, traffic, health of their dwellers, and other factors with sensors and must predict capacity for thousands of locations. Forecasting competitions were created in order to open the room for everyone to innovate to address that increasing demand for forecasting at scale and at different levels of granularity. An example is the M5 Forecasting competition [81] of the International Institute of Forecasters [83], done in Kaggle, where thousands of Walmart product items were forecasted. Other M5 competitions were similarly focused on the forecast of multiple time series.

Visiting websites of forecasting software companies such as Forecast Pro and Auto-box (Automatic Box–Jenkins ARIMA modeling and forecasting package), among others, for example, the reader will get a good flavor of how competitive the industry is, in addition to learning a lot about the state of the art in automatic forecasting [50]. Reading the magazine *OR/MS Today* [140], the reader will see occasionally surveys of the forecasting software suppliers to offer an overview of the latest features and market trends. For an extensive survey of the state of the art in commercial software the reader is referred to summaries by Fry and Mehrotra [50] and Fildes and colleagues [48].

In the rest of this case study we will explore how we could do automatic forecasting for multiple time series. The methods will be illustrated with one time series. Extra programming, not done here, would do the scaling of the model fitting and forecasting when more time series need forecasting.

6.11.2 Automatic Forecasting in R

The preceding discussion brings to the front the question: how can we do automatic forecasting in R? To start with, what we do using R must be open source, since in-house automatic forecasting tools are not usually shared by organizations. And, certainly, not just anyone can forecast with R, since R must be learned first. In this case study, we will visit two well-known automatic forecasting functions: prophet

---

[9]The reader should realize that, unlike in an academic course, or a textbook, in the workplace the type of time series that must be analyzed, albeit huge, is pretty homogenous in nature. A specific industry, such as, for example, the electricity industry, will specialize in time series analysis for that industry. So the reader should approach the study of time series with that mindset.

and `forecast`. The reader will recall that we used `forecast` in the case study of Chapter 5, without getting into details there.

Thus, what we intend to do in this case study cannot be replicated just by any employee of a high-tech company and is certainly not what an in-house automatic forecasting tool looks like in a high-tech company.

6.11.3   Facebook's Prophet

In 2017, open-source Facebook's automatic forecaster Prophet [46], created by Sean J. Taylor and Ben Letham [186], became very popular, giving rise to numerous applications and case studies. R's `prophet` package [187] is an open-source package that contains the automatic forecaster, the `Prophet forecaster`, which can account for missing data and outliers, level changes, human scale seasonalities and irregularly spaced holidays, data sets at various granularity levels, such as minute-by-minute or hourly data and trends that indicate nonlinear growth [132].

The method implements a time series model that is flexible enough for a wide range of business time series, yet configurable by nonexperts who may have domain knowledge about the data generating process but little knowledge about time series models and methods. In short, `prophet` is an analyst-in-the-loop approach to forecasting.

The Prophet model is called a "decomposable time series model," with three adjustable components: trend, seasonality and holidays. Polynomials in time are sometimes added. Thus, it is appropriate for time series where the trend and/or seasonality and special features such as holidays or special events are the only things defining the time series, namely models like

$$y_t = g_t + s_t + h_t + w_t,$$

where g_t is the trend, s_t is the seasonality, h_t are the holidays, and w_t is the random term. The model can be adjusted as the data changes. This is a curve fitting approach, that is, the requirement that measurements be regularly spaced is not needed; missing values do not require imputation.

The forecasting approach of `prophet` is driven by the nature of the time series that Facebook forecasts, so it could be inappropriate for other types of time series. The model was originally intended for forecasting daily data with weekly and yearly seasonality plus holidays effects. It works best if there is strong seasonality [78].

The R version of `prophet` consists of a single function that can be used for many types of time series.

```
prophet(
    df = NULL,
    growth = "linear",
    changepoints = NULL,
    n.changepoints = 25,
    changepoint.range = 0.8,
    yearly.seasonality = "auto",
```

```
      weekly.seasonality = "auto",
      daily.seasonality = "auto",
      holidays = NULL,
      seasonality.mode = "additive",
      seasonality.prior.scale = 10,
      holidays.prior.scale = 10,
      changepoint.prior.scale = 0.05,
      mcmc.samples = 0,
      interval.width = 0.8,
      uncertainty.samples = 1000,
      fit = TRUE,

      ...
)
```

This function has now become one of the methods embedded in the most recent `tidyverts` package, which, at the time of writing this book, had not yet been incorporated into CRAN, but is available in github (https://tidyverts.org/). None of the preliminary analysis, model identification or any of the other routines that we have learned in Chapter 6 are necessary, thus an analyst without training in time series can run it. However, exploratory preliminary analysis is recommended to fine-tune the arguments inputted in the function. One more characteristic of `prophet` is that the time series does not have to be a `ts` object. A data frame with a date and the variables is all that is needed. The date can be a character class item and the variables must be numeric.

A person using this function and all the routines in the package must be able to run an R program, access a package, and most importantly, understand what the arguments of the functions in the package require. Probably, Facebook has a user-friendly interface that allows the analyst-in-the-loop to adjust the arguments of the function. But that was not shared by Facebook.

6.11.4   The `forecast` Package

In the context of models that we have already seen in this book, the `forecast` [77] package in R contains an automatic stepwise algorithm for forecasting with ARIMA models. Hyndman and Khandakar [78] summarize attempts to automate ARIMA modeling in the 25 years before 2008, both academically and commercially, and describe the package. Hyndman and Athanasopoulos [75] use the package in their book, where the reader will find several examples of its use.

The `forecast` package was built by Robert Hyndman and coauthors for time series data structures like the one we have so far been using in this book, namely `ts`, which stores the time index using three time series parameters (`start`, `end` and `frequency`). According to Hyndman, for most time series tools such as `arima`, `ets` and `stl` this structural information is sufficient.

Despite the alleged ease of use by non–time-series experts claimed for automated forecasting methods, tuning the method requires a thorough understanding of how the underlying time series models work. Consider, for example, the description of the theory behind the `forecast` package in [78], Section 3, which the reader should be able to understand after having studied and practiced the examples provided in Chapter 6. As indicated in [78], the first input parameters to automated ARIMA are the maximum orders of differencing, the autoregressive components, and the moving average components, the p, d, q, P, D, Q. Running all possible combinations of those letter values could be problematic. So the user must put some thought into what to feed the software. A typical analyst will not know how to adjust these orders without having at least some introductory notion of time series analysis. This type of expertise is hard to scale [186].

It should be noticed that the `forecast` package assumes that the time series is homoscedastic. If a time series, upon preliminary inspection of the data, is found not to be variance stationary, it must be transformed before the `forecast` package is applied.

The reader should be aware that the `forecast` package is no longer supported. This package is now retired in favor of the `fable` package, which is part of the previously mentioned `tidyverts` package. The forecast package will remain in its current state and will be maintained with bug fixes only according to one of its authors' github site [70]. For the latest features and development, that site recommends forecasting with the `fable` package.

6.11.5 The `tidyverts` with `tsibble` Data Structure

The `ts` data structure lacks details that are present in modern time series data sets: multiple seasonality, irregular observations, exogenous information, many time series (that differ in length). For that reason, Hyndman, in collaboration with Earo Wang and Mitchell O'Hara-Wild, built the `tydiverts` suite of packages, available at tidyverts.com, to work with a new data structure, `tsibble`, the latter being the first package of the `tidyverts` suite to make it to CRAN. With this package, R has come a step closer to approaching time series analysis and forecasting at a scale of the kind done by high-tech companies.

The learning curve for `tidyverts` is a little steeper than for `prophet` and `forecast`. The reader will find tutorials on this newer package at Hyndman's website for seminars https://robjhyndman.com/seminars/. In particular, the seminar at https://robjhyndman.com/seminars/tidyverts/ [72] is a reasonable introduction. More recently, the author, in a presentation at the R annual conference [73], stated that `tidyverts` is for those who:

- already use the `tidyverse` packages in R such as `dplyr, tidyr, tibble` and `ggplot2`,
- need to analyze large collections of related time series,

- would like to learn how to use some new tidy tools for time series analysis including visualization, decomposition and forecasting [67].

6.11.6   Prophet Forecast of Blood Glucose

Since 2020 closed-loop system technology adjusts insulin levels via an insulin pump and provides readings of interstitial blood glucose continuously via a sensor for Type I diabetic patients. The pump acts based on predictions of blood glucose for the next half hour based on past sensor readings. By its insulin delivery the pump in turn affects the blood sugar readings. When the patient is sleeping, that works very well. When awake, with unexpected events taking place, body reactions, food, and so on, it gets complicated. Health care managers rely on the technology to manage the diabetes of patients. A large amount of data is continuously being stored. This study is based on Sanchez [174].

In this section, we will be looking at 16,017 readings of blood glucose between May 1 and June 26, 2021, read every five minutes by the sensor. The data was donated by a Type I diabetic patient and is used with permission. The following snapshot of the raw data is shown in order to give an indication of the time-stamp format.

```
               date_time                    cgmreadings
1  2021-05-01T00:02:00             122
2  2021-05-01T00:07:00             125
3  2021-05-01T00:12:00             127
4  2021-05-01T00:17:00             129
5  2021-05-01T00:22:00             132
6  2021-05-01T00:27:00             134
```

We will inspect the data first in order to see what the important features are. Figures 6.17, 6.18, and 6.19 describe the most important features.

Based on the description of the data, and given the granularity of the observations (every five minutes a reading is observed), we explore fitting and forecasting blood glucose reading with Prophet. Figure 6.20 shows the fitted curve and the forecast of 168 five-minute unknown readings based on that curve. The features used for that forecast are the day, hour, week of the observation and a nonlinear trend. That is a total of 14 hours ahead forecasted.

As we indicated earlier in this book, Prophet is a curve-fitting method that uses the components of the time series. The decomposition made is shown in Figure 6.21.

The blood glucose reading analysis presented here was done with program *JSM-cgmprogram.R*, and the data set is *cgmdataonly.csv*.

6.11.7   Exercises

Exercise 6.19   Use program *JSM-cgmprogram.R* and repeat the analysis done to predict blood glucose readings, but this time leave out some hours worth of data as a test

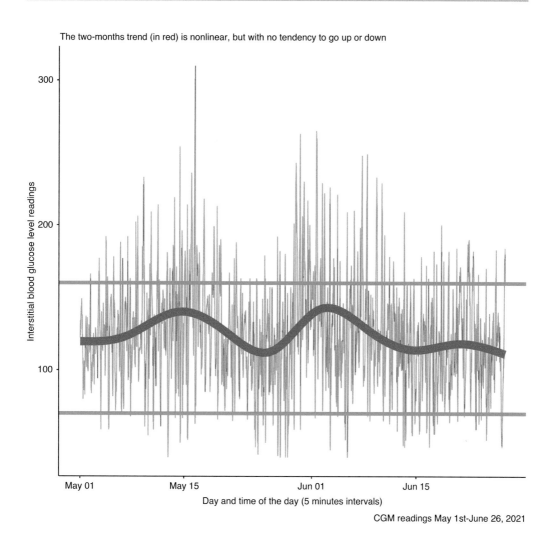

The two-months trend (in red) is nonlinear, but with no tendency to go up or down

CGM readings May 1st-June 26, 2021

Figure 6.17 The green lines indicate the healthy range recommended to this patient by the health care manager. Going outside this range is reason to be alarmed. The red curve is a smoother that indicates the trend. We can see that the trend stays within the healthy range. The variability around that trend, however, is substantial.

set. After fitting the curve and forecasting, compare the forecast with the actual data values. Create a data frame that contains the test data, the forecast, and the prediction intervals. Calculate the RMSE.

Exercise 6.20 Use the `forecast` package and discover what model is automatically selected for the same training set interval that you chose in Exercise 6.19. Compare the RMS of the forecasted values obtained with the model chosen by `forecast` and the one obtained with Prophet.

There is clearly an hour-of-the-day effect in the CGM blood glucose readings

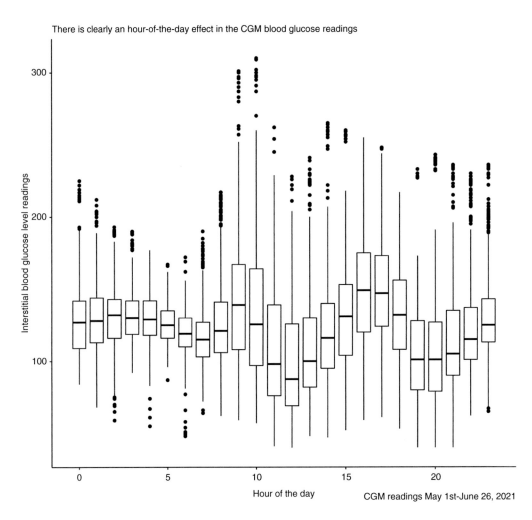

CGM readings May 1st-June 26, 2021

Figure 6.18 Seasonal box plots by hour of the day. Each box represents measurements at that time of the day. We can see that there is some periodicity during the day, with blood glucose getting lower after breakfast and after dinner. There is daily seasonality.

Exercise 6.21 Program *automaticforecast.R* contains code to fit an automatic model for one of the unemployment rate variables already viewed in Chapter 1. The data is *unemploymentall.csv*. This is a challenging time series, because it includes the pandemic years. Compare the fit and forecast for the last two years obtained with Prophet and `forecast`.

6.12   Appendix: ARIMA(p, d, q)(P, D, Q)$_F$ Notations

It is always convenient to consult other time series books or articles. The reader will find, however, that the notation is sometimes very similar in most books and slightly different in others.

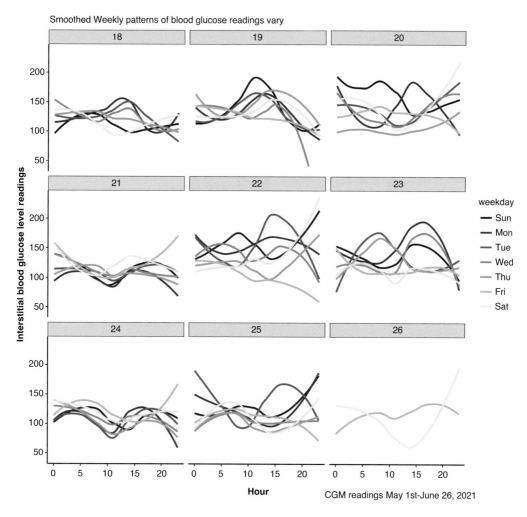

Figure 6.19 There are three features represented in this graph: the hour of the day, the week of the year, and the day of the week. The weekly data has been sliced into pieces, each displaying a day. Some weeks show more variability than others, and the magnitude of the daily seasonality varies across weeks.

Recall that a time series is integrated of order d, denoted as $I(d)$, if the dth difference is stationary. A time series $\{x_t\}$ follows an ARIMA(p, d, q)($0, 0, 0$) process if the dth regular differences of the $\{x_t\}$ series is an AR(p) or an MA(q) or an ARMA(p,q) process. If we use backshift operator notation, $y_t = (1 - B)^d x_t$, the reader will often find a model of this sort expressed as

$$\theta_p(B)y_t = \phi_q(B)w_t$$
$$\theta_p(B)(1 - B)^d x_t = \phi_q(B)w_t,$$

where θ_p and ϕ_q are polynomials in the backshift operator of orders p and q, respectively, with coefficients of the polynomial being the coefficients of the ARMA(p, q)

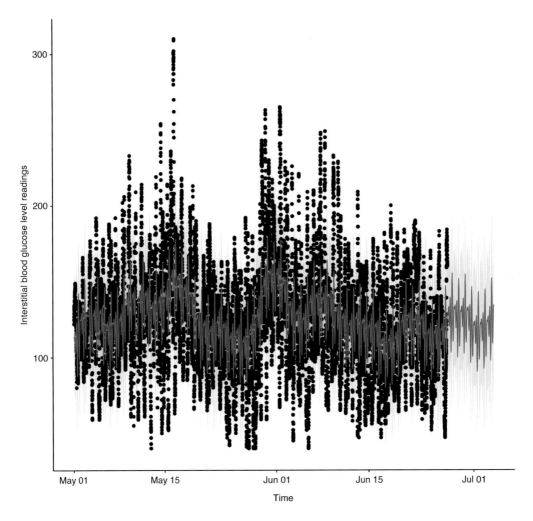

Figure 6.20 Curve fitted to the readings of blood glucose obtained from the sensor and forecast of readings in the next 14 hours. The prediction intervals around the forecast, in lighter blue, indicate our uncertainty.

process. The roots of the θ_p must be larger than one for the autoregressive part to be stationary. The roots of the ϕ_q must be larger than one for the MA part to be invertible.

As we indicated earlier in this chapter, an $ARIMA(p, d, q)(P, D, Q)_F$ is sometimes called a Seasonal Autoregressive Integrated Moving Average Process or $SARIMA(p, d, q)(P, D, Q)_F$. A model like this is expressed in the literature as

$$\Theta_P(B^s)\theta_p(B)(1 - B^s)^D(1 - B)^d x_t = \Phi_Q(B^s)\phi_q(B)w_t,$$

where Θ_P, θ_p, Φ_Q, and ϕ_q are polynomials of orders P, p, Q, and q, respectively.

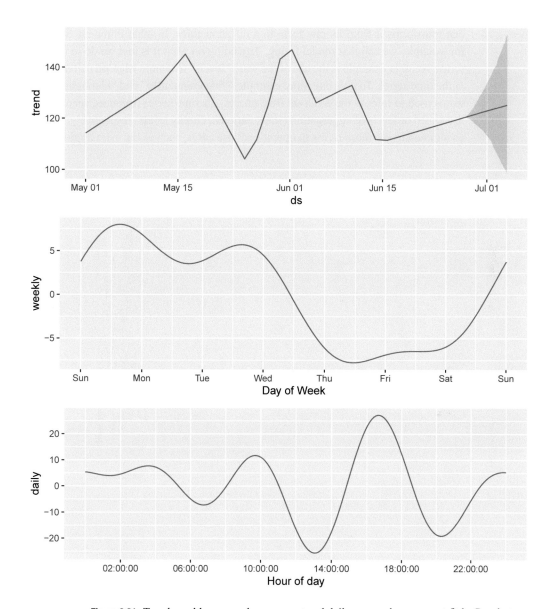

Figure 6.21 Trend, weekly seasonal component and daily seasonal component fit by Prophet.

In general, the model is nonstationary, although if $D = d = 0$ and the roots of the characteristic equation (polynomial terms on the left-hand side) all exceeded unity in absolute value), the resulting model would be stationary.

The polynomial notation is not the only thing where a flag needs to be raised. In browsing literature related to ARIMA modeling, the reader will encounter different languages that are saying the same thing. The word "filtering" is an example. Filtering is a word well known in Engineering. It just means doing something to the data that transforms it from its raw form to a different form. In other domains, Statistics for example, we call that model fitting, depending on what it is that we do to the time series. Another example is the use of the word "levels" to denote what we call the value of the time series. To give one more example, "fine granularity" and "hierarchical" are words used to refer to the different frequencies of a time series (minutes, second, milliseconds) or levels of aggregation (state, national, county levels) respectively; it is not uncommon to see those terms in operational use cases, such as business applications.

Part III

Multivariate Modeling and Forecasting

Part III of this book is about the modeling of the relations among variables. With ARIMA models, we could forecast thousands of time series without the intention of investigating the relationship among those time series (forecasting at scale). But most time series data analysis projects involve more than learning and forecasting at scale. Sometimes the objective is to study how the change in the value of one variable, might change the course or the forecast of another completely. Hidden Markov Models (HMM) are an example of such models and they are discussed in Chapter 7. Other times there is interest in investigating the relationships, perhaps to learn about causal effects, or simply to investigate the propagation through a system of a particular stimulus to a given variable. Macroeconomic models of the economy or ecosystems fall under this last category. Vector Autoregressive (VAR) models are an example. We study models of this type and the things we can do with them in Chapter 8. Regression models that study causality in just one direction from independent to dependent variables fall under this category as well, and they are studied in Chapter 9. Finally, in Chapter 10 we survey Machine Learning regression, including Random Forest, Gradient Boosting and Neural Networks.

Studying relations between two time series requires understanding of the cross-correlation function which is the generalization of the ACF to two variables. Thus, in Chapter 8 we introduce the cross-correlogram, which has a lot of uses in the study of complex systems.

Issues that have been investigated when studying relations among multiple time series are introduced in this part of the book, but the reader is made aware of the fact that each discipline has its own concerns regarding relations. For example, economists might be very interested in unit roots and cointegration and volatility, but these issues may not be so important to someone studying rain and temperature and other of nature's variables that do not usually have long-term stochastic trends or volatility. It is for this reason that those issues are touched lightly and references to more detailed sources are provided.

7 Latent Process Models for Time Series

7.1 Summary Statistics for Binary and Categorical Data

It is very possible that it is not realistic to assume that a time series is generated all along by the same model, a model that has fixed coefficients. Instead, it is very possible that the model changes with time, depending on some states. Consider, for example, the Type I diabetic individual. During periods of stress, the interaction of insulin and blood glucose is probably very different from the interaction occurring during periods of relaxed exercise. It would be unrealistic to assume the same model for both states. In the statistician's frame of mind, the model will vary depending on the state. That is the assumption of "hidden Markov models" for time series, which we study in Chapter 7.

Acknowledging a changing model implies that we need to collect data on the state at which the diabetic person is at each time point. The state could be binary (stress, no stress) or could be multinomial, a categorical variable with several values. Then summaries of this variable have to be obtained. In fact, for hidden Markov models we will need to obtain summaries of the transition proportions of the Markov matrix. When we talk about these models, we will discuss the kind of summaries needed for them. For now, suffice it to say that time series analysis has ways to model all sorts of time series data. But for all types of modeling and analysis, the descriptive statistics, smoothing, and all the concepts learned in Part I of this book are required knowledge. We cannot move to more sophistication without having learned the building blocks of time series analysis first.

7.2 Introduction

Most statistical analysis (theory and practice) is concerned with static, time-invariant models: models with one set of parameters whose values are fixed across all experimental or observational units. This is true of regression models. It is also the case in classical time series models (ARIMA, GARCH, Exponential smoothing) that we have studied in the earlier chapters of this book, and it will be true of VAR models and of many machine learning approaches to time series.

A static model implicitly assumes that the quantified relationships remain the same across the design space of the data. Every observation provides information on each parameter. In some situations, this is a perfectly reasonable approximation and is quite valid. In other situations, it is not. In time series, it is a very dangerous assumption. The passage of time alone always brings changed circumstances, new situations, fresh considerations [156]. A way of compensating for the limitation of static models has traditionally consisted of updating the static model as new observations are obtained, that is, estimating a new static model using the old and the newly obtained observations. This is the approach by many analyst-in-the-loop machine learning models for time series. An alternative solution, based on latent process models, was proposed long time ago: assume that a latent process is generating the observed data and model the data with a latent process model.

In general, the latent process model is characterized by two principles.

- First, there is a hidden or latent, autocorrelated process called the *state process*, represented by a system equation, also known as state equation, which describes how the state evolves through time.
- Second, the observations are independent given the states [179]. The observed is represented by the *observation or measurement equation*, which describes how time series observations are produced from the state process.

Forecasting with latent process models involves first predicting the state process and then predict the variable of interest assuming that state. The model for the variable of interest will change as the state changes.

There are latent process models of two kinds, distinguished by the nature of the state equation:

- *Models with finite number of states that follow a discrete-valued Markov chain over time*, called Hidden Markov Models (HMM). Hidden Markov models are also known as "Hidden Markov process," "Markov-dependent mixture," "Markov-switching model," "Markov mixture model," or "models subject to Markov regime." Within HMM we could have:

  - HMM with *observed process following a discrete distribution,* for example, Poisson.
  - HMM with *observed process following a continuous distribution*, for example, Gaussian.

  Knowledge of the state implies knowledge of the stochastic distribution of the observed random variable. Each state corresponds to a different distribution for the observed.
- *Models that allow for any state to be theoretically possible* (the state process is continuous-valued), which are called State Space Models (SSM). Both the observed and the state process are usually assumed to be Gaussian.

Naturally, the computing details of each type of models are different.Computation for SSM is more developed than HMM in the sense that SSM have a common mathematical framework to be used for model development that allows for common computer software to be employed for making forecasts from a variety of techniques [125, 60]. This common framework derives from systems theory and can be used even with static models. The framework is widely employed nowadays in many of the automated forecasting methods used in Machine Learning because of its flexibility, allowing for missing observations, time series with values recorded at unequally spaced times, multiple components and covariates [13]. The computational workhorse behind state space models is the Kalman filter algorithm.

In contrast to SSM, HMM have less developed computing methods, and each case scenario must be taken as a separate case. However, there are some common models widely used for which software has been created.

Both types of models are widely used. Zucchini and MacDonald [211] illustrate several applications of HMM to modeling epileptic seizures, Drosophila speed, wind direction, births at a hospital, homicides and suicides and animal behavior. Several authors studied those examples before them, but they make their modeling accessible to a wider audience. Shumway [179] presents an example where a process that we will simulate in Section 7.3.3 is a reasonable modeling assumption to make. Shumway talks about a time series of annual counts of major earthquakes. The author found that the mean and the variance of the earthquake counts are very different, with the variance much larger than the mean. This is called overdispersion. A feature of a Poisson distribution is that the mean equals the variance. If we observe overdispersion, that is a signal that the data are not Poisson. A solution for overdispersion of counts is to model the data as a mixture of Poissons. A mixture per se gives independent observations, as we will see in Section 7.3, but Shumway found that the counts are autocorrelated. So a mixture of Poisson random variables alone would not be an appropriate model. Shumway suggests that we assume a Poisson HMM, which assumes that the earthquake counts per year are generated by a process like the one that we will simulate in 7.3.2, but with only two states. Monitoring epidemiologic surveillance data using HMM assumes a baseline stochastic time series model for "normal" behavior of disease, and a different model for when there are epidemics. Surveillance data consists of time series counts of incident disease cases, aggregated weekly, daily, or more frequently. Ozonoff and colleagues [143] fitted a HMM to 312 weeks of influenza data for the years 1990 to 1996 and found that it fitted the data better than Serfling and periodic autoregressive models. Strat and Carrat [185] also found advantages in using HMM with influenza and with poliomyelitis data. Over the years, HMM have become a common tool to detect flu epidemics. An R program written by Lytras and colleagues [110] can raise an alert when the probability that an epidemic has started in Greece exceeds a given threshold. Raghavan [160] demonstrated the superiority of HMM in predicting terrorist attacks.

Before we proceed to learning how to estimate and forecast with some models, we explain, using simulation, how data is generated by a latent process model. We will use the HMM case, as it is easier to visualize. The SSM are based on the same idea, but the continuous nature of the state makes it harder to demonstrate.

7.3   Autocorrelated Time Series from Mixtures

Latent process models are extensions of mixture models to allow for correlated observations. Mixture models have been used to model unknown distributional shapes, data with group structure or, in unsupervised learning methods, such as cluster analysis [118]. For a monograph on finite mixture modeling, the reader is referred to McLachlan and Peel [118]. In this section, we demonstrate with simulation how we go from a simple finite mixture of independent observations to an autocorrelated one, and we call that *data generated by an HMM*. We then estimate the parameters of the model that generated the data, to demonstrate that estimation methods are doing their job of giving us the model that generated the data, as they should.

To illustrate latent process models in simple terms then, we will start by considering an autocorrelated mixture of three Poisson distributions for an observed discrete random variable Y, also known as Poisson HMM. Before doing that, we will first familiarize the reader with the notion of having data generated from a finite mixture of three Poisson distributions for a discrete random variable. This gives a set of independent observations, like mixture models do. In order to generate the autocorrelated mixture we extend the mixture model by assuming a simple first-order, three-state discrete Markov process deciding which distribution in the mixture is generating the data. Generating data from the mixture under the assumption that the states follow a first-order Markov process is assuming that the data is generated by an HMM. The Markov process introduces the autocorrelation.

After the reader sees the evolution of states and the corresponding levels of the variable at each state, we take the data generated and estimate the transition probabilities, the parameters of the Poisson models, and the estimated hidden states using code from [211]. We will do that in Section 7.3.3.

7.3.1   A Poisson Mixture without Autocorrelation

A mixture of three Poisson distributions for discrete random variable Y is a probability mass function defined as

$$P(y) = \alpha_1 P(Y = y \mid \lambda_1) + \alpha_2 P(Y = y \mid \lambda_2) + \alpha_3 P(Y = y \mid \lambda_3),$$

where α_1 is the probability that the observation is generated by distribution $P(Y = y \mid \lambda_1)$, α_2 is the probability that the observation is generated by distribution $P(Y = y \mid \lambda_2)$ and $\lambda_3 = 1 - \lambda_1 - \lambda_2$. The parameters $\lambda_1, \lambda_2, \lambda_3$ are the expected value of the variable Y under the different state distributions, respectively.[1]

We generate a sequence of independent observations from a mixture of three Poisson distributions using *Part I* of program *ch7Poisson-HMM-estimation.R*. If the state

---

[1] As indicated in many probability books [173, 175, 17], a Poisson probability mass function is an appropriate model when the random variable measured is a count, a nonnegative number representing the number of events occurring in an interval (of time or space), assuming that the events are equally likely to happen at any point within the interval that is, assuming a Poisson process determines the event occurrence. For example, the number of COVID-19 cases per day in a given location.

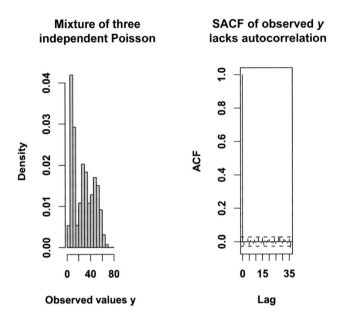

Figure 7.1 Data generated from a mixture of three Poisson distributions. The sample autocorrelation function (SACF) plot of the generated data show that the random numbers generated from a mixture of distributions are not autocorrelated, which makes sense since they are independent.

is 1, which happens with probability 0.4, then the observation comes from a Poisson with mean 10; if state is 2, which happens with probability 0.3, then the observation comes from a Poisson with mean 50; and if state 3, which happens with probability 0.3, then the observations comes from a Poisson with mean 30.

As we can see in Figure 7.1, the sample ACF of the data generated from such a mixture of three Poisson distributions lacks autocorrelation. The sample ACF is the sample ACF of white noise. This confirms what we said earlier that a simple mixture distribution will produce independent observations. As we have learned in this book, if observations are independent there is no time series analysis needed.

7.3.2   Introducing Autocorrelation in a Poisson Mixture

The data generated from the mixture in Section 7.3.1 consists of independent observations. But quite often it is the case that the state at which a system is today affects the state at which it will be tomorrow. This introduces autocorrelation in the observed data. A common assumption to account for that fact is that the state is generated by a first-order Markov process. The reader should consult the Appendix in Section 7.8 for a refresher or introduction to a Markov process.

Consider now that random variable Y, coming from the same mixture of Poisson distributions as in 7.3.1, takes values depending on the state generated by a first-order

Markov process with transition probability matrix

$$P = \begin{pmatrix} 0.6 & 0.3 & 0.1 \\ 0.3 & 0.3 & 0.4 \\ 0.1 & 0.6 & 0.3 \end{pmatrix}.$$

That is, the probability of moving from state 1 to state 2 is 0.3; the probability of moving from state 2 to state 1 is 0.3, and so on.

Then Y will be a stochastic process with autocorrelation. We demonstrate that with the simulation found in Part II of program *ch7Poisson-HMM-estimation.R*, which is like Program A.2.1, page 242 in [211]. Figure 7.2 shows a time plot of the series generated. This is one of the many possible members of the ensemble of all possible time plots. The histogram of the data generated indicates that there is some multimodality. The sample ACF of the mixture data shows autocorrelation and dies away sinusoidally. The sample PACF cuts off after lag 1. This is a typical correlation pattern of an AR(1), as indicated in Chapter 5. Thus, although we have obtained data from a mixture of three Poisson distributions, this new mixture is in sharp contrast with the mixture in Section 7.3.1, which showed no autocorrelation because, in this new mixture, the states are determined by a hidden first-order Markov model, which results in autocorrelated observations. To get an idea of what the R program is doing, we show a few steps here:

- Assume we start at state 1 as initial value. Then we draw an observed random number from a Poisson distribution with mean value 10.
- To determine which state to go next, a random draw is made from a multinomial distribution,[2] with parameter vector $p = (0.6, 0.3, 0.1)$ and number of observations $n = 1$. This makes the next state (1,0,0), the next state is 1; or (0,1,0), the next state is 2; or (0,0,1), the next state is 3.
- If the next state is 1, again, we draw an observed value from a Poisson with mean 10. If it is 2, we draw from a Poisson with mean 50, and if we draw a 3, we generate an observed value from a Poisson with mean 30.
- Suppose we ended at state 2. Thus, we draw an observed value from a Poisson with mean 50. Then, to determine what state to go next, we draw a random number from the multinomial distribution with parameter vector $p = (0.3, 0.3, 0.4)$, and so on; we repeat the process many times.

The point of this simulation is that we are generating each value of the time series y_t according to the "hidden (in practice)" state of the system at time t, sometimes state 1, and sometimes state 2, and sometimes state 3. But the state at time t depends on what the state was at time $t - 1$ because of the Markov nature of the state. This fact is what creates the autocorrelation in the observed time series. The state is what is autocorrelated.

---

[2] A multinomial distribution is a generalization of the binomial distribution to measure the probability of observing a distribution of observations among k mutually exclusive classes [173, 175].

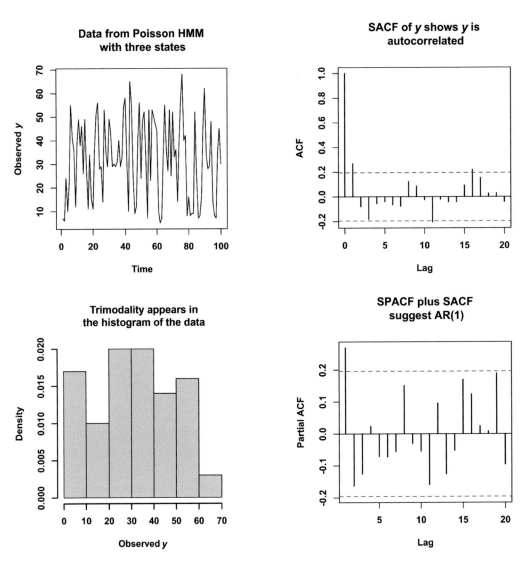

Figure 7.2 Generating HMM from mixtures of distributions according to a Markov process introduces autocorrelation in the data. The observations of this Poisson HMM process at time t conditional on the state at t are independent of the observations at time $t-1$ conditional on the state of the system at $t-1$, but the observations are autocorrelated as indicated by the sample ACF (SACF).

Notice that the R program calls the transition matrix gamma, and the vector of Poisson means lambda, the state state and the observed values x. We preserve the same notation used in [211] to facilitate the location of the program in that source.

Notice also that although the sample ACF and the sample PACF identify the autocorrelations of an AR(1) process, we would not model this time series as an AR(1) static model, because that would ignore the fact that the values depend on the Markovian state of the system, and also the fact that the observations are not Gaussian. Rather we would consider fitting to it a Poisson HMM. The reason is that the sample ACF that we observe is really the sample ACF of the Markov chain that defined the states of the system at each t. As we saw in Section 7.3.1, without the Markov process, the data would not be autocorrelated.

In the simulation example just done, we have created a time series of autocorrelated observations where the autocorrelation is created by the dependence between the states generated by a Markov process. We have generated a Poisson Hidden Markov process. Poisson HMM would be estimated using methods appropriate for them.

7.3.3 Fitting a Poisson HMM to the Autocorrelated Mixture

Assuming an HMM for Poisson counts is just the first step. To be able to predict with it, we first must estimate the parameters of the model. Parameter estimation in HMM usually relies on maximum likelihood or Bayesian methods. Mixture models for independent observations already encounter difficulties in estimation. The dependency structure that a Markov assumption adds to mixture models only exacerbates the difficulties encountered by the former [118]. Shumway [179], Chapter 6, describes how to set up the likelihood function to be maximized. The parameter estimation will produce the estimated states behind each observed count. The model assumed is

$$P(y_t \mid y_{1:t-1}) = \sum_{j=1}^{3} [p(state_{j,t} \mid state_{t-1})P_j(y_t)].$$

The log likelihood is

$$\sum_{t=1}^{n} \log \left(\sum_{j=1}^{3} [p(state_{j,t} \mid state_{t-1})P_j(y_t)] \right),$$

where n is the number of observations. Using algorithms for nonlinear optimization, we can obtain the maximum likelihood estimators of the transition matrix and the parameters of the Poisson distributions of the mixture. From these, a smoothing distribution formed by the probabilities of being at each state is formed. Shumway [179] found that a model similar to this model fits very well the earthquake data.

Part III of program *ch7Poisson-HMM-estimation.R* uses code from [211] to fit an HMM to the data that we generated in Section 7.3.2. The results are described next.

MLE Estimate of the Poisson Means

This is what the R program calls `lambda`, a vector containing the estimates of the parameters of the Poisson distributions in the mixture:

$$\widehat{\lambda_1} = 10.6; \quad \widehat{\lambda_2} = 51.2; \quad \widehat{\lambda_3} = 31.84.$$

Compare that with the true values that we used to simulate the data: 10, 50, 30, respectively.

MLE Estimate of the Transition Matrix

Recall that the program in [211] calls the transition matrix gamma. We called it P.

$$\widehat{P} = \begin{pmatrix} 0.69 & 0.081 & 0.23 \\ 0.125 & 0.295 & 0.58 \\ 0.35 & 0.50 & 0.15 \end{pmatrix} \quad \text{compare with} \quad P = \begin{pmatrix} 0.6 & 0.3 & 0.1 \\ 0.3 & 0.3 & 0.4 \\ 0.1 & 0.6 & 0.3 \end{pmatrix}.$$

Estimate of Equilibrium Distribution

The process has a stationary or equilibrium distribution that gives the probabilities of its being in the respective states after many transitions have evolved. This is what the R program calls delta:

$$\pi = (0.44, 0.26, 0.30).$$

Notice that these are close to the initial probabilities of being in each state assumed in the mixture that generated the data.

Other Output

The maximum likelihood value is 375.0259, the Akaike Information Criterion (AIC) is 768.05 and the Bayesian Information Criterion (BIC) is 791.4983. We use the latter two for model selection when more than one model is being considered. We would naturally try to fit other models, with perhaps different state processes, if the latent process is unknown. In a simulation like the one we did, our objective is just to show that the estimation method works. Showing this is always better done with a simulation.

Evaluating the Output

It may seem disconcerting to the reader that the estimates are not closer to the parameters that we used to generate the data. However it should not be surprising. We generated only 100 observations. Were we to generate many more observations, say 1,000, the estimates would be much closer to the true parameters. We leave that as an exercise.

In-Sample State Prediction

The R programs also estimated the state in the in-sample period. This is the state.hat output in the program.

```
1 3 2 2 3 1 1 3 1 1 2 2 3 2 3 1 1 1 1
1 1 3 3 1 1 1 1 1 1 2 3 1 1 3 3 2 2 2
1 1 2 3 3 2 3 2 3 2 3 2 2 3 2 2 3 3 2
3 2 1 1 1 3 1 1 1 1 3 2 3 1 1 1 3 1 1
1 1 1 3 3 1 1 1 1 1 3 1 1 3 2 2 1 2 3
2 3 2 3 2
```

When compared with the true states, we obtain

```
state.hat   1    2    3
        1   43   1    0
        2    0   1   25
        3    0  30    0
```

which shows that 43 times that the state was state 1 the program predicted that it is state 1. But the program made a lot of errors estimating states 2 and 3: 25 states 3 were estimated as 2, and 30 states 2 were estimated as 3. Again, the small number of observations used for the estimation is the reason for that departure from the true values.

7.3.4   Exercises

Exercise 7.1   Generate 1,000 observations from the HMM assumed in the simulation of Section 7.3.2 and do the following:

(a) Plot graphs like those in Figure 7.2. Describe what you observe in the plots.
(b) Use the data generated to fit to it an HMM with the same initial values as those used in program *ch7Poisson-HMM-estimation.R*. Compare the estimates to the true parameters assumed to simulate the data.
(c) Estimate the states and do a cross-tabulation of the estimated states and the states simulated. Discuss the errors made. Use part III of program *ch7Poisson-HMM-estimation.R* for that.
(d) Forecast six steps ahead.

7.4   Autocorrelated Gaussian Mixtures

The observed values generated by a Hidden Markov process do not have to be Poisson. In other words, assuming a finite number of states and a discrete Markov chain does not imply that we are bound to also have discrete random variables generating the observed data. The observed data can be continuous. One common assumption is that of a mixture of Gaussian distributions. Instead of simulating an autocorrelated Gaussian mixture, we will leave the simulation for an exercise and focus on a real case where a Gaussian HMM would make sense.

Yellowstone National Park in Wyoming, United States, is home to two-thirds of the number of geysers in the entire world. A very famous geyser is Old Faithful, located in Yellowstone's Upper Geyser Basin in the southwest section of Yellowstone National Park. Visitors from all over the world stand in the geyser-viewing area to wait for eruptions of Old Faithful. A ranger station tracks the time, height and length of an eruption to predict the next eruption. Although the geyser is faithful in the sense that it erupts on average 20 times a day, the time, length and height of an eruption is never known with complete certainty. Thus, waiting time is uncertain. Sometimes the Geyser erupts for a long time and forcibly, other times the eruption is short and weak. Thus,

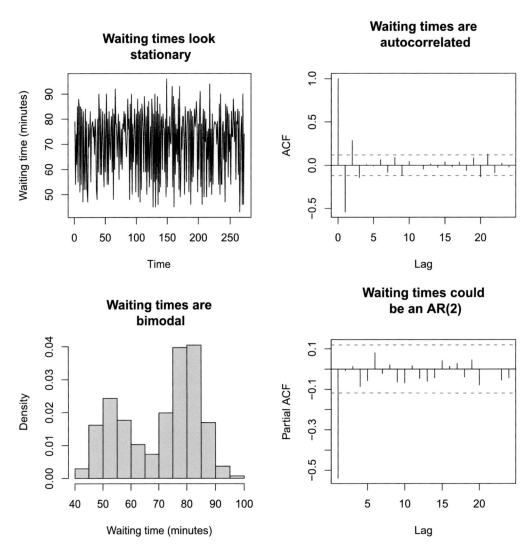

Figure 7.3 Some think that the bimodal distribution displayed in the histogram of waiting times is associated with how long the eruption was prior to the waiting time.

eruption duration is also random. A typical question posed by visitors after seeing an eruption is, "How much would we have to wait for the next eruption?" [101] Data certainly can help answer that question. We have seen that we can predict the future with models, and certainly the park rangers use all the information they can to make that prediction.

Figure 7.3 shows the time plot, the histogram, and sample ACF and sample PACF of a time series of waiting times to see Old Faithful. The data behind those plots is R's `faithful` data set. The data contains time series of 272 waiting times and eruption duration. In view of the information in Figure 7.3, how would the reader answer the visitor's question (i.e., how long is the time between eruptions, the waiting time)?

Why isn't the time between eruptions clear-cut? Because a geyser is part of a complex Earth system. According to National Geographic, "Geysers are made from a tube-like hole in the Earth's surface that runs deep into the crust. The tube is filled with water. Near the bottom of the tube is molten rock called magma, which heats the water in the tube. Water in the lower part of the tube, close to the magma, becomes super hot. Gradually, it begins to boil. Some of the water is forced upward. The boiling water begins to steam, or turn to gas. The steam jets toward the surface. Its powerful jet of steam ejects the column of water above it. The water rushes through the tube and into the air. The process continues until all the water is forced out of the tube... After the eruption, water slowly seeps back into the tube. The process begins again" [129]. The whole process can take from a few minutes to days.

The visitors and the Park Ranger's office do not observe the process going on inside the Earth. The indicators of what that process might be are the observed duration of the eruptions, their height and the time elapsing between eruptions or waiting time. These are all time series that are observed, but not at equidistant times.

Obviously, judging by Figure 7.3, there are multimodality features in the histogram of waiting times. That is, one single distribution does not seem to fit the data well. The waiting times are stationary and autocorrelated according to the sample ACF. The multimodality would not allow us to use a "one distribution" model like the ARIMA models we have seen. More useful would be to have a model that allows us to capture the changing distributional behavior, which is probably determined by the unseen process going on inside the Earth. A Gaussian HMM might be worth trying with this data set, and that is what Zucchini and MacDonald [211] fitted to the `oldfaithful` data.

Figure 7.4 illustrates a simple HMM for the Old Faithful geyser data set. The unobserved process (what is going on under the ground) is latent but autocorrelated. Conditionally on the state under the ground, the observed waiting times are independent.

A Gaussian HMM can be fit to the `faithful` data with program *ch7faithful.R*, the same program used to produce Figure 7.3. The program fits a Gaussian HMM, with continuous probability models for the levels, and first-order Markov states for the `faithful` data set.

7.4.1 Exercises

Exercise 7.2 HMM could be made with mixtures of continuous random variables. With the same transition probability matrix and R program as used in Section 7.3.2, generate data from an autocorrelated mixture of three Gaussian distributions,

$$f(y) = \alpha_1 f(y \mid \mu_1, \sigma_1) + \alpha_2 f(y \mid \mu_2, \sigma_2) + \alpha_3 f(y \mid \mu_3, \sigma_3),$$

where α_i is the probability that the observation is generated by distribution $f(y \mid \mu_i, \sigma_i)$, $i = 1, 2, 3$, and $\mu_1 \neq \mu_2 \neq \mu_3$, $\sigma_1 \neq \sigma_2 \neq \sigma_3$ are the models' parameters.

(a) First generate a sequence of 100 independent observations from a mixture of three normal distributions using program *Poisson-HMM-estimation.R* but replacing the

Poisson generation with Gaussian random number generation: with probability 0.5 a number comes from a normal with mean 10 and standard deviation 1; with probability 0.3, the number comes from a normal with mean 17 and standard deviation 1; and with probability 0.2 a number comes from a normal with mean 28 and standard deviation 3. This will produce a sequence of independent observations. Plot the time plot, histogram and sample ACF. Comment on what you discovered in the plots.

(b) Now introduce autocorrelation in the mixture using the same program. Assume that the state at which a system is today affects the state at which it will be tomorrow. Consider that the member of the mixture from which we will generate data will be decided by a first-order Markov process with transition probability matrix as given in the program. Do plots like in Figure 7.2 and comment on what you discovered from the plots.

(c) Estimate the parameters of the model, the transition probability matrix and the states. Use part III of program *Poisson-HMM-estimation.R* for that.

(d) Forecast with the estimated model six steps ahead.

7.5   Univariate State Space Models (SSM)

SSM, in contrast to HMM, assume a continuous state space [188]. The main idea of an SSM can be easily understood using an autoregressive model of order 1, now that the reader is familiar with these models. Figure 7.5 shows this model.

Let Y_t be a time series. Then the standard (static and time invariant) AR(1) (see Section 5.4.1 and Chapter 6) model is written usually as

$$Y_t = \alpha Y_{t-1} + v_t,$$

where v_t plays the same role as W_t in previous chapters. As we see, in the usual model, the index t only appears in the Y's and the v_t.

In the dynamic, or time-varying context, the model is expressed in a form similar to the standard form but with a parameter α that changes with time, which we indicate by putting the subscript t in the α:

$$Y_t = \alpha_t y_{t-1} + v_t.$$

In the context of state-space models, this is known as the *observation equation*, also known as *measurement equation*. Everything else is as in the standard form seen in earlier chapters.

The t subscript accompanying the parameter α would pose a lot of problems for estimation were it not for the addition of the assumption that the parameters are distinct but stochastically related through what is called a *system equation*, also known as *state equation*. The state equation describes the model dynamic, the mechanism of parametric evolution of α through time. This evolution has the form of a first-order continuous Markov process, for example:

$$\alpha_t = g\alpha_{t-1} + W_t,$$

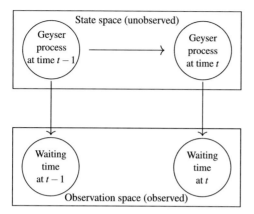

Figure 7.4 A discrete-state Hidden Markov Model assumes discrete states of the system that may be represented by a one-state Markov model, such as, for example, the behavior of waiting time for the Old Faithful geyser.

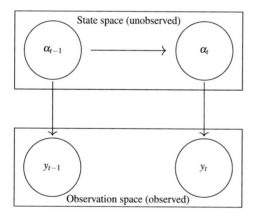

Figure 7.5 A state space model has a continuous state space.

where g_t is a known coefficient and W_t is an unobservable stochastic term, normal white noise usually being assumed. Figure 7.5 illustrates this idea.

The *(hidden, latent) state of the system* is represented by the system equation. The latent variables are in the sequence of the α parameter over time, which is found by means of probabilistic routines. The Markov property followed by the hidden state guarantees that (i) the state of the process summarizes all the information from the past that is necessary to predict the future (independence of the future of a process from its past, given the present state) and (ii) that the observed values of Y_t are independent, given the state α_t.

The computations used to obtain a fitted model and forecasts rely on algorithms that are naturally different from those used for the static models. But the end result of an SSM is similar to that of a static model: a fitted model and parameter estimates, and forecasts. Naturally, because of the assumptions made, Bayesian statistics plays a very important role in dynamic linear models and is a natural habitat for this type of model.

In Section 7.5, we introduce simple state space models with Gaussian states and Gaussian levels. Base R code with the function StructTS is then used to fit state space models to simulated data and to the Nile data set, widely studied in the state space literature.

7.5.1   The General Univariate SSM

The general univariate SSM is written as [150]:

Observation equation:

$$Y_t = F_t\theta_t + v_t, \qquad v_t \sim N(0, V_t)$$

System equation:

$$\theta_t = G_t\theta_{t-1} + W_t, \qquad W_t \sim N(0, W_t),$$

where Y_t denotes the observation series at time t; F_t is a vector of known constants; θ_t is the latent vector of model state parameters; v_t is a stochastic term having a normal distribution with zero mean and variance V_t; G_t is a matrix of known coefficients that defines the systematic evolution of the state vector across time; and W_t is a stochastic term having a normal distribution with zero mean and covariance matrix W_t. The v_t and the W_t are temporarily independent and mutually independent. To complete the model specification, an initial value for θ must be specified.

Since there are many physical systems for which the state of nature θ_t changes over time according to a relationship prescribed by engineering or scientific principles, the ability to include a knowledge of the system behavior in the statistical model is an apparent source of attractiveness of the model and the Kalman Filter algorithm used to estimate it [121]. Notice that F_t and G_t could be such that they do not change with time, as is also true of the variances V_t and W_t. We are just using the most general notation.

Example 7.1   To give some context to the parameters, we will refer to an example presented by Meinhold and Singpurwalla [121] about tracking a satellite's orbit around the Earth. In their example, the parameter θ_t is the position and speed of the satellite at time t, with respect to a spherical coordinate system with origin at the center of the Earth. These quantities cannot be measured directly. From tracking stations around the Earth, we obtain measurements of distance to the satellite and the accompanying angles of measurement; these are the Y_t. The principles of geometry, mapping Y_t onto θ_t, would be incorporated in F_t, while v_t would reflect measurement error; G_t prescribes how the position and speed change in time according to the physical laws governing orbiting bodies, while W_t would allow for deviations from these laws owing to such factors of nonuniformity of the Earth's gravitational fields, and so on [121].   □

Example 7.2   Local level models

To add more context to the equations, we will also refer to an example presented by West and Harrison [202] in the context of production planning and stock control, which are always very important for business to succeed. A first-order polynomial model is suitable to forecast demand in order to know how much to produce and how much to stock. The authors give the example of a pharmaceutical company which sells an average of 100 units per month. Medical advice leads to a change in drug formulation that is expected to lead to wider market demand for the product. It is agreed that from January, $t = 1$, the new formulation with the new packaging will replace the current product, but the price and brand name will remain unchanged. In order to plan production, stocks and raw material supplies, short-term forecasts of future demand are required. The model that has been working with the old drug formulation will be used for short-term forecasting. Let y_t denote observed demand. The following observation and system equations describe the behavior of the system as usual. But then there is some extra initial information that needs to be incorporated. At $t = 0$, the view of the market for the new product is that demand is most likely to have expanded by 30 percent to 130 units average per month and it is unlikely that demand has fallen by more than 10 units or increased by more than 70. The model then needs to be fitted to historical data using this new information if good forecasts are to be made.

Observation equation: $y_t = \mu_t + \varepsilon_t,$ $\varepsilon_t \sim N(0, 100)$
System equation: $\mu_t = \mu_{t-1} + \xi_t,$ $\xi_t \sim N(0, 5)$
Initial information: $(\mu_0 \mid D_0) \sim N(130, 400)$

To compare with the general notation introduced earlier, $\theta_t = \mu_t$, $100 = V$, $5 = W$, $F_t = G_t = 1$. The only parameters we need to know are the V and the W, which are estimated using maximum likelihood or Bayesian methods. In this example, they are given.

This model assumes that the average demand is locally constant. The variances for ε_t and ξ_t are saying that the observed variation around the mean will be larger than the variation of the mean itself. The signal-to-noise ratio will thus be 0.05. The initial information combines with the information in the data to update the μ_t.

Later on, at time $t = 10$, it is discovered that a competing drug is withdrawn from the market. This leads to updating the prior initial information, the assumption for W_t and updating the prior as well at time $t = 10$ to reflect what is known.

The model just described is known as a *first-order closed constant polynomial model*. It is constant because all the matrices defining its dynamics are constant. The model is also called *random walk with noise model* by the authors of R's `dlm` package [149], *structural equation model* and *local level* model in base R.

The major applications of the constant model are in short-term forecasting and control with a signal-to-noise ratio between 0.2 and 0.001, where the main benefit is derived from data smoothing. Mathematical derivations to be found elsewhere can show that this model is equivalent to using a simple exponential smoother.

A local-level process

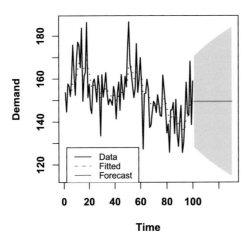

Figure 7.6 A simple state space model, a local-level model, fit to the pharmaceutical company data. The shaded region represents the 95 percent confidence interval for the expected value of the test data.

State space models, under different model assumptions, reduce to ARIMA, exponential smoothing and other classical regression models, but with varying coefficients determined by the state. Thus, when machine learning methods use them, they are implicitly generalizing the classical models to the more prevalent situations generated by the velocity, volume and variety of time series data that we face nowadays.

The function StructTS in R is an easy way to fit a structural equation state space local-level model. The R package dlm [149] is also a very complete array of functions to estimate and forecast with state space models.

We simulate data from the Harrison and West model using program *ch7locallevel.R*. Then we use the function StructTS and KalmanRun to fit the model and to forecast with it 30 steps ahead (which is not practical, since this type of model is only used for short-term forecasting).

Figure 7.6 shows the values fitted in-sample using the Kalman filter and the forecast.

The residuals from fitting this model are white noise.

But the usefulness of these models is in their ability to add components that change over time to the model, such as trends, seasonality and other variables. For that reason, we discuss some of these models next and then we will fit a trend model to a well-known data set, the Nile data set.

The reader should notice that the notations change a lot across literature resources, even though all of them are doing the same.   ☐

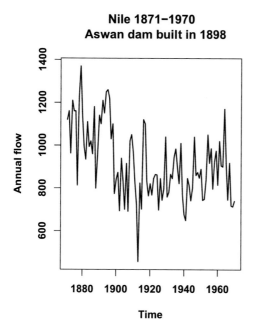

Nile 1871–1970
Aswan dam built in 1898

Figure 7.7 The Nile river data set, widely used as example in the state space literature, has a jump in trend after the construction of the Aswan dam in 1898.

Example 7.3   SSM with local linear trend

The local trend model adds a time-varying trend to the level models.

Observation equation: $y_t = \mu_t + \epsilon_t,$ $\epsilon_t \sim N(0, 100)$
System or transition equation: $\mu_t = \mu_{t-1} + v_{t-1} + \xi_t,$ $\xi_t \sim N(0, 5)$
System equation: $v_t = v_{t-1} + \xi_t.$

The data set `Nile` embedded in R contains measurements of the annual flow of the river Nile at Aswan (formerly Assuan), 1871–1970, in $10^8 m^3$ with apparent change-point near 1898 [28]. It is a classic example of data in the state space literature. Looking at Figure 7.7 we notice a jump in the flow of the river following the construction of Aswan dam in 1898. Some authors suggested that the Nile river volume experienced a permanent decline in the mean of its distribution in 1899 [156]. This can be done by inflating the system variance in 1899 using a multiplier bigger than one [150].

A local linear trend might be appropriate. So we proceed to fit one using the `dlm` package [149]. We illustrate the results in a series of graphs. The estimation only yields estimates of the variances and also estimates of the fitted values, a forecast and confidence intervals. It is easy to implement.

We used program *ch7Nilesriver.R* to obtain the graphs in Figures 7.7 , 7.8, 7.9 and 7.10 but program *ch7StructTS-Nile.R* will do the same work.   □

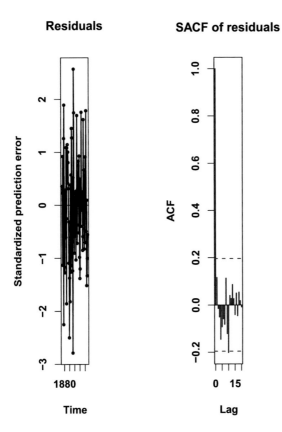

Figure 7.8 The residual of the model fitted to the Niles data set and the sample ACF of the residuals indicate that the residuals are white noise.

7.5.2   General State Space Structural Model

This is a local trend model with an additional seasonal component:

Observation equation: $y_t = \mu_t + s_t + \varepsilon_t,$ $\varepsilon_t \sim N(0, 100)$
System equation for level: $\mu_t = \mu_{t-1} + v_{t-1} + \xi_t,$ $\xi_t \sim N(0, 5)$
System equation for trend: $v_t = v_{t-1} + \eta_t.$
System equation for seasonals: $s_t = s_{t-1} - \ .... - s_{t-s+2} + \eta_t$

Other covariates believed to be related to the dependent variable could be added to this model. The fitting of this model to the Airline Passenger data set will be left as an exercise.

7.5.3   Exercises

Exercise 7.3   The pharmaceutical data set to which we fit a local level SSM model could be split into a training set and a test set. Do that, and then reestimate the model using only the training data. Compare the out-of-sample forecasts to the test data.

Filtered and smoothed

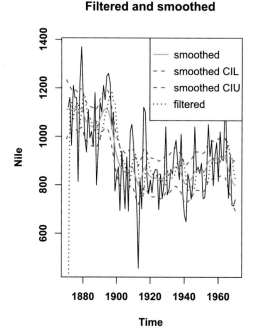

Figure 7.9 The smoothed levels of the Nile time series, smoothed confidence intervals (C) and the actual filtered values of the level.

Obtain the RMSE of the forecast. Do a plot that contains the training data, the test data, the predicted values in the training set and the forecast and confidence intervals.

Exercise 7.4 The Niles data set to which we fit an SSM with local linear trend could be split into a training set and a test set. Do that, and then reestimate the model using only the training data. Compare the out-of-sample forecasts to the test data. Obtain the RMSE of the forecast. Do a plot that contains the training data, the test data, the predicted values in the training set and the forecast and confidence intervals.

Exercise 7.5 Fit a local trend model with seasonality to a training set of the `Air-Passenger` data set that comes with R. Forecast the test period and obtain the root mean square error of the forecast.

7.5.4   Recursive Estimation

This section will make more sense to the reader that has some knowledge of Bayesian statistics and some introductory knowledge of numerical computing in Statistics.

We have used a program that has provided a pretty good fit to the Niles data set, has estimated the variances, and has provided in-sample and out-of-sample forecasts. The reader interested in understanding what the program is doing, can read this section, or alternatively, may read [150]. As indicated in Section 7.2, the state vector at time t summarizes the information from the past that is necessary to predict the future. Thus,

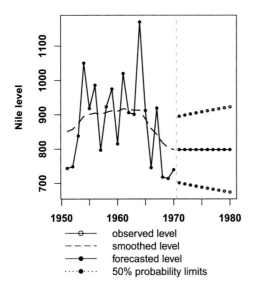

Nile level

—○—   observed level
– – –   smoothed level
—●—   forecasted level
····●····   50% probability limits

Figure 7.10 The out-of-sample forecast for the Nile level.

before forecasts of future observations can be calculated, it is necessary to make inferences about the state vector θ_t. To do that, software that does state-space estimation
uses the Kalman filter, an algorithm that became famous many decades ago and is still
used today, or other algorithms.

To specify the distribution of the state vector θ_t, it is necessary to start with a distribution of the state vector at time zero (θ_0). This, together with the system equation,
determines the distribution of the state vector at $t = 0, 1, 2, \ldots$. We refer to these distributions as the prior distributions, since they represent our "belief" prior to having
observed the data $y_t, y_{t-1}, \ldots$.

After the history of the process $Y_t = \{y_t, y_{t-1}, \ldots\}$ has been observed, we want to
revise our prior distribution of the unknown vector θ_t. The new or revised information
about θ_t is expressed by the conditional distribution $p(\theta_t \mid Y_t)$. This is also called the
posterior distribution, since it measures the information about θ_t after the observations
in Y_t have become available. From the state-space model given at the beginning of the
chapter and the prior distribution, we can derive recursive equations that propagate
the conditional distributions $P(\theta_t \mid Y_t) \rightarrow P(\theta_{t+1} \mid Y_t) \rightarrow P(\theta_{t+2} \mid Y_{t+1}), \ldots$. Those
probability distributions are also normally distributed under the assumptions made for
v_t, w_t and θ_0. The Kalman filter just consists of updating the estimate of the expected
value and the variance-covariance matrices of those updating distributions. Those are
the parameters that we estimate.

As pointed out by Petris and colleagues [150], the Kalman filter can suffer from
instability and give nonpositive definite matrices. Singular value-based algorithms are
more robust, and that is the algorithm used by dlm [149] for filtering and smoothing.
It does sequential singular value decomposition of the variance-covariance matrices.

First Step: Estimate the V and W by maximum likelihood

Then to estimate by maximum likelihood estimation, dlm [149] relies on the optim() function in R, an optimizer that we use to calculate maximum likelihood estimators of the V and W matrices. We have demonstrated in Example 7.3 how to conduct estimation using the famous Nile river data, a univariate time series of measurements of the annual flow of the river Nile.

The variances could have been parametrized in terms of their natural logs to avoid having to set lower and upper values for the variances (to avoid the algorithm to visit negative values of variance), but we did not do that and used instead the *lower=* argument. The parm argument is a placeholder to put the final parameter estimates.

Get the Fitted θ_t

If convergence $= 0$, then we can proceed to set up the model for estimation of the θ_t at each t. It is this step that uses the Kalman filter, given the likelihood function that we estimated. The Kalman filter can be run by calling the function dlmFilter [149]. This produces an object of class dlmFiltered, a list containing, most notably, predicted states and filtered estimates of state vectors together with their variances. Method functions residuals and tsdiag can be used to compute the one-step forecast errors and plot several diagnostics based on them.

The residuals are all within three standard deviations with no apparent pattern and they have the ACF of white noise.

Smoothing the Thetas

We may also obtain smoothed values of θ_t, which seems to give better estimates.

After we obtained all of the above, we plotted the filtered and smoothed values of the θ's.

Forecasting

The standard practice of fitting a state-space model by maximum likelihood, plugging the MLEs into the model, and considering state estimates and forecasts, together with their variances, obtained from the Kalman filter, does not account in any way for the uncertainty about the true values of the model parameters. This may be an issue, since precise parameter estimates in state-space models are the exception rather than the rule, even in simple models like the local-level model (recall the large standard errors of the variances obtained from the Nile river data). Moreover, maximum likelihood estimation in state-space models can be numerically hard. The Bayesian approach offers a natural way of dealing with parameter uncertainty in a state-space model.

7.6   Problems

Problem 7.1   Think of two data sets to which you think an HMM or SSM should be fitted instead of a static model with constant coefficients. Justify your choice.

Problem 7.2   Use the Italian COVID-19 data set studied in the case study of Chapter 5, difference the data to obtain the new number of cases per day, and fit to it an appropriate Poisson HMM using the program presented in this chapter for that type of model. Justify your choice of number of states.

Problem 7.3   Do you think that the Birmingham data set, for a given parking structure such as Broad Street, could be well modeled by a Gaussian hidden Markov model? Fit such a model to the data from that parking structure. Before you do, justify the number of states that you are assuming.

7.7   Quiz

Question 7.1   Which of the following is always true?

(a) Observations from a mixture of Poisson distributions are not autocorrelated with themselves, but if they are from a mixture of Gaussian distributions, they are.
(b) Observations from a mixture of Poisson distributions are not autocorrelated with themselves, and observations from a mixture of Gaussian distributions are not autocorrelated with themselves.
(c) Observations from a mixture of Poisson distributions are autocorrelated with themselves, and observations from a mixture of Gaussian distributions are autocorrelated with themselves.

Question 7.2   If an HMM is appropriate, then there is (check all that apply)

(a) a discrete state space and a continuous observation space.
(b) a continuous state space and a discrete observation space.
(c) a discrete state space and a discrete observation space.
(d) a continuous state space and a continuous observation space.

Question 7.3   If an SSM model is appropriate, then there is (check all that apply)

(a) a discrete state space and a continuous observation space.
(b) a continuous state space and a discrete observation space.
(c) a discrete state space and a discrete observation space.
(d) a continuous state space and a continuous observation space.

Question 7.4   It makes sense to fit to data an HMM and/or an SSM to a data set if which of the following is true?

(a) During the period in which the data is observed, conditions are very similar in the context in which the data is collected.
(b) During the period in which the data is observed, conditions vary in the context in which the data is collected.
(c) During the period in which the data is observed, the only thing observed is white noise.

(d) During the period in which the data is observed, the only thing observed is a constant trend.

Question 7.5 What is the difference between the `Nile` data set and the `faithful` data set regarding the appropriateness of fitting a HMM to each of them?

(a) The Nile data set has changes in their state and the faithful data set does not.
(b) The faithful data set has changes in their state and the Nile data set does not.
(c) There is no difference between the two data sets. They both have changes in state.
(d) The Nile data set's observation space is continuous, and the faithful data set's observation space is discrete.

Question 7.6 The pharmaceutical data set was fit a constant-level SSM. Which model does the SSM fit resemble?

(a) The random walk model
(b) The simple exponential smoothing model
(c) An AR(1) model
(d) White noise

Question 7.7 When fitting a Poisson HMM to a data set set, we must first make sure of which of the following?

(a) There is no overdispersion in the data set.
(b) The states are not autocorrelated.
(c) The observations are generated by a uniform distribution.
(d) The states are Poisson distributed.

Question 7.8 In HMM and in SSM for autocorrelated data, which of the following apply?

(a) The data is autocorrelated because the state space is autocorrelated.
(b) The data is autocorrelated because the observation space given the state is autocorrelated.
(c) The trace of the states over time cannot be estimated.
(d) The trace of the states over time can be estimated.

7.8 Appendix: First-Order Markov Process

To introduce a Markov process generating the state, consider a system that can be in any of a finite number of states, and assume that it moves from state to state according to some prescribed probability law. The system, for example, could be the visitors to a web page and the states the web page categories visited (frontpage, news, sports, etc.) [172, 153]. Observing visitors over a long period of time would allow one to find the probability of visiting a particular web page given that the visitor just finished visiting another web page, for example, the probability that the visitor goes to the News page given that the visitor was in the sports page.

Let Z_t denote the state of the system at time point t, and let the possible states be denoted by $1, 2, , \ldots, m$ for a finite integer m. We are interested not in the elapsed time between transitions from one state to another, but only in the states and the probabilities of going from one state to another, that is, the transition probabilities.

We will denote the probability of moving to state k given that the last state was j by

$$P(Z_t = k \mid Z_{t-1} = j) = p_{jk},$$

where p_{jk} is the transition probability from state j to state k; and this probability is independent of t. So the transition probabilities do not depend on the time points; they only depend on the states. The event $\{(Z_t = k \mid Z_{t-1} = j)\}$ is assumed to be independent of the past history of the process conditional on the present. Such a process is called a Markov process with stationary transition probabilities. The transition probabilities can conveniently be displayed in a matrix:

$$P = \begin{bmatrix} p_{11} & p_{12} & \cdots\cdots & p_{1m} \\ p_{21} & p_{22} & \cdots\cdots & p_{2m} \\ .. & \cdots & \cdots & \cdots \\ .. & \cdots & \cdots & \cdots \\ .. & \cdots & \cdots & \cdots \\ p_{m1} & p_{m2} & \cdots\cdots & p_{mm} \end{bmatrix},$$

where the probabilities in a given row add up to 1. The transition probability p_{12} is the probability of moving from state 1 to state 2, for example.

If matrix P is regular (i.e., P^n has all positive entries for some power of P), then the process has a stationary or equilibrium distribution that gives the probabilities of its being in the respective states after many transitions have evolved. Suppose that limit exists and denote it by $\pi = (\pi_1, \ldots, \pi_m)$. Then it must satisfy the following condition:

$$\pi = \pi P.$$

Markov chains can be found explained in many introductory probability books, such as Scheaffer [175].

8   Vector Autoregression

8.1   Introduction

Prey breed and get eaten in proportion to predators; predators reproduce in proportion to prey eaten and die. Understanding the dynamics of organisms and population densities within an ecological system is essential for successful environmental management and sustainability. Issues arise as to how organisms will respond to the introduction of additional plants or animal species, natural and/or human-made environmental disruptions, as well as other perturbations and unexpected changes in population densities. *Vector Autoregression Models (VAR)* allow capturing the essence of a predator–prey interaction like that represented in predator–prey models studied in Biology and Ecology. For example, Ewing and colleagues [45] examined data on two freshwater organisms: the predator, Didinium species, and its prey, Paramecium species. They were interested in the response of the predator and prey growth rates to *shocks* (unexpected change) inflicted on their environment. To that end, they estimated a VAR model with two equations, one for the model of the growth rate of the predator and another model for the growth rate of the prey. In the absence of change, the dynamics of the system remain the same and the interrelation can be estimated using the data. However, when an unexpected exogenous event occurs such that a process is set into motion, and absent other exogenous events, the system is altered for some time and eventually returns to its usual dynamics. Compared to other methods, the VAR models fitted by the authors are the best to study the impulse response to an unexpected change (also known as innovation, impulse or shocks) in the ecological system.

> VAR models are multiequation models that contain an equation for each variable, and each variable depends on past values of itself and past values of the other variables. In other words, all variables are endogenous[109].

Figure 8.2 shows the nature and the contrast of VAR models with the univariate models studied earlier in this book. Whereas in univariate ARIMA models values of a variable X at time t are affected by past values of itself, in the VAR multivariate case past values of other variables, for example, Y, R, could be affecting variable X. Figure 8.2 shows a multivariate model with two model equations, one for density of

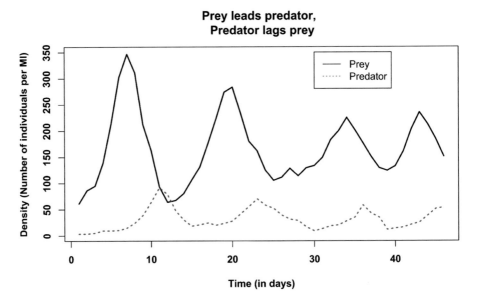

Figure 8.1 Prey and predator density over time help illustrate the concept of leading indicator, as prey density leads predator density, meaning that there has to be prey for predators to be present, but not vice versa.

prey and one for density of predator. It was fitted to the data shown in Figure 8.1, which is the data used by [45]. The data can be found in file *predatorprey.csv* and the results obtained running the program *ch8predatorpreyprog.R* indicate that prey density at time t is affected by prey density at $t-1$ and $t-2$ and predator density at $t-1$ and $t-2$. Predator at time t is affected by prey and predator at both $t-1, t-2$.

The system in Figure 8.2 will follow that model over time as far as it is not disturbed by external shocks. If external shocks disturb it, there will be a chain reaction that will eventually end and bring the system back to the dynamics represented by the model in Figure 8.2. That chain reaction is demonstrated in the Impulse Response Functions (IRF). Figure 8.3 shows the impulse response of both prey and predator density to a shock resulting in more predators. As we can see, prey responds by decreasing, which leads to predators decreasing, but then as prey recovers predators come back; and as time goes on these chain reactions are less severe, until there is a point, approximately 46 days later, where the system goes back to its usual dynamics, seen in Figure 8.2.

VAR is also appropriate for studying the dynamic interaction between wolf and deer populations when both are affected by a natural hazard such as forest fire, human-made hazards, climatic change, sudden eradication of disease, or even environmental regulation [45]. In addition to that application, VAR is appropriate for the modeling of the communication among regions of the brain that comprise a neural network of specific cognitive/perceptual conditions or the resting state and also follows dynamics that are much in need of being understood [22]. Other areas of application of VAR include immune dynamics, where viruses replicate, viral presence stimulates T-cell

$$\widehat{Prey_t} = +66.42^* + 1.42Prey_{t-1}^* + 0.07Pred_{t-1} - 0.75Prey_{t-2}^* - 0.39Pred_{t-2}$$

$$\widehat{Pred_t} = 1.07 - 0.09Prey_{t-1}^* + 0.92Pred_{t-1}^* + 0.17Prey_{t-2}^* - 0.33Pred_{t-2}^*$$

Figure 8.2 A VAR model assumes that all variables are affected by their own past values, but they can also be affected by and affect other variables [45].

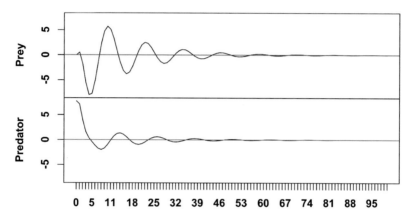

Orthogonal Impulse Response from Predator

Figure 8.3 Impulse response function of both prey and predator density to a shock resulting in more predators. After the initial shock, there is a chain reaction of predator and prey growth until no more growth than the equilibrium one takes place (until reactions settle at 0 change).

production and T-cells destroy a proportion of viruses they encounter. VAR is also appropriate for the analysis of engineering or climate complex systems. In particular, VAR has also been a tool in macroeconometrics since the mid 1990s, due to its ability to capture complex dynamic interrelationships among macroeconomic variables in a relatively parsimonious econometric framework. More recently, VAR models have proved to be a powerful tool for monitoring macroeconomic conditions in real time, or nowcasting [24], while at the same time retaining their proficiency in the tasks that they have been routinely used for, namely, structural analysis, forecasting and scenario analysis [94, 10]. Financial time series can also be modeled with VAR. For example, stock prices in Germany depend on stock prices in the United States, the United Kingdom, Japan and other countries. At the same time, the stock prices of these countries depend on Germany's and other countries' stock prices. Each country's stock prices depend on the prices of the other countries. Forecasting stock prices in the countries accounting for their interrelation can improve forecasting performance. Moreover, a shock in one of these countries propagates to the others and the effect is felt in all of them until the system rests.

In this chapter, we learn to apply simple VAR models with R. First, in Section 8.2.1 we will introduce the concept of cross-correlations among time series and use it to tentatively identify a VAR model. Modeling with VAR will be discussed in Sections 8.3 and 8.4. Then in Section 8.7 we revisit the IRF. Section 8.8 discusses how to distinguish between spurious and real relations and briefly talks about unit roots and cointegration. Software currently available for conducting VAR range from generic to those specialized for brain imaging data, ecological data, economic data and other types of data. Some software packages are proprietary and others are open-source. R allows us to introduce the main ideas without the methodological complexities that accompany these multiequation models. More specialized or mature commercial software to conduct VAR is RATS (Regression Analysis of Time Series) [161].

8.2   Cross-Correlation between Two Time Series

A VAR model captures the mean and the variance structure of the individual time series and the pairwise *cross-correlation* between each other at different time lags. For that reason, the *Cross-Correlation Function (CCF)* plays a very important role in multivariate time series models.

Consider two stationary time series, Y_t and X_t, for which values have been collected at regular intervals in the past. We define the CCF of X_t and Y_{t-k} as

$$\rho_k(X, Y) = \frac{E[(X_t - \mu_X)(Y_{t-k} - \mu_Y)}{\sigma_X \sigma_Y},$$

and we say that the combined process is second-order stationary if the cross-correlations depend only on the lag k and not on time.

The *sample cross-correlation function* or sample CCF of the observed x_t and y_{t-k}, which is used to estimate $\rho_k(X, Y)$, is defined by

$$r_k(x, y) = \frac{\sum_{t=1}^{n-k}[(x_t - \bar{x})(y_{t-k} - \bar{y})]}{\sqrt{\sum_{t=1}^{n}(x_t - \bar{x})^2}\sqrt{\sum_{t=1}^{n}(y_t - \bar{y})^2}}.$$

The following are some important features of the CCF to keep in mind:

- Usually, $r_{x,y} \neq r_{y,x}$ at any lag k.
- A plot of the sample cross-correlation function against lag k is called *cross-correlogram*.
- As in the univariate autocorrelation, the CCF is defined for stationary random processes. If stationarity is not satisfied, the CCF will die away slowly and will have significant cross-correlations for very large values of k, thus depriving the CCF of any use for multivariate model identification and modeling. Differencing is needed before the identification step if the individual time series are nonstationary. If the CCF dies away quickly or cuts off, then there is stationarity in the relation between the two series. Of course, the sample CCF will have a lot of other distracting cross-correlations due to sampling, the expected 5 percent, as in the univariate correlograms.
- We need at least 50 pairs of observations to obtain a useful estimate of the CCF.

- Sample cross-correlations at lag k larger than the two standard errors line when viewed in the cross-correlogram plot reveal that the corresponding $\rho_k(X, Y)$ is significantly different from 0.
  The hypothesis test underlying that conclusion at each lag k is:

$$H_0 : \rho_k(X, Y) = 0$$
$$H_a : \rho_k(X, Y) \neq 0.$$

  An $r_k(x, y)$ outside the two standard errors band shown in a cross-correlogram means that the p-value of the test is less than 0.05 and therefore we reject the null hypothesis, leading to the conclusion that there is statistically significant evidence that $\rho_k(X, Y)$ is different from zero, that is, there is correlation between X_t and Y_{t-k}.

- We may obtain the sample CCF or cross-correlogram by typing in R `acf(X,Y)`, where x, y would be replaced by the name of the observed time series, that is, a very complete plot that will give the sample ACF and the sample CCF. By typing `print(acf(x,y))`, we can see the values of the autocorrelations and cross-correlations. Base R has another command to obtain the sample CCF, `ccf(x,y)`, but this one just gives the sample cross-correlation function.

Example 8.1   Cowpertwait and Metcalfe [30] studied the relation between building permit application approvals and building construction activity in Australia and found that past approvals lead (affect) construction activities but past construction activity does not affect approvals. In this example, we investigate whether that is the case in the United States. Let X_t denote quarterly building permits (PERMIT, aggregated to quarterly in FRED's database) from 1974Q1 to 2020Q3, and let Y_t denote quarterly housing starts (HOUST in FRED) for the same times.[1] Figure 8.4 shows the time plot of the data before differencing.

The sample ACF and CCF of the time series (not shown) reveal nonstationarity. Since we must have stationary series in order to interpret the CCF, we differenced. The CCF of the differenced time series can be seen in Figure 8.5. The program to obtain them is program *ch8USApprovActiv.R*, which uses the command `acf`. This command, when applied to an object of class `mts` will produce both the sample ACF and the sample CCF.

The plots in the diagonal in Figure 8.5 are the sample ACFs. other plots are the cross-correlograms or sample CCFs. The titles of the cross-correlograms should be interpreted as indicating how the second variable listed in the title affects the first variable.

The reader should remember that once we have let R know that we have a quarterly time series, the label 1 in the horizontal axis represents a whole cycle, that is, a year.

---

[1] PERMIT is Thousands of New Private Housing Units authorized by building permits, and it is a monthly, not seasonally adjusted time series that was extracted from 1960:1 to 2020:11 and converted to quarterly; STARTS are New Privately Owned Housing Starts-Total one-family units, in thousands, and it is quarterly, seasonally adjusted 1974Q1 to 2020Q3. The data sets are, respectively, *PERMIT-quarterly.csv* and *HOUST-Q.csv*.

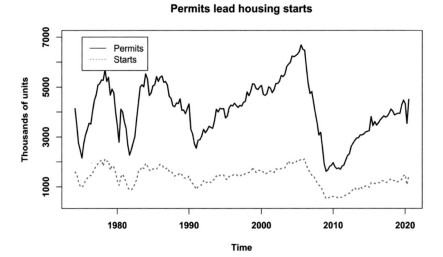

Figure 8.4 Building permits and housing starts help illustrate the concept of leading indicator, as permits lead activity, meaning that approvals happen first, then comes the activity, but not vice versa.

Thus, there are five lags represented between 0 and 1 (one for lag 0), and four lags between 1 and 2, and so on.   □

8.2.1   Identifying a Multivariate Model

The correlogram and the cross-correlogram plots help us identify a multivariate, multiequation model for the change in permits and the change in housing starts. Let's illustrate how identification works with the data in Example 8.1.

Before identification, looking at Figure 8.5, we can see that:

- The only significant cross-correlation in the "Permits & Starts" plot in Figure 8.5 is at lag 0, meaning that change in permits at time t is contemporaneously correlated with change in housing starts at time t.
- There is one significant cross-correlation in the "Starts & Permits" plot, other than at lag 0, indicating that past change in permits affects change in housing starts at time t.

Notice that interpretations are software-dependent. So, reading the documentation is recommended if not using R (and when using R).

To identify a model, we do the following:

- Identify the lag where the first significant spike appears (exclude 0 lag). Looking at the "Starts & Permits" plot in Figure 8.5, our first significant spike appears at lag 1, so we know that change in housing starts at time t depends on change in permits at time $t - 1$.

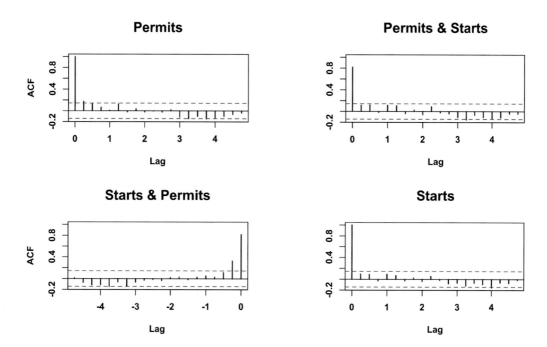

Figure 8.5 Correlation structure of change in housing starts and permits. The sample correlograms of each differenced variable are in the diagonal plots. The cross-correlograms are in the off-diagonals. The "Permits & Starts" plot indicates how the second variable (Starts) affects the first (Permits). The "Starts & Permits" plot indicates how Permits affects Starts.

- To determine other past lags in approvals, look at how many lags there are between the first significant spike and the lag at which decay starts. Decay starts right after lag 1. So we know now that change in housing starts at time t does not depend on change in permits at time $t - 2$.
- To determine the dependence of change in housing starts on past values of itself, we look at what happens in the cross-correlation function after lag 1. If the sample cross-correlation dies away in a damped exponential fashion, we set the lag dependence at 1, that is, change in housing starts at time t depends on change in housing starts at time $t - 1$. If it dies away in a damped sine-wave fashion, we set the lag at 2, that is, we would decide that change in housing starts at time t depends on change in housing starts at times $t - 1$ and $t - 2$. We settle tentatively for the latter.

Thus, if we let $Y_t^* = $ change in housing starts and $X_t^* = $ change in permits, the model identified from Figure 8.5 is

$$Y_t^* = \alpha_1 X_{t-1}^* + \beta_1 Y_{t-1}^* + \beta_2 Y_{t-2}^* + W_{t,Y} \tag{8.1}$$

$$X_t^* = W_{t,X}. \tag{8.2}$$

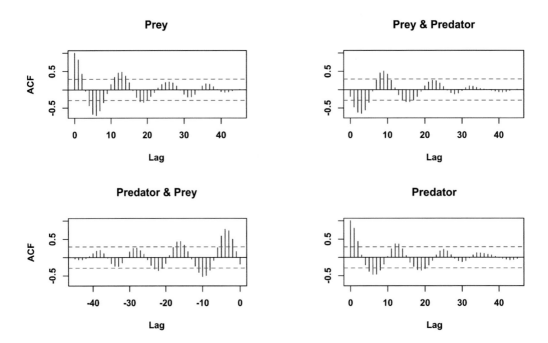

Figure 8.6 The correlograms and cross-correlograms of predator and prey densities reveal a model very different from that revealed in Figure 8.5.

Because there are no significant cross-correlations at positive lags, X_t^* is just represented by its variance.

To offer another example, we look at the cross-correlations of the predator and prey time series seen in Section 8.1. Applying the identification rules to the image in Figure 8.6, we can tentatively identify a VAR model with two equations in which each variable depends on the past three values of the other, and the past two values of itself. Tentatively, a good initial model could be a VAR(p=2). If such a model produces *white noise bivariate residuals*, then we could be satisfied with such a model and might not need a VAR(p=3).

To be able to determine whether the residuals of a VAR model are multivariate white noise, we need first to understand what that is.

8.2.2   CCF of Bivariate White Noise

Two time series X_t and Y_t are bivariate white noise if their individual ACFs are those of white noise and their CCFs have a significant $\rho_k(X, Y)$ value at only lag $k = 0$.

Figure 8.7 indicates what the CCF of white noise would look like in theory. This definition generalizes to more than two variables, in which case we contemplate multivariate white noise.

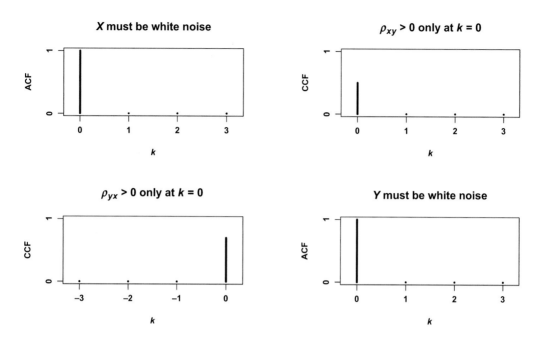

Figure 8.7 Correlation structure of bivariate white noise in theory.

In the case of multivariate and bivariate white noise, the series can be correlated contemporaneously at time t (lag 0), but the cross-correlations are 0 at any other lag.

As is the case with a single time series, the CCF of a sample will look slightly different from Figure 8.7 because of sampling. To demonstrate that, we generated bivariate white noise using the code at the bottom of program *ch8ApprovActiv.R*. Figure 8.8 shows that there are some nonzero and nonsignificant autocorrelations, and even possibly 5 percent significant autocorrelations and cross-correlations in a random sample of bivariate white noise. We can however discern that the variables are contemporaneously correlated and they do not have have any other significant auto or cross-correlations.

Bivariate white noise CCF is the desired outcome for the residuals obtained after fitting a VAR model, for example. If the VAR model fits a multivariate time series well, then the bivariate residuals should have a cross-correlogram resembling that of bivariate white noise, like that seen in Figure 8.8.

8.2.3   Exercises

Exercise 8.1   If they are not, two time series X_t and Y_t must be made stationary before interpreting the information in the cross-correlogram. It is common to use the differencing operations that we have learned in Chapter 4 to achieve univariate stationarity. In the time series used by Cowpertwait and Metcalfe's example [30], with Approvals and Activity in Australia, one of the variables has an upward trend and it is possible that the cross-correlogram plot is masked by the smoothness of the trend. So it is worth

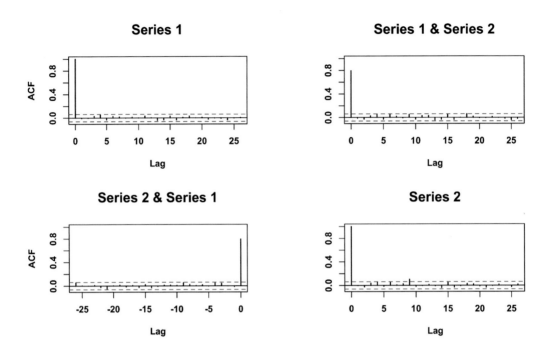

Figure 8.8 Correlation structure of sampled bivariate white noise.

looking at the differenced data, that is, the data after being made stationary by taking the first regular difference, to determine if the model that those authors identified will be the same as the model identified after differencing. Program *ch8ApprovActiv.R* and data *ch8ApprovActiv.dat* are needed for this exercise.

(a) Do a time plot of the first difference of the time series and comment.
(b) Plot the cross-correlogram and identify a model for the first difference.
(c) Cowpertwait and Metcalfe [30] do not difference the data. They look at the decomposed random component. Thus their cross-correlogram looks a little different. Obtain the cross-correlogram of the random component of the decomposed time series and identify a model. Compare the cross-correlogram and the identified models with those obtained for the differenced data. Which one do you think would be easier to handle?

Exercise 8.2   Simulate bivariate white noise using the code given at the bottom of program *ch8ApprovActiv.R*. Do not use set.seed().

(a) Why isn't the image of the cross-correlogram that you obtain identical to that in Figure 8.7? Do the simulated series have cross-correlation at lag 0? Explain.
(b) Repeat the simulation again. Compare the plot of the cross-correlogram that you get now with the ones already plotted and Figure 8.8. Explain the difference if any.
(c) Repeat the simulation several times. What are the differences in the images obtained? What are those differences due to?

Exercise 8.3 We have observed time series y_t denoting total sales (in thousand of cases per month) and x_t, which denotes advertising expenditures (in thousands of dollars per month). There is no need to pretransform the data. The data set is *ch8salesandexpenditures.txt* and can be read into R using the code to do so in program *ch8ApprovActiv.R*. Conduct a cross-correlation analysis of these time series and identify a model for their relationship. If deemed necessary, difference the time series, justifying your choice using the appropriate visual displays.

8.3 Vector Autoregression Models (VAR)

As indicated in Section 8.1, a VAR model is a very particular type of time series model that contains as many equations as variables, and each variable depends on past values of itself and past values of the other variables in the system. All included variables are assumed to be jointly endogenous. The goal in VAR is to model the relationship among all of the variables over time. Forecasting where the system will be in the future is aided by the mutual interrelation assumed by the VAR model. Scenario analysis of the effect of a shock to the system with the impulse response functions can help gain insight into how long the effect of a shock will last. In this section, we will introduce the theoretical representation of VAR models.

Starting from the autoregressive representation of weakly stationary processes, VAR is an extension of the AR models, capturing the historical patterns of each variable and its relationship to the others. Christopher Sims [181] developed VAR as an alternative to traditional simultaneous equations system, an approach to modeling vector time series that has been and still is used by economists.

In a VAR of order p (VAR(p)), each component of the vector variable X depends linearly on its own lagged values up to p periods as well as on the lagged values of all the other variables up to order p. The number of observations available limits the number of lags and the number of variables that can be used, that is, the order p and the number of variables. However, we always seek to identify the smallest possible number of lags, and that is why before specifying a VAR(p) we first observe the cross-correlation function.

With VAR, we can answer how new information that appears at a certain point in time in one variable is processed in the system and what impact it has over time not only for this particular variable but for the other variables of the system. In this context, we study the impulse response function and variance decomposition.

8.3.1 VAR(1) Model Specification, Two Variables

We will consider the more general notation now. Two time series, X_t and Y_t, follow a vector autoregressive process of order 1 (denoted $VAR(p = 1)$) if

$$X_t = \theta_{11}X_{t-1} + \theta_{12}Y_{t-1} + W_{X,t}$$
$$Y_t = \theta_{21}X_{t-1} + \theta_{22}Y_{t-1} + W_{Y,t},$$

where $W_{X,t}$ and $W_{Y,t}$ are bivariate white noise and θ_{ij} are model parameters.

A VAR model with two variables will have two equations, but it can have any order p that we identify. The number of equations in a VAR model is determined by the number of variables that we want to forecast. The order p does not depend on the number of variables.

The model can be written in matrix notation as

$$\mathbf{Z}_t = \boldsymbol{\Theta}\mathbf{Z}_{t-1} + \mathbf{W}_t,$$

where

$$\mathbf{Z}_t = \begin{pmatrix} X_t \\ Y_t \end{pmatrix}, \quad \boldsymbol{\Theta} = \begin{pmatrix} \theta_{11} & \theta_{12} \\ \theta_{21} & \theta_{22} \end{pmatrix}, \quad \mathbf{W}_t = \begin{pmatrix} W_{X,t} \\ W_{Y,t} \end{pmatrix}.$$

Using the backward shift operator, this equation can also be written in reverse form as

$$(\mathbf{I} - \boldsymbol{\Theta}\mathbf{B})\,\mathbf{Z}_t = \mathbf{W}_t,$$

where $\mathbf{I}$ is the 2×2 identity matrix and $\mathbf{B} = \begin{pmatrix} B & 0 \\ 0 & B \end{pmatrix}$. Alternatively,

$$X_t - \theta_{11}X_{t-1} - \theta_{12}Y_{t-1} = W_{X,t}$$
$$Y_t - \theta_{21}X_{t-1} - \theta_{22}Y_{t-1} = W_{Y,t},$$

and then

$$\begin{pmatrix} 1 - \theta_{11}B & -\theta_{12}B \\ -\theta_{21}B & 1 - \theta_{22}B \end{pmatrix} \begin{pmatrix} X_t \\ Y_t \end{pmatrix} = \begin{pmatrix} W_{x,t} \\ W_{y,t} \end{pmatrix}.$$

We can see that

$$\begin{vmatrix} 1 - \theta_{11}B & -\theta_{12}B \\ -\theta_{21}B & 1 - \theta_{22}B \end{vmatrix} = (1 - \theta_{11}B)(1 - \theta_{22}B) - \theta_{12}\theta_{21}B^2.$$

This is a polynomial in B. Analogous to AR models, a $VAR(p)$ model is stationary (stable) if that polynomial has no roots in or on the complex circle. All $|\theta(B)|$ must exceed unity in absolute value. The absolute value of the roots of this or any polynomial is given by the function `polyroot()` with `Mod` in R. But determining the characteristic polynomial can be very cumbersome if the order p of the VAR model is large. Thus, in practice we rely on the equivalent condition that all the eigenvalues of the coefficient matrix obtained when estimating the VAR(p) model have modulus less than 1. We will see in future examples that R produces the eigenvalues when printing a summary of the estimated model.

Example 8.2 Consider the VAR(1) example given in Cowpertwait and Metcalfe's book [30].

$$X_t = 0.4X_{t-1} + 0.3Y_{t-1} + W_{X,t}$$
$$Y_t = 0.2X_{t-1} + 0.1Y_{t-1} + W_{Y,t}.$$

Rearranging,

$$X_t - 0.4X_{t-1} - 0.3Y_{t-1} = W_{X,t}$$
$$Y_t - 0.2X_{t-1} - 0.1Y_{t-1} = W_{Y,t}.$$

Simplifying,

$$\begin{pmatrix} 1 - 0.4B & -0.3B \\ -0.2B & 1 - 0.1B \end{pmatrix} \begin{pmatrix} X_t \\ Y_t \end{pmatrix} = \begin{pmatrix} W_{x,t} \\ W_{x,t} \end{pmatrix}.$$

Then the characteristic equation is given by

$$\begin{vmatrix} 1 - 0.4B & -0.3B \\ -0.2B & 1 - 0.1B \end{vmatrix} = 1 - 0.5B - 0.02B^2.$$

The absolute values of the roots of the polynomial

$$(1 - 0.4B)(1 - 0.1B) - (-0.3B)(-0.2B) = 1 - 0.5B - 0.02B^2 = 0$$

are found as follows:

```
Mod(polyroot(c(1, -0.5,-0.02))),
```

which gives $B_1 = 26.861407$ and $B_2 = 1.861407$, larger than 1 in modulus. From this, we can deduce that the VAR(1) model is stationary or stable since both roots exceed unity in absolute value. This is equivalent to saying that the eigenvalues of the coefficient matrix

$$A = \begin{pmatrix} 0.4 & 0.3 \\ 0.2 & 0.1 \end{pmatrix}$$

are less than 1. The absolute values of the eigenvalues of A are 0.53722813 and 0.03722813. Notice that they are the reciprocals of the roots of the backshift polynomial. ☐

Example 8.3 Suppose we observe vector random variable $\mathbf{X}_t = \begin{bmatrix} X_{1,t} \\ X_{2,t} \end{bmatrix}$, which we want to forecast, and let $\mathbf{W}_t = \begin{pmatrix} W_{1,t} \\ W_{2,t} \end{pmatrix}$ be bivariate white noise with variance-covariance matrix

$$\Sigma_{WW} = \begin{pmatrix} 1 & 0.7 \\ 0.7 & 1.49 \end{pmatrix}.$$

Suppose the model is VAR(p=1):

$$\begin{pmatrix} X_{1,t} \\ X_{2,t} \end{pmatrix} = \begin{pmatrix} 0.6 & -0.3 \\ -0.3 & 0.6 \end{pmatrix} \begin{pmatrix} X_{1,t-1} \\ X_{2,t-1} \end{pmatrix} + \begin{pmatrix} W_{1,t} \\ W_{2,t} \end{pmatrix}.$$

According to the model, X_1 at time t is explained by lagged values of itself and lagged values of X_2 plus a random error term, W_t. Note that neither variable depends on contemporaneous values of the other variable, unlike in structural equation models. The random errors W_1 and W_2 represent everything else that is outside the two-equation system. Since many factors would affect both variables, it is possible that the two errors are correlated.

We can rewrite the model as follows:

$$X_{1,t} = 0.6X_{1,t-1} - 0.3X_{2,t-1} + W_{1t}$$
$$X_{2t} = -0.3X_{1,t-1} + 0.6X_{2,t-1} + W_{2t}.$$

We can rearrange in reverse form:

$$X_{1,t} - 0.6X_{1,t-1} + 0.3X_{2,t-1} = W_{1t}$$
$$X_{2t} + 0.3X_{1,t-1} - 0.6X_{2,t-1} = W_{2t},$$

or

$$\begin{pmatrix} 1 - 0.6B & 0.3B \\ 0.3B & 1 - 0.6B \end{pmatrix} \begin{pmatrix} X_{1,t} \\ X_{2,t} \end{pmatrix} = \begin{pmatrix} W_{1,t} \\ W_{2,t} \end{pmatrix}.$$

More compactly,

$$(\mathbf{I} - \mathbf{AB})X_t = W_t,$$

where I is the identity matrix, $\mathbf{A} = \begin{pmatrix} 0.6 & -0.3 \\ -0.3 & 0.6 \end{pmatrix}$, and B is the backshift polynomial.

To check whether the system is stable, the roots of $|\mathbf{I} - \mathbf{AB}| = 0$ have to be calculated, that is, we have to solve the system:

$$\begin{vmatrix} 1 - 0.6B & 0.3B \\ 0.3B & 1 - 0.6B \end{vmatrix} = 0.$$

The determinant is a polynomial in B:

$$(1 - 0.6B)^2 - (0.3B)^2 = 1 - 1.2B + 0.27B^2 = 0.$$

Alternatively, we could just find the eigenvalues of the matrix A.

> A system is stationary, and therefore stable, if, after a shock, the system eventually returns to the dynamics prevailing before the shock. That will happen if the roots of the characteristic equation are larger than one in absolute value. An equivalent condition is that the absolute value of the eigenvalues of the model's coefficients are all less than one.

We find the roots with R's `polyroot()` function:

```
Mod(polyroot(c(1,-1.2,0.27)))
```

which gives the two roots $10/9$ and $10/3$. These roots are larger than 1 in modulus and therefore the system is stable. Alternatively, we look at the eigenvalues of the matrix **A** using

```
A=matrix(c(0.6,-0.3,-0.3,0.6), ncol=2)
Mod(eigen(A)$values)
```

which gives 0.9 and 0.3 as the absolute values of the eigenvalues. Notice that it is just a coincidence that the A matrix is symmetric. This does not need to be the case.

Notice that when we estimate VAR models for observed time series, R will give us the eigenvalues printed in the summary of the output, but it calls them "roots of the characteristic polynomial," which may lead to confusion. We will revisit this point in the examples where we fit VAR models to data.

The textbook by Cowpertwait and Metcalfe [30], chapter 11, shows how to simulate VAR processes with examples using the function $\mathtt{ar}(\ )$. □

8.3.2   VAR(2) Model Specification, Two Variables

A VAR(1) process can be extended to a VAR(2) or a VAR of any order p larger than two. A VAR model can have any number of variables and any order.

Example 8.4   For example, a VAR(2) can be written as

$$X_t = \theta_{11}X_{t-1} + \theta_{12}X_{t-2} + \theta_{13}Y_{t-1} + \theta_{14}Y_{t-2} + W_{x,t}$$
$$Y_t = \theta_{21}X_{t-1} + \theta_{22}X_{t-2} + \theta_{23}Y_{t-1} + \theta_{24}Y_{t-2} + W_{y,t}.$$

Which can be written as:

$$\begin{bmatrix} 1 - \theta_{11}B - \theta_{12}B^2 & -\theta_{13}B - \theta_{14}B^2 \\ -\theta_{21}B - \theta_{22}B^2 & 1 - \theta_{23}B - \theta_{24}B^2 \end{bmatrix} \begin{pmatrix} X_t \\ Y_t \end{pmatrix} = \begin{pmatrix} W_{x,t} \\ W_{y,t} \end{pmatrix}.$$

The roots of the determinant

$$\begin{vmatrix} 1 - \theta_{11}B - \theta_{12}B^2 & -\theta_{13}B - \theta_{14}B^2 \\ -\theta_{21}B - \theta_{22}B^2 & 1 - \theta_{23}B - \theta_{24}B^2 \end{vmatrix}$$

help determine whether this AR(2) is stationary.

Alternatively, the four eigenvalues of the matrices

$$\begin{pmatrix} \theta_{11} & \theta_{13} \\ \theta_{21} & \theta_{23} \end{pmatrix}, \quad \begin{pmatrix} \theta_{12} & \theta_{14} \\ \theta_{22} & \theta_{24} \end{pmatrix}$$

would have to be less than one in absolute value. □

The time series in a VAR(p) model must be autocorrelated but also cross-correlated for a VAR(p) model to make sense. To identify the VAR(p) model, we use the correlogram and the cross-correlogram.

8.3.3   MA Representation of a VAR(p)

The MA representation of $(I - AB)X_t = W_t$ is

$$\mathbf{X_t} = \frac{I}{I - AB} \mathbf{W_t} = (1 + AB + A^2 B^2 + \ldots\ldots) W_t$$

and it is known as Multivariate Wold Decomposition.

Example 8.5   We refer again to Example 8.3 and will now write its MA representation,

$$\begin{pmatrix} X_{1,t} \\ X_{2,t} \end{pmatrix} = \begin{pmatrix} 0.6 & -0.3 \\ -0.3 & 0.6 \end{pmatrix} \begin{pmatrix} X_{1,t-1} \\ X_{2,t-1} \end{pmatrix} + \begin{pmatrix} W_{1,t} \\ W_{2,t} \end{pmatrix}$$

or, in vector form,

$$\mathbf{X_t} = \mathbf{AX}_{t-1} + \mathbf{W}_t.$$

The Wold decomposition similar to that of a univariate system takes form in this multivariate system as follows (vector notation follows):

$$\begin{pmatrix} X_{1,t} \\ X_{2,t} \end{pmatrix} = \begin{pmatrix} W_{1,t} \\ W_{2,t} \end{pmatrix} + \begin{pmatrix} 0.6 & -0.3 \\ -0.3 & 0.6 \end{pmatrix} \begin{pmatrix} W_{1,t-1} \\ W_{2,t-1} \end{pmatrix}$$

$$+ \begin{pmatrix} 0.45 & -0.36 \\ -0.36 & 0.45 \end{pmatrix} \begin{pmatrix} W_{1,t-2} \\ W_{2,t-2} \end{pmatrix}$$

$$+ \begin{pmatrix} 0.378 & -0.351 \\ -0.358 & 0.378 \end{pmatrix} \begin{pmatrix} W_{1,t-3} \\ W_{2,t-3} \end{pmatrix}$$

$$+ \ldots$$

We can check that $\mathbf{A}^2 = \begin{pmatrix} 0.45 & -0.36 \\ -0.36 & 0.45 \end{pmatrix}$ and $\mathbf{A}^3 = \begin{pmatrix} 0.378 & -0.351 \\ -0.358 & 0.378 \end{pmatrix}$.

$$\mathbf{A}^2 = \mathbf{A}^T \mathbf{A} = \begin{pmatrix} 0.6 & -0.3 \\ -0.3 & 0.6 \end{pmatrix} \begin{pmatrix} 0.6 & -0.3 \\ -0.3 & 0.6 \end{pmatrix} = \begin{pmatrix} 0.6^2 + 0.3^2 & -2(0.6)(0.3) \\ -2(0.6)(0.3) & 0.6^2 + 0.3^2 \end{pmatrix}$$

$$= \begin{pmatrix} 0.45 & -0.36 \\ -0.36 & 0.45 \end{pmatrix}.$$

$$\mathbf{A}^3 = (\mathbf{A}^2)^T \mathbf{A} = \begin{pmatrix} 0.45 & -0.36 \\ -0.36 & 0.45 \end{pmatrix} \begin{pmatrix} 0.6 & -0.3 \\ -0.3 & 0.6 \end{pmatrix}$$

$$= \begin{pmatrix} 0.45(0.6) + 0.36(0.3) & 0.45(-0.3) - 0.36(0.6) \\ 0.45(-0.3) - 0.36(0.6) & 0.45(0.6) + 0.36(0.3) \end{pmatrix} = \begin{pmatrix} 0.378 & -0.351 \\ -0.351 & 0.378 \end{pmatrix}.$$

The interpretation of the final matrix obtained is as follows: the upper left diagonal entry is the effect on X_1 of a shock to X_1; the lower right diagonal entry is the effect on X_2 of a shock to X_2; the upper right off-diagonal entry is the effect on X_2 of a shock to X_1; and the low left off-diagonal entry is the effect on X_1 of a shock on X_2.   □

8.3.4   Exercises

Exercise 8.4   Consider the VAR(1) model

$$X_t = 0.6X_{t-1} - 0.3Y_{t-1} + W_{x,t}$$
$$Y_t = -0.3X_{t-1} + 0.6Y_{t-1} + W_{y,t}.$$

(a) Find the roots of the backshift polynomial.
(b) Find the eigenvalues of the coefficient matrix.
(c) Is this VAR(1) stationary?
(d) What is the relation between the solutions to (a) and (b)?

Exercise 8.5   Consider the following AR(3) model:

$$X_t = \theta_{11}X_{t-1} + \theta_{12}X_{t-2} + \theta_{13}X_{t-3} + \theta_{14}Y_{t-1} + \theta_{15}Y_{t-2} + \theta_{16}Y_{t-3} + W_{x,t}$$
$$Y_t = \theta_{21}X_{t-1} + \theta_{22}X_{t-2} + \theta_{23}X_{t-3} + \theta_{24}Y_{t-1} + \theta_{25}Y_{t-2} + \theta_{26}Y_{t-3} + W_{y,t}.$$

Set up the model in matrix form and write the reverse form, leaving only the W terms on the right. Indicate the matrix whose determinant will help us find the roots of the characteristic polynomial. Indicate also the matrices for which we would have to find the eigenvalues as an alternative to determining the stationarity or stability of the model.

8.4   Applying VAR Models

In this section, we will discuss how to estimate VAR models of order p in practice with R and how to forecast and obtain impulse response functions with them.

8.4.1   Estimation

One thing to notice in VAR models is that since all equations have the same independent variables, we can estimate the parameters independently by Ordinary Least Squares (OLS), one equation at a time, although other methods have been used [94, 60].

We will continue our practice of fitting the model to training data and compare the forecast to test data. The goal is to identify a stationary model that fits the training data well and gives forecasts close to the test data. Validating the model with the test data is necessary in order to feel more confident that the forecasts of the future for which we do not have data are going to be accurate.

We will use the library `vars` in R [151, 152]. The reader should follow the code as we go over the first example.

Example 8.6   For this example, we will use the quarterly data set *USeconomic*, which comes with the `tseries` package of R. This data set has four seasonally adjusted time series recorded for the period 1954–87, but we will be interested in the log of Gross National Product, log(GNP), and the log of Money Supply, log(M1), and will

The data appear nonstationary

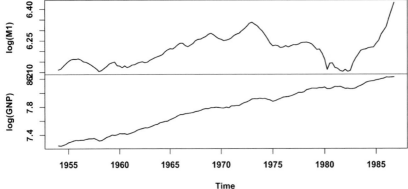

Figure 8.9 These time series will be fitted a VAR(p=2) model

Table 8.1 Output of the VAR model of change in log GNP (GNP) and change in log M1 (M1). Significance codes: $0^{***}, 0.001^{**}, 0.01^{*}, 0.05., 0.11$

GNP				
	Estimate	Std. Error	t-value	Pr($> |t|$)
GNP.l1	0.305	0.088	3.449	0.000767***
M1.l1	0.231	0.088	2.630	0.009606**
GNP.l2	0.240	0.085	2.821	0.00556**
M1.l2	0.056	0.095	0.596	$7.031e - 01$

M1				
	Estimate	Std. Error	t-value	Pr($> |t|$)
GNP.l1	−0.125	0.092	−1.362	0.176
M1.l1	0.552	0.091	6.057	0.000***
GNP.l2	0.093	0.088	1.056	0.293
M1.l2	0.139	0.098	1.416	0.159

fit a VAR(p=2) model using the VAR () function from the library vars. Because we have two variables, the VAR model will have two equations. Program *chapter8var1.R* part I contains the program to estimate the parameters of the model. We split the data into a training set containing the first quarter of 1954 to the fourth quarter of 1986 and a test set containing data for the four quarters of 1987. We will validate the model with the test set. The time plot of the training set can be viewed in Figure 8.9. Both time series are nonstationary and have been differenced before estimating the model. Thus, the name GNP after differencing is the "change in log GNP," and the name M1, after differencing, "means change in the log of M1."

The estimated model coefficients are given in Table 8.1. Notice that the order in which they are presented here could be different from the order of the results when running the R program.

The variable labeled GNP is the change in log of GNP. GNP.l1 means change in log of GNP at time $t - 1$; M1.l1 means change in log M1 at $t - 1$, and so on. The R output is a block of results for the first equation and a block of results for the second equation. A summary of the estimation will also write the linear models estimated, each on top of its model coefficients. Translating the R output into a regression-like set of equations, where a coefficient with a star exponent means that the coefficient is statistically significant at the 5 percent level of significance (p-value less than 0.05), or lower, we get

$$\widehat{GNP}_t = 0.305GNP^*_{t-1} + 0.231GNP^*_{t-2} + 0.240M1^*_{t-1} + 0.055M1_{t-2}$$
$$\widehat{M1}_t = -0.125GNP_{t-1} + 0.093GNP_{t-2} + 0.552M1^*_{t-1} + 0.139M1^*_{t-2}.$$

According to this model, the change in log of GNP at quarter t is significantly affected by itself at $t - 1$ and $t - 2$ and by change in log M1 at quarter $t - 1$. But the change in the log of M1 is not significantly affected by past change in log GNP at all. Rather, change in log M1 is significantly affected only by the last quarter values of itself. The coefficient results lead one to conclude, if we were to adopt this model, that a change in log M1 will affect change in log GNP, making M1 a leading indicator, a term used for macroeconomic variables that trigger a reaction in the economy when they change. For example, if housing construction starts going up, that is a leading indicator of improvement in the economy. In fact, M1 increases are used by the Federal Reserve in the United States (the Fed) as a signal that interest rates must be decreased to boost investment and that way improve the economy. So the results make sense.

The residuals from this model are bivariate white noise. Thus, we tentatively adopt this model for forecasting and simulating response to external shocks. □

8.4.2 Determining Whether the Model is Stable

As indicated in Section 8.3, it is important to determine whether the model is stable or stationary. In order to do that, we presented there two alternative ways to check: on the one hand, we could look at solutions for the backshift polynomial for the model; on the other hand, we could check the eigenvalues of the coefficient matrix. The backshift polynomial for the estimated model of change in log GNP and change in log M1 is a little tedious to find and it is not produced by the R output. Instead, the R output gives the eigenvalues of the coefficients matrix. Those need to be less than 1 and they are, indicating that the model is stable.

Eigenvalues are called roots of the characteristic equation used to find them, hence the name "roots," not to be confused with the roots of the backshift polynomial, which the R results lead to deduce are larger than 1.

A direct way to obtain the eigenvalues of the coefficients matrix resulting from a VAR(p) model fitting is roots(VARobject), where VARobject is the name

Table 8.2 R output for the forecast obtained with the model of Example 8.6.

GNP

fcst	lower	upper	CI	test data
0.0121	−0.0076	0.0317	0.0197	0.01118
0.0123	−0.0091	0.0337	0.0214	0.0122
0.0121	−0.011	0.0351	0.0231	0.0110
0.010	−0.0134	0.0345	0.024	0.01481

M1

fcst	lower	upper	CI	test data
0.0247	0.0043	0.0452	0.0204	0.01901
0.0175	−0.0057	0.0407	0.0232	0.0046
0.0127	−0.0120	0.0374	0.0247	−0.0074
0.0090	−0.0163	0.0345	0.0254	0.0006

given to the VAR model fitting results. The eigenvalues can also be seen at the top of the output when using summary (VARobject).

Example 8.7 The R part of the output for Example 8.6 obtained by using sum-mary (US.var) that indicates the eigenvalues of the coefficients matrix is:

Roots of the characteristic polynomial:

0.8346   0.7868   0.5939   0.5939   0.4435   0.05525

Those values are a little rounded. The same values can be obtained using roots (US.var). The reader may want to take another look at program *chapter8var1.R* to see the sequence of R statements leading to this. □

8.4.3   Forecasting

Forecasting with a VAR model should be done if the residuals of the fitted model are multivariate white noise. If that is the case, the predict () function when applied to our VAR object will give the forecasts. After applying the function to the model estimated in Example 8.6, we obtain the forecast for the time interval of the test data in that example seen in Table 8.2. In order to compare the forecast with the test data, the test data is also included as a column (rounded).

In Table 8.2, the forecast (out-of-sample) is denoted by fcst, and the lower and upper bands of the CI for the future quarter are given. The values under the column headed by CI are the margin of error of the forecast. The interval can be interpreted as saying that we are 95 percent confident that the true future value of the time series lies within the interval. Comparing the forecast with the actual values of the test data

Orthogonal Impulse Response from GNP

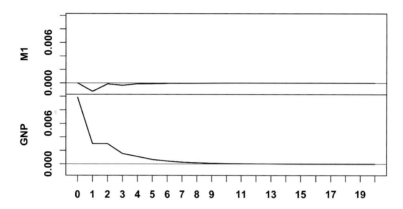

Figure 8.10 Reaction of the system to a shock to GNP. The system restores its usual dynamics after approximately seven quarters, with M1 restoring first.

and looking at the intervals, we can see that all future actual values of the test data fall within the intervals. That is a good sign.

8.4.4   Impulse Response Functions

As indicated earlier, the impulse response functions allow us to see how the estimated VAR system would react to external shocks. In practice, we like to look at the reaction of the system to a shock of each variable separately and see how long it takes for the system to settle back to its original functioning, as expressed by the estimated VAR model. We will study the impulse response functions of the system estimated in Example 8.6.

Example 8.8   Figures 8.10 and 8.11 show the response to such a shock to either variable.

We can see in those images that a shock to the money supply, such as, for example, some monetary policy, has a lasting effect on GNP and the M1 itself, taking approximately 11 quarters to restore the usual dynamics, an indication of which is the curves that lie on the zero line (no more reaction to the initial change). □

We will now look at another example, with more variables, to indicate to the reader the potential of VAR models for more than two variables.

8.5   VAR Models for More than Two Time Series

VAR models are not limited to bivariate time series. They can be extended to higher-dimensional time series of three, four or more variables. Having more variables means having more model equations. We illustrate that with another example.

Orthogonal Impulse Response from M1

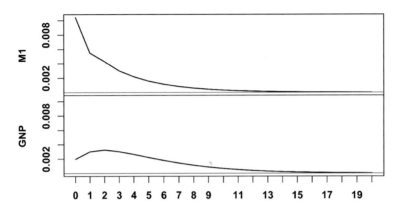

Figure 8.11 Reaction of the system to a shock to M1. The system restores its usual dynamics after approximately 11 quarters.

Example 8.9 The R code for this example is in program *chapter8var2.R*. The example uses the Canada time series data set (which can be loaded from within the vars [151] package with the command data(Canada). There are four quarterly time series ranging from the first quarter of 1980 to the last quarter of 2000: labor productivity (*prod*) is defined as the log difference between GDP and employment; *e* is the log of employment; *U* is the unemployment rate and *rw* is real wages, defined as the log of the real wage index. The data is taken from the OECD data base and spans from the first quarter 1980 until the fourth quarter of 2000.

We split the data into a training set ranging from 1980, quarter 1, to 1995, quarter 4. The test data is from first quarter of 1996 to last quarter of 2000. This leaves five years or 20 quarters.

Using the VAR function in the vars package, we fit a multivariate VAR model to the four economic variables in the training set of the Canadian data. A time plot of the four time series in the training set can be seen in Figure 8.12. We realize, looking at the time plot, that the variables have a trend or long-term cyclical behavior, all signals of nonstationarity. The sample ACFs and CCFs shown in Figure 8.13 confirm the nonstationarity in some variables. We do a first regular difference of all the time series in order to be able to see some features that might help us identify a model. Differencing economic variables and measuring their change instead of their level is standard, as often we are more interested in whether there is change than in the actual volume, which is always affected by population growth. Figure 8.14 indicates that the first-order regular difference makes all the time series stationary.

We tentatively identified a VAR(3) and fit that model. The residuals of this model are multivariate white noise. However, values at lag $t-3$ are not statistically significant, indicating that the lag 3 variables are not really needed. Since parsimonious models are always more desirable than large models, we opted for a VAR(2) model for the

Time series in the Canada data set

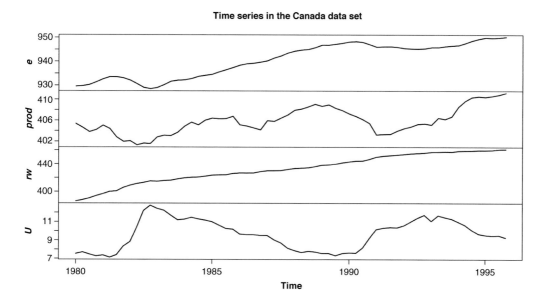

Figure 8.12 The time plot of the Canadian economic data reveals nonstationarity due to long-term trends or long-term cyclical behavior.

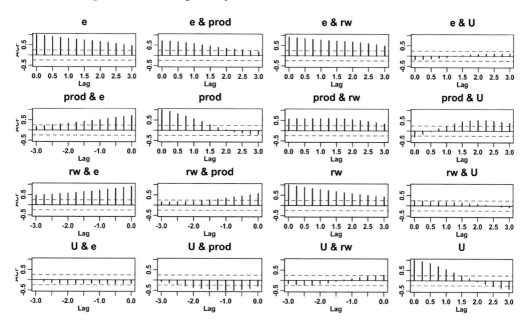

Figure 8.13 The cross-correlation and autocorrelation plots reveal nonstationarity

change in the variables (i.e., the regularly differenced variables) and fitted this model to the training set.

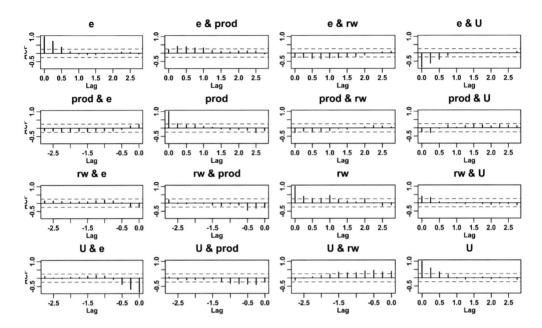

Figure 8.14 The sample cross-correlations and autocorrelation functions of the differenced time series in the Canada data set. We can see that the first difference, the change in the variables, results in stationary time series.

The VAR(2) model gives white noise residuals as indicated by Figure 8.15, so we look at it more closely. The VAR(2) model has four equations and is:

$$\hat{e}_t = -1.36^* + 1.65e^*_{t-1} + 0.167prod^*_{t-1} - 0.631rw_{t-1} + 0.26U_{t-1}$$
$$- 0.49e^*_{t-2} - 0.1prod_{t-2} + 0.003rw_{t-2} + 0.1U_{t-2}$$
$$\widehat{prod}_t = -166 - 0.17e_{t-1} + 1.15prod^*_{t-1} - 0.05rw_{t-1} - 0.47U_{t-1} + 0.38e_{t-2}$$
$$- 0.17prod_{t-2} - 0.11rw_{t-2} + 1.01U^*_{t-2}$$
$$\widehat{rw} = -33 - 0.26e_{t-1} - 0.08prod_{t-1} + 0.9rw^*_{t-1} + 0.02U_{t-1} + 0.36e_{t-2}$$
$$- 0.005prod_{t-2} + 0.05rw_{t-2} - 0.12U_{t-2}$$
$$\hat{U} = 149 - 0.5e^*_{t-1} - 0.07prod_{t-1} + 0.01rw_{t-1} + 0.4U^*_{t-1} + 0.05e^*_{t-2}$$
$$+ 0.04prod_{t-2} + 0.04rw_{t-2} - 0.07U_{t-2},$$

where the $*$ indicates whether the variables significantly affect the contemporaneous value of the variable on the left-hand side of the equal sign.

As we can see, e is affected by e and $prod$, $prod$ is affected by itself and U (so $prod$ leads e), rw is affected by rw only and U is affected by e and U (e leads U).

Using the fitted VAR model we forecast the values in the test set period. Adding these predictions to a time series plot of each variable, and visualizing the forecast and the test data together, we can see what the out-of-sample predictions are. Figure 8.16 shows the forecasts obtained for each of the four differenced series with the VAR(2) model. We can see that whereas in the short term the forecast seems reasonable, the

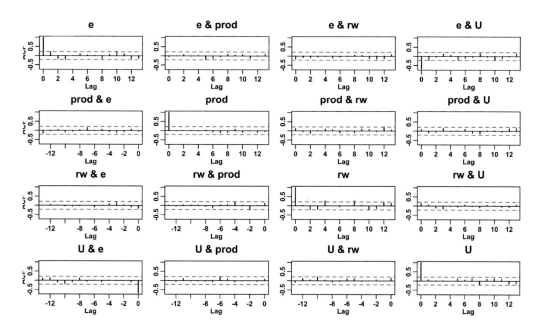

Figure 8.15 The residuals of the VAR(2) model fitted to the Canada data set are the residuals of multivariate white noise.

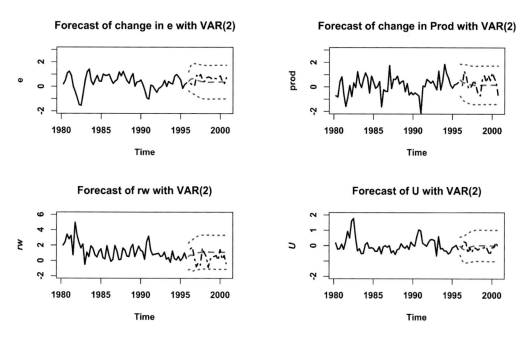

Figure 8.16 Forecasts (in red color) and prediction intervals (in blue color) of the differenced variables are close to the true values (in black color) of the differenced series in the test set when the forecast is short term, but the forecasts tend to be biased the longer the forecasting period is.

forecast departs from the true data as the step ahead gets farther. Although the data is inside the prediction intervals, the forecast tends to be biased as it gets farther from the training set. Short-term forecasts are better than long-term forecasts.

Of course, we could keep playing with this and compute the root mean square error of the forecast for the VAR(2) model. When comparing with other models, that is important.   □

8.5.1   Impulse Response Functions

When there are more than two time series, we have the possibility of more than two sources of shocks to a system. There could be a shock coming via any of the time series. Regardless of where the shock comes from, we will have one impulse response function for each of the variables in the model. We will demonstrate this with the impulse response functions for the VAR model fitted to the Canada data set.

Example 8.10   Besides forecasting, one of the main uses of a VAR(p) is to study the dynamic reaction of the whole estimated system to a shock in one of the variables with IRF. This is a dynamic view of how all the variables in the VAR model change over time as a shock is inflicted on one of them, under the estimated model assumption. The `irf()` function in R makes that very easy to find.   □

Example 8.11   With the model that we fitted to the Canadian data in Example 8.9, we find the impulse response functions for all the variables after exerting a shock to variable e. Then we apply a shock to each of the other variables, one at a time, and study the response of the other variables. We will also describe what we find out and sketch the impulse response functions. The R code to do this is in program *chapter8var2.R*

As we can see in Figure 8.17, a shock in the change in employment leads in the short term to a decrease in the change in productivity, a rise in change of real wages, simultaneous with a decrease in change of unemployment and eventually reverses movement of all of the variables. It takes about 40 quarters (10 years) for the effect of the shock to disappear completely in the four series. This is consistent with our earlier finding that *prod* leads e.

An increase in the change in real wages leads very quickly to an increase in the change in unemployment, a decrease in the change in employment, decrease in change of productivity, and within a year, decrease in real wages. In about four years (20 quarters), the effect of the shock notably disappears. See Figure 8.18.

We will leave it as an exercise for the reader to plot the impulse response functions corresponding to a change in the other two variables and to interpret them.   □

8.6   Automatic Fittings

Suppose we observe two time series, x_t, y_t. The R command `xy.ar = ar(cbind (x, y))` will fit a best VAR(p) model decided by the AIC criterion. For example, when fitted to some data, we obtained this output from xy.ar:

```
        x      y
   x  0.399  0.321
   y  0.208  0.104
```

Letting the function `ar( )` select your model automatically for you results in very large models usually. It is always advisable to check the cross-correlogram before fitting a well-identified model without the help of the automatic selectors. This practice will result in a a smaller model if we take into account that 5 percent of the auto- and cross-correlations will not be real.

However, when we use this automatic model selection procedure with the differenced Canada's training set, we are told that the order is p=2, which is what we selected upon using the sample CCF.

8.6.1   VAR Model Selection with `tsibble` Data Class

Hyndman and Athanasopoulos [76] contains a section that introduces the execution of VAR estimation using time series data of class `tsibble`. They bring up the important point that model selection using the AIC will result in larger than needed model order p. For that reason, they created a function that will make the model selection using the BIC.

The reader might want to reproduce their analysis in Section 12.3 of their book.

8.6.2   Exercises

Exercise 8.6   Using program *chapter8var2.R*, select automatically a VAR model for the `Canada` data set that was studied in Section 8.5 without differencing it. Investigate the white noise property of the residuals and show the impulse response functions and interpret them if applicable.

Exercise 8.7   In Example 8.6, we differenced the time series and we deduced by analyzing the IRF that the estimated system was stable. Fit a VAR(2) model to the undifferenced time series, check the residuals and obtain the impulse response functions after a shock in M1 and after a shock in GNP. Compare the IRF obtained with the ones obtained with the differenced time series.

Exercise 8.8   Convert the Canada data set from `mts` data class to `tsibble` data class and use the two functions provided by Hyndman and Athanasopoulos [76] to select automatically a VAR model for the four variables. Is there any difference in the order p of the model proposed by the AIC and the BIC? Check the multivariate residual's ACF and CCF in each case and determine whether they are multivariate white noise in both case.

Orthogonal Impulse Response from e

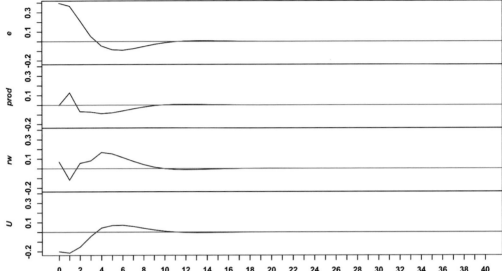

Figure 8.17 Dynamics of the Canadian economy after a one time shock to change in employment.

8.7   Impulse Response Functions (IRF)

The understanding of the dynamic response of a system like a VAR to a small shock is of considerable interest in Engineering, Biology, Climate analysis, Economics and many other areas. In this section, we introduce more formally this very important tool called impulse response functions of which we have seen several examples throughout this chapter.

The IRF of a system helps us to visualize and gain intuitive understanding of that dynamic response. We will now give an intuitive understanding of how IRFs arise by using a one-equation model. We will then see how to produce IRF curves for multiple variables using the `var` package in R.

8.7.1   Wold Decomposition Theorem

According to Wold Decomposition Theorem, any mean 0, covariance-stationary time series process can be written as an infinite sum of weighted random shocks. Let $\{X_t\}$ be a time series from which we subtracted the mean. Then,

$$X_t = W_t + \psi_1 W_{t-1} + \psi_2 W_{t-2} + \cdots = W_t + \sum_{j=1}^{\infty} \psi_j W_{t-j}.$$

For example, the decomposition of an AR(1):

$$Y_t = \alpha Y_{t-1} + W_t = \alpha[\alpha Y_{t-2} + W_{t-1}] + W_t = \alpha^2 Y_{t-2} + \alpha W_{t-1} + W_t.$$

Orthogonal Impulse Response from *rw*

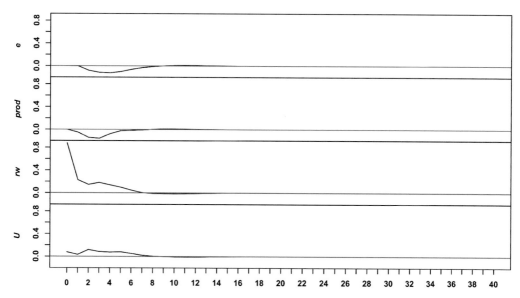

Figure 8.18 Effect of a shock on real wages on the Canadian economy.

If we keep substituting for the *y* terms using the AR(1) model,

$$Y_t = \alpha^2[\alpha Y_{t-3} + W_{t-2}] + \alpha Y_{t-1} + W_t = \alpha^3 Y_{t-3} + \alpha^2 W_{t-2} + \alpha W_{t-1} + W_t,$$

and so on. Substituting, we would end up with an infinite sum of random shocks weighted by functions of the model coefficient α. The weights near each W term are the impulse responses $\psi_j, \ j = 1, \ldots, \infty$.

We can consider any stationary time series model as a linear filter subject to a sequence of random shocks.

8.7.2   IRF for One-Equation Model

How a process reacts to a single shock at time *t* provides us with important information about how the shock propagates throughout the system, and what effect it has, over time.

Example 8.12   Assume an AR(2) univariate stationary process modeled as

$$X_t = 0.9824X_{t-1} - 0.3722X_{t-2} + W_t,$$

where W_t is where random shocks will be inflicted at a given time *t*, and X_t has the mean removed. If we inject a random shock of size one standard error at time *t*, the system will react, but after some times it will eventually return to its original dynamics represented by the model. By how much will it react and for how long? Knowing the answer to that helps us understand the effects of shocks on an otherwise very stable system.

Table 8.3 Impulse response function of a one-time shock to an AR(2) stochastic process.

time(t)	w_t	ψ_j
-2	0	0.0000
-1	0	0.0000
0	1	1.0000
1	0	0.9824
2	0	0.59291
3	0	0.21683
4	0	-0.00767
5	0	-0.08824
6	0	-0.08383
7	0	-0.04951
8	0	-0.01744
9	0	0.00130
10	0	0.00776
11	0	0.00715
12	0	0.00413
13	0	0.00140
14	0	-0.00016
15	0	-0.00068

The answer can be known by studying the ψ's mentioned in Section 8.7.1, also known as the impulse response function.

The IRF can be computed easily with a spreadsheet program directly from the autoregressive model given.

Suppose we want to simulate that our system is hit by a simple shock at time $t = 0$, $w_0 = 1$ and $w_t = 0$ for $t > 0$. Suppose also that the system is at rest before time $t = 0$. That means that $x^*_{t-2} = 0$ and $x^*_{t-1} = 0$, where x^*_t denotes change in the system at time t. Then,

$$x^*_0 = 0.9824x^*_{0-1} - 0.3722x^*_{0-2} + w_t = 0.9824(0) - 0.3722(0) + 1 = 1$$
$$x^*_1 = 0.9824x^*_{1-1} - 0.3722x^*_{1-2} + w_t = 0.9824(1) - 0.3722(0) + 0 = 0.9824$$
$$x^*_2 = 0.9824x^*_{2-1} - 0.3722x^*_{2-2} + w_t = 0.9824(0.9824) - 0.3722(1) + 0 = 0.59291$$

. . . .

The numbers obtained on the extreme right-hand side are the impulse responses at time t, the ψ's. Continuing the computations, we obtain Table 8.3. $\square$

To show the dynamics plotted, we simulate the process, or just plot the numbers given against the lags. We do that with program *chapter8manirf.R*, part I.

Looking at Figure 8.19, we can see that the shock at time $t = 0$ of size $w_0 = 1$ standard error causes a response by the system that lasts 10 time periods. At around $t = 10$ the system stops propagating the initial shock.

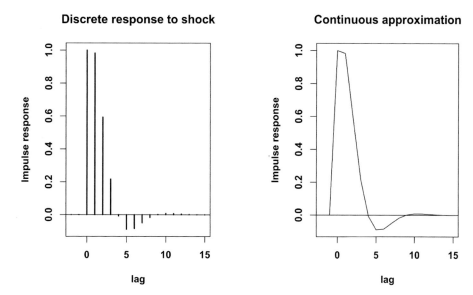

Figure 8.19 Impulse response function of $x_t = 0.9824x_{t-1} - 0.3722x_{t-2} + w_t$, after a shock of one unit at time $t = 0$.

8.7.3 Exercises

Exercise 8.9 Change the parameters of the model in Example 8.12 to $\alpha_1 = 0.2$ and $\alpha_2 = -0.8$ and observe the new dynamic reaction to a shock injected on w_t at time $t = 0$. Suppose $x_{t-1} = x_{t-2} = 0$. Complete Table 8.4 with the new impulse response function, showing work at each line. Plot the impulse response function, as in Figure 8.19: a plot for the discrete response and a separate plot for the continuous approximation. Compare the new impulse response table with the one given in Table 8.3 and comment on how long it takes the new system to rest after the shock. Include your table, your plot and your comments. What property of the process, or some reasonable explanation, is explaining the behavior of the IRF?

Exercise 8.10 What is the difference between forecasts, simulation and impulse response function? Explain using the following AR(2) model:

$$x_t = 0.9824x_{t-1} - 0.3722x_{t-2} + w_t.$$

8.7.4 Implementing More Than One Shock to a System

It is possible for a system to be subject to more than one shock, each shock happening at different points in time. Suppose we estimated the following model:

$$x_t = 0.2x_{t-1} - 0.8x_{t-2} + w_t.$$

We implement a shock of size one standard error at time $t = 0$, another shock of size one standard error at time $t = 5$ and a shock of size one standard error at time $t = 8$. We do the computation of the table and the plot of the two impulse response functions.

Table 8.4 Complete this table according to the instructions given in Exercise 8.9.

time(t)	w_t	ψ_j
−2	0	
−1	0	
0	1	
1	0	
2	0	
3	0	
4	0	
5	0	
6	0	
7	0	
8	0	
9	0	
10	0	
11	0	
12	0	
13	0	
14	0	
15	0	

We can see the IRF of the aggregate three shocks to the system in Figure 8.20, and the calculations up to $t = 15$ are in Table 8.5. The calculations for $t = 16 - 20$ are left as an exercise. Notice how the amplitude of the oscillations has increased, going beyond values of 1 and −1 and how the additional shocks have delayed the tendency to resting behavior beyond lag 15. We do the plot with program *chapter8manirf.R*, part II.

8.7.5   Exercises

Exercise 8.11   Add a few more steps to Table 8.5 by completing the empty lines for $t = 16, 17, 18, 19, 20$.

8.8   Spurious Relations, Stochastic Trends, Unit Roots and Cointegration

As in multiple regression, in multivariate time series models we also approach our results with caution and question whether causality is warranted or not.

> The exception to having to difference the time series in a VAR if they are nonstationary is when they are cointegrated series.

Table 8.5 Impulse response function of an AR(2) model with three shocks.

time(t)	w_t	ψ_{j1}	w_t	ψ_{j2}	w_t	ψ_{j3}	ψ_{total}
-2	0	0	0	0	0	0	0
-1	0	0	0	0	0	0	0
0	1	1	0	0	0	0	1
1	0	$0.2(1) - 0.8(0) = 0.2$	0	0	0	0	0.2
2	0	$0.2(0.2) - 0.8(1) = -0.76$	0	0	0	0	-0.76
3	0	-0.312	0	0	0	0	-0.312
4	0	0.5456	0	0	0	0	0.5456
5	0	0.35872	1	1	0	0	1.35872
6	0	-0.3647	0	0.2	0	0	-0.1647
7	0	-0.3599	0	-0.76	0	0	-1.1199
8	0	0.2198	0	-0.312	1	1	0.9078
9	0	0.3319	0	0.5456	0	0.2	1.0775
10	0	-0.1095	0	0.3587	0	-0.76	-0.5107
11	0	-0.2874	0	-0.3674	0	-0.312	-0.9641
12	0	0.0301	0	-0.3599	0	0.5456	0.2158
13	0	0.2359	0	0.2198	0	0.3587	0.8145
14	0	0.02312	0	0.3319	0	-0.3674	-0.0097
15	0	-0.1841	0	-0.1095	0	-0.3599	-0.6535
16							
17							
18							
19							
20							

When doing vector autoregression, how can we be sure that we are not fitting spurious regressions, that is, concluding that there are relations between variables that do not really exist? After all, if two variables have trends, it is not uncommon that they will appear to be highly correlated even when we know that they cannot possibly be related. Conducting policy changes and interventions based on conclusions of a VAR model when the relations are not real could have devastating consequences. For example, if it was not true that global warming causes the rise of the waters in the Netherlands, moving the seaside towns inward might cause unnecessary hardship to people in the Netherlands. This problem prompted statisticians to diagnose spurious relations using several well-known tests. These tests do not have much power, but they are still widely used.

The term spurious regression is used to characterize regression models where the variables are not in fact related but the regression model or correlation coefficients say they are.

The spurious regression could be due to confounders common to both time series, or to both having stochastic trends.

> The only time that we could do a VAR of nonstationary time series without previously differencing variables that have trends in mean or in seasonality is when the variables are cointegrated. That is, when the time series share common stochastic trends, VAR can be fitted without prior differencing.

In time series data, spurious relations could appear under two very different disguises.

- **Spurious regression due to coincident stochastic trends just by chance**
  Many time series, in particular financial time series, can be approximated by a random walk model, and although not being related at all, appear very correlated among themselves because they have coincident "walks," namely coincident stochastic trends. Thus, even though the variables may not be related, they appear spuriously related. Program *chapter8manirf.R* Part III allows the reader to confirm that by doing Exercise 8.12.
  Stochastic trends are a feature of an *ARIMA* process with a unit root (i.e., $B = 1$ is a solution of the characteristic equation), which is the case in a simple random walk process, but could also be the case in a very complicated ARIMA process. Cowpertwait and Metcalfe [30] illustrate this by simulating two independent random walks. The values of two independent simulated random walks plotted against each other are correlated because each of them has a stochastic trend and the trends of the two could be coincident by chance. Stochastic trends are common in economic series, and so the phenomenon described could be present.

- **Spurious relation due to confounders**
  Time series that are not random walks, may also appear to be related, when in fact they are not. They both could be related due to a confounder, for example, population growth, which affects both of them and gives rise to an underlying trend. Or similar seasonal behaviors could give rise to an underlying seasonal trend. This is to be watched for if the time series have trends or seasonality. For example, the Australian electricity and chocolate production series, which have correlation of 0.96, are correlated because they share an increasing trend due to an increasing Australian population [30]. The high correlation of 0.96 does not imply that the electricity and chocolate production variables are causally related. Rather, it is only due to the population growth. Although we can fit a regression of one variable as a linear function of the other, with added random variation, the regression model is a spurious regression because of the lack of any causal relationship. In this case, it would be far better to regress the variables on the Australian population.

Caution is needed before concluding that two series with stochastic trends present a spurious relation. For example, consider the daily exchange rate series for UK pounds, the Euro, and New Zealand dollars, given for the period January 2004 to December 2007 (all per US dollar). Cowpertwait and Metcalfe's book, Fig. 11.4, shows these data in a scatterplot of the UK and EU exchange rates. Both rates are per US dollar. As their book shows, the correlogram plots of the differenced UK and EU series indicate that

IRF of 3 shocks to $x_t = 0.2\,x_{t-1} -$
$0.8x_{t-2} + w_t$

Continuous approximation

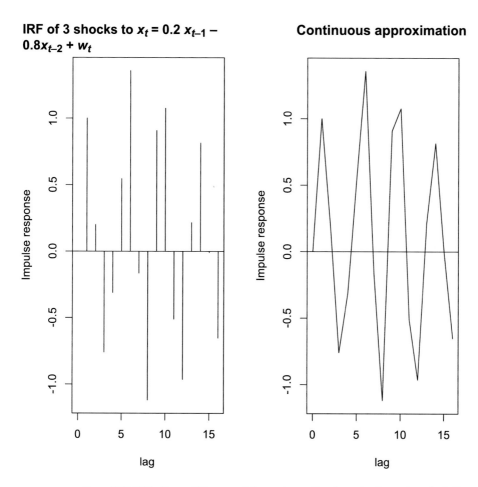

Figure 8.20 IRF of $x_t = 0.2x_{t-1} - 0.8x_{t-2} + w_t$ after three shocks at times 1, 5, 8.

both exchange rates can be well approximated by random walks, while the scatterplot of the rates shows a strong linear relationship, which is supported by a high correlation of 0.95. It is possible that the proximity of the two economic entities geographically and financially make them actually related. In that case, we say that those time series with unit roots are cointegrated. Thus time series models with unit roots could very well be reflecting a causal relation between the variables.

8.8.1 Exercises

Exercise 8.12 Use program *chapter8manirf* part 2, to generate two random walks.

(a) Compile the function and run it 20 times without using a seed. How many of the 20 times did you get a correlation coefficient larger than 0.7 in absolute value? Write the values you obtained.

(b) Do you think that the variables are really correlated? Add code to the program to find that out.

(c) Save the last x and y generated.

8.8.2   Unit Roots Checking

What to do before doing vector autoregression of nonstationary series to guarantee that we are not just getting a spurious regression? A common practice in financial and economic circles is to

- Check for unit roots.
- Check common stochastic trends (cointegration).
- If the time series have unit roots, it only makes sense to do a VAR of unit root series that are cointegrated. If they are not cointegrated, then we should make the time series stationary before we apply VAR.

Unit Root Tests

There are some well-known tests to determine if there are unit roots in time series.

- Dickey and Fuller [36] developed a test of the null hypothesis that $\alpha = 1$ against an alternative hypothesis that $\alpha < 1$ for the model $X_t = \alpha X_{t-1} + W_t$ in which W_t is white noise.
- A more general test, known as the Augmented Dickey–Fuller (ADF) test and developed by Said and Dickey [168], allows the differenced series to be any stationary process rather than white noise, and approximates the stationary process with an AR model.
- The method is implemented in the adf.test function of the tseries library [191].
- If the observed time series x_t is well modeled by a random walk, the null hypothesis of a unit root will not be rejected. The unit root test consists of testing the following null and alternative hypotheses:

$$H_o : \alpha = 1$$
$$H_a : \alpha < 1.$$

The unit root test is for an AR(1) model. The more general Augmented Dickey–Fuller test is for more general models with a unit root. Program *chapter8manirf.R* contains the code to do the augmented test. A p-value of 0.99 would lead to the conclusion that we cannot reject the null hypothesis that $\alpha = 1$, that there is a unit root and the time series has random walk behavior.

Example 8.13   An Augmented Dickey–Fuller test was performed for a time series of precipitation. The results were displayed as follows by R:

```
data: prec.ts
Dickey-Fuller= -7.3586, Lag order=6, p-value=0.01
alternative hypothesis: stationary.
```

We may deduce from this output that we reject the null hypothesis that there is a unit root in the process assumed for the time series and conclude that there is evidence that the time process is stationary. This makes sense since the time series is seasonal but otherwise mean stationary. Unit roots refer to stochastic trends.   □

8.8.3   Exercises

Exercise 8.13   Generate two random walks x_t and y_t again, using program *chapter8manirf.R*, part III.

(a) Conduct a unit root test to determine whether the time series is stationary or not.
(b) Find the correlogram of the first difference of x_t and of y_t. What is your conclusion after seeing the correlogram of the differenced series?
(c) Which would you trust more, the ACF of the differenced data or the unit root test? Explain.

8.8.4   Cointegration

Two nonstationary time series $\{X_t\}$ and $\{Y_t\}$ are cointegrated if some linear combination $aX_t + bY_t$, with a and b constant, is a stationary series.

Example 8.14   Consider a random walk μ_t given by $\mu_t = \mu_{t-1} + W_t$, where W_t is white noise with zero mean, and two series X_t and Y_t given by $X_t = \mu_t + W_{X,t}$ and $Y_t = \mu_t + W_{Y,t}$, where $\{W_{X,t}\}$ and $\{W_{Y,t}\}$ are independent white noise series with zero mean.

Both series are nonstationary, but their difference $\{X_t - Y_t\}$ is stationary since it is a finite linear combination of independent white noise terms. Thus the linear combination of $\{X_t\}$ and $\{Y_t\}$, with $a = 1$ and $b = -1$, produced a stationary series, $\{W_{X,t} - W_{Y,t}\}$. Hence $\{X_t\}$ and $\{Y_t\}$ are cointegrated and share the underlying stochastic trend $\{\mu_t\}$.   □

Example 8.15   Suppose we are interested in the relationship between total consumption (C) and Disposable Income (DI),

$$C_t = \beta_0 + \beta_1 DI_t + \epsilon_t.$$

Suppose that we find that both C_t and DI_t are both nonstationary, I(1), series. Then

$$\epsilon_t = C_t - \beta_0 - \beta_1 DI_t$$

could also be nonstationary, I(1).

Should we proceed with the least squares estimation of this model? The answer is NO.

$$\varepsilon_t = \varepsilon_{t-1} + w_t,$$

where w_t is stationary. There is a unit root in the error model, and least squares will not be good. It will give highly significant relation between C_t and DI_t even when $\beta_1 = 0$. This is spurious regression.

But C_t and DI_t being I(1) does not always imply that ε_t is I(1). It could be I(0), it could be stationary. In this last case, we say that C_t and DI_t are cointegrated.

Cointegrated means that the two variables have similar stochastic trends. They have a long-term equilibrium relation $C_t = \beta_0 + \beta_1 DI_t + \varepsilon_t$ and ε_t is just a short-term deviation from that equilibrium.

If C_t and DI_t are cointegrated, least squares gives estimators with good statistical properties. □

8.8.5   Tests for Cointegration

Two series can be tested for cointegration using the Phillips–Ouliaris test implemented by the po.test function in the tseries library. The function requires that the series be given in matrix form and produces the results for a test of the null hypothesis that the two series are not cointegrated:

$$H_0 : x, y \text{ not cointegrated}$$
$$H_a : x, y \quad \text{cointegrated.}$$

The test for cointegration is the po.test.

Example 8.16   Sometimes, it is easier to see what a concept means if a simulation is used to illustrate it. We will generate two time series that share a common stochastic trend and therefore are cointegrated.

```
### program to generate co-integrated ts
x <- y <- mu <- rep(0,1000)

## generate a stochastic trend
for(i in 2: 1000) {mu[i] = mu[i-1]+rnorm(1) }

## generate data with common trend mu
x1=mu+rnorm(1000)
y1=mu+rnorm(1000)

## unit root test
## because po test makes sense with unit root data
adf.test(x1)
adf.test(y1)

##confirm that x1, y1 is random walk
```

```
acf(diff(x1))
acf(diff(y1))

## cointegration test
po.test(cbind(x1,y1))
```

We see that, when applied to two random walks, the conclusion of the `adf.test` is to retain the null hypothesis that the series have unit roots. The `po.test` then provides evidence that the series are cointegrated since the null hypothesis is rejected at the 1 percent level.   □

We now give an example of the result of the Phillips–Ouliaris Cointegration test for precipitation and temperature in the same location. The test results are

```
data:cbind(prec.ts, temp.ts)
Phillips-Ouliaris demeaned=-217.0705,
Truncation lag parameter=2,
p-value=0.01
```

Thus we reject the null hypothesis that the series are not cointegrated. They share a common trend but they do not have a unit root. Thus the cointegration is ignored.

8.8.6 Exercises

Exercise 8.14   Generate two random walks like we did earlier in Exercise 8.13 and do a unit root test and a cointegration test of them. Write the conclusions.

Exercise 8.15   The data set *bluebirdlite* is in package TSA [19] of R, is of class data frame and can be accessed simply by typing its name in R, after installing the package TSA. It contains weekly unit sales (log-transformed) of Bluebird Lite potato chips (New Zealand) and their price for 104 weeks. Make the data set a multiple time series object. Visually inspect the time plot and the ACF and CCF of the data. Determine whether sales and price have a unit root and whether the two time series are cointegrated.

Exercise 8.16   Check whether the Canada vector time series has a unit root and if it does, do a cointegration test. If the time series are cointegrated, perform a VAR of the undifferenced time series. Compare the estimates, forecasts and IRF.

8.9   VAR Models Software

In this chapter, we have just touched the surface of VAR and issues related to VAR. The software dedicated to identification and estimation of VAR models is RATS, which stands for Regression Analysis of Time Series. Developed and sold by Estima, the software started to be developed when VAR first appeared in the 1980s, and has since been the specialized software for VAR. Being a commercial package means that it is not open source.

Other commercial software packages run VAR, in particular, SAS ETS, which is specialized in time series.

Over the years, several authors have developed packages that provide estimation and other functions for VAR modeling. The ones we have used in this chapter are the most user friendly, for simple models, in order to introduce the methodology.

8.10   Problems

Problem 8.1   Consider the following VAR(1) process:

$$x_{1,t} = 0.6x_{1,t-1} - 0.5x_{2,t-1} + W_{1,t}$$
$$x_{2,t} = 0.4x_{1,t-1} + 0.5x_{2,t-1} + W_{2,t}.$$

(a) If we do not have the data, that is, this model was given to us as the true model, how can we show that the VAR(1) process is stationary?
(b) If the model is the best model we could assume using the sample ACF, and cross-correlogram before model fitting, what must those functions have looked like to conclude that the preceding was the right model? Sketch the cross-correlogram.
(c) If the process is stationary, forecast $x_{1,t+1}$ and $x_{2,t+1}$, that is, forecast both time series one step ahead out-of-sample.

Problem 8.2   Let Y_t be the log of the area planted with sugar cane in Bangladesh in year t and let X_t be the log of the price of sugar cane in year t. We fitted a simple linear regression model:

$$\hat{y}_t = \quad 6.111 + 0.974x_t$$
$$s.e. = \quad (0.169) \quad (0.111).$$

The ACF of the residuals shows that there is an AR structure left in the residuals, $w_t = 0.8w_{t-1} + v_t$, where v_t is white noise.

Do we have enough information to say that the log area planted and log price are cointegrated? Explain.

Problem 8.3   The following VAR model was fitted to cointegrated observed time series x_t (London stock) and y_t (New York stock):

$$\hat{x}_t = 8.386 + 0.985x_{t-1} + 0.0749y_{t-1}$$
$$(P - value) = (0.000) \quad (0.000) \quad (0.000)$$
$$\hat{y}_t = 0.078 + -0.0003x_{t-1} + 1.002y_{t-1}$$
$$(P - value) = (0.830) \quad (0.477) \quad (0.000).$$

(a) According to the fitted model, which variable "leads"? Which "lags"? Describe how.

(b) According to Figure 8.21, should we believe what this model is suggesting about the relation between X and Y? Why ?

Note that the image in Figure 8.21 corresponding to the cross-correlogram was obtained using the `ccf ()` function with the multiple time series object containing both the London and NY time series. The cross-correlogram thus contains, on the right of 0, the one that we could have obtained in the upper right-hand corner had we used `acf ()` with the same multiple time series object. On the left of 0, we can see the cross-correlogram that we would have obtained had we used the `acf ()` function. The `ccf ()` provides less information than we would have obtained with the `acf ()`, because the latter gives not only the cross-correlograms but also the correlograms. Figure 8.21 required that we used the `acf ()` on the individual series and the `ccf ()` on the multiple time series object, twice the work.

Problem 8.4   Use program *ch8stockmarket.R* to do the following problem.

The data set *stockmarket.dat* contains stock market data for seven cities for the period January 6, 1986 to December 31, 1997.

(a) Use an appropriate statistical test to test whether the London and/or the New York series have unit roots. Does the evidence from the statistical tests suggest that the series are stationary or nonstationary? Why?
(b) Let x_t represent the London series (Lond) and y_t the New York series (NY). Fit a VAR(1) model, giving a summary output containing the fitted parameters and any appropriate statistical tests (including the residuals' ACF).
(c) Which series influences the other the most? Why might this happen?
(d) Test the London and New York series for cointegration.
(e) Fit the following model, giving a summary of the model parameters and any appropriate statistical tests:

$$x_t = a_0 + a_1 y_t + w_t.$$

(f) Test the residuals of the fitted model for unit roots. Does the test support or contradict the result in part (d)? Explain your answer.

Problem 8.5   A data set consists of 100 observations of Y_t (total sales in thousand of cases in month t) and X_t (advertising expenditures in thousands of dollars) in month t. The data set is *ch8salesandexpenditures.txt* and a program you may use to do this problem is at the bottom of program *chapter8manirf.R*. You will need to add code to it, using as examples the codes of other problems that we have introduced in this chapter.

(a) Do a time plot of the two variables and plots of the sample correlograms and cross-correlograms. Comment on the results and determine whether differencing is necessary. Before you decide, recall that AR processes sometimes look like nonstationary data.
(b) Identify a model for the two variables.
(c) Would a VAR model be an appropriate model for this data set ? Why or why not? Justify your answer.

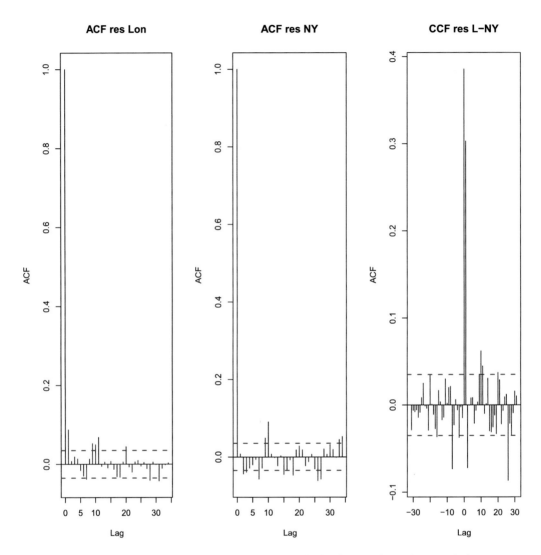

Figure 8.21 Analysis of residuals from VAR model fitted to London and NY stock data.

Problem 8.6   Consider the predator and prey density time series seen in Figure 8.1. The data file is *predatorprey.csv* and the program *ch8predatorpreyprog.R* was used to obtain it, as well as the impulse response functions in Figure 8.3. The latter require estimating first the VAR model. That VAR model is for the density. Ewing and colleagues [45] studied the growth of the species instead of the density itself.

(a) Convert the two time series to growth rates.
(b) Obtain a time plot of the two growth rates over time, labeling and giving a legend, as in Figure 8.1.
(c) Plot the correlograms and cross-correlograms using the acf() function of Base R.

(d) Identify a tentative first multivariate model for the growth rate and decide what VAR model to fit first, based on that information.

(e) Estimate the VAR model, determine whether the residuals are multivariate white noise or not. If your residuals are not white noise, then propose an alternative model.

(f) Once a satisfactory model is obtained, write the equations of the model. Determine which coefficients are statistically significant and interpret the findings. Check the stability conditions as well.

(g) Produce the impulse response functions resulting from a shock to prey first and then those resulting from a shock to predator.

(h) Forecast with the final estimated mode 12 steps ahead and calculate the RMSE.

8.11   Quiz

Question 8.1   The following VAR model was fitted to observed time series x_t and y_t:

$$\hat{x}_t = 8.386^* + 0.985^* x_{t-1} + 0.0749^* y_{t-1}$$
$$\hat{y}_t = 0.078 - 0.0003 x_{t-1} + 1.002^* y_{t-1}.$$

The star over a number means that the coefficient is statistically significant at

$$\alpha = 0.05.$$

According to this model

(a) Y leads X
(b) X leads Y
(c) Y lags X
(d) X lags Y

Question 8.2   When simulating the effect of an unexpected change in the size of a predator population, researchers usually start by making the change equal to

(a) 1 at time 1 and no change at $t \neq 1$
(b) one standard error at time 1 and no change at $t \neq 1$
(c) 0 at time 1 and 1 standard error at every $t \neq 1$
(d) three standard errors at time 1 and no change at $t \neq 1$

Question 8.3   When we fitted a VAR(2) to the Canada data set, we found that the eigenvalues of the coefficient matrix are reported by R as follows:

```
Roots of the characteristic polynomial:
0.7141 0.7141 0.5754 0.5754 0.473 0.473 0.3278 0.3064
```

What can we deduce from this?

(a) The VAR process is stable.
(b) The VAR process is unstable.

(c) The roots of the backshift polynomial are larger than 1 in absolute value.
(d) The residuals are not white noise.

Question 8.4   A VAR model allows us to

(a) estimate the effect of immediate past values of a variable on today's value of another
(b) simulate, after estimation, the response of one variable to a sudden change in the value of another variable
(c) forecast, after estimation, future values of variables that are interrelated, assuming that there is no drastic change in any variable
(d) ignore the stationarity of the time series involved

Question 8.5   If several related time series are not stationary, a sudden change to one of them will

(a) cause the system to continue wandering around without ever returning to its normal steady state
(b) cause the system to change but will eventually return the system to its steady state
(c) cause the variance of the changes over time to increase without bounds
(d) not have any effect in the system

Question 8.6   Interpret what Figure 1 in Ewing, Riggs and Ewing (2006)'s article [45] is saying about a shock to the predator–prey system studied:

(a) It will cause the system to continue wandering around without ever returning to its normal steady state.
(b) It will cause the system to change but will eventually return the system to its steady state.
(c) It will cause the variance of the changes over time to increase without bounds.
(d) It will not have any effect in the system.

Question 8.7   Ewing, Riggs and Ewing (2006)'s article [45] estimated a VAR model of order 3 for the predator prey population density growth rate. According to this model:

(a) The response of growth rate in prey on growth in predator is more immediate and has less persistence than the other way around.
(b) The response of growth rate in predator and growth rate in prey in the past has a current effect in current growth rate.
(c) There is no relationship between the predator and prey growth rates.

Question 8.8   According to the VAR model of order 2 in the article by Sanchez and Wang (2007) [171], what came first, the chicken or the egg?

(a) the chicken
(b) the egg
(c) we cannot tell. The effect goes both ways.

Question 8.9   In the article by Sanchez and Wang [171], the results are

(a) the results of impulse response analysis
(b) the estimated VAR model without any drastic changes imposed
(c) the forecasts obtained from the VAR model
(d) cross-correlation analysis is missing

Question 8.10   Suppose we have the following stationary univariate process:

$$y_t = 0.9824y_{t-1} - 0.3722y_{t-2} + W_t.$$

Suppose we give a one-unit shock to the system at time $t = 0$. What would be the value of the impulse response at time $t = 9$?

(a) 0.00130
(b) −0.00776
(c) 0.00715
(d) −0.00767

8.12   Case Study

VAR have been used often to study leading or lagging indicators. A leading indicator is a variable that helps anticipate what will happen to other variables on a regular basis. For example, construction is believed to be a leading indicator, because when construction starts rising in a country with a market economy, soon afterward the economy starts improving. It is said that construction leads in the sense that it signals recovery.

Sanchez and Wang [171], in an article published in *Stats, the Magazine for Students of Statistics* (now discontinued), applied this concept to the old dilemma of which came first, the chicken or the egg? This case study contains the contents of that article, reproduced here with permission from the American Statistical Association.

The full view of the original article may be seen in the `timeseriestime.org` website for the book. The data analysis supporting that article was originally done with SAS. The reader may access the original data used in the article and import it into R (the data set is called `chicken-egg-data.sas7bdat`), then attempt to reproduce what is described in the narrative using R code learned in this chapter.

8.12.1   Which Came First, the Chicken or the Egg?

Chickens play a huge role in our lives. As well as sometimes acting as pets or ending up sitting on plates, chickens show up in pop culture. They had feature roles in the hit movie *Chicken Run*; they frequently guest star in popular games such as *The Legend of Zelda* and *Final Fantasy*; and they take center stage in the classic dilemma of causality: *Which came first, the chicken or the egg?* Philosophical issues regarding this question were presented in a CBS News Video, *Was the Chicken or Egg 1st?* in 2006. Sanchez and Wang use time series analysis of historical data from the United States chicken industry to shed some light on the direction of causality.

In their study, the question of causality can be rephrased as *Does the chicken depend on the egg, or does the egg depend on the chicken?* In other words, does the egg lead or does the chicken lead? They use a VAR model of chicken and egg time series to answer that question.

8.12.2   The United States Chicken Industry

The main products of the poultry industry in the United States are broilers (chicken for meat) and table eggs (eggs for cooking). Being the world's largest producer of poultry meat, 14 percent of the total United States' annual poultry production is exported. The United States is also the second-largest egg producer in the world.

Although it has become a highly specialized agricultural business nowadays, the commercial poultry industry was made up of millions of small backyard farms before the 1950s, when meat was a byproduct of egg production. At the time of Sanchez and Wang's article, poultry products accounted for about 10 percent of all farm revenue, and the industry had been transformed almost completely from a fragmented, home-owned industry to a highly organized, vertically integrated industry linking all production decisions from farm to market.

8.12.3   Chicken and Egg Time Series Data

We can analyze this data set with program *chicken.R*. One of the two variables studied by Sanchez and Wang was monthly chickens hatched with the intended purpose of becoming broilers. This variable is called hatched in the program. The other variable is broiler eggs, called eggs in the program. (The other variable studied was *eggs*, but not table eggs; rather, broiler eggs.) The data span the years between 1975 and 2002. Because the values of the time series display an upward trend with increasing variability, the time series are not stationary.

Differencing

As we have seen since Chapter 4, differencing is an easy and effective method to help stabilize a nonstationary time series. Simple differences are differences taken one period apart. Seasonal differences are differences taken 12 periods apart. In some time series, we need to do both simple and seasonal differencing.

After regular and seasonal differencing of hatched and eggs, the time series fluctuate around a constant mean of zero and the variance looks relatively stable with a few extreme values.

But in time series analysis, we do not rely on only our eyes. We also look at two plots that play a prominent role in understanding a time series: the ACF and the PACF.

As we saw in Chapter 3, autocorrelation is the association between values of the same variable over time. The autocorrelation coefficient (ρ_k) measures the autocorrelation between two values in a time series k time periods apart, and the sample autocorrelation function (ACF) plots the autocorrelation coefficients' values for k from 0 and up.

Partial autocorrelation is the association between times series values separated by k time periods with the effects of the intermediate observations eliminated. A plot of these values is the partial autocorrelation function PACF.

8.12.4   Autocorrelation Functions

If we want to know whether the memory of a time series goes as far back as k months, that is, whether the value of `hatched` this month depends on what happened k months ago, then we can test:

$$H_0 : \rho_k = 0$$
$$H_a : \rho_k \neq 0$$

where ρ_k is the autocorrelation between the value of the series at time t and its value k periods before. If a sample autocorrelation, r_k, is larger than two standard errors, this means that the p-value for the test at that lag k is smaller than 0.05, and, therefore, we can reject the null hypothesis. If so, autocorrelation is statistically significant and we say there is evidence in the data supporting the conclusion that there is memory or correlation between values ofthe variable at time t and at time $t - k$.

The ACF of the differenced series has a significant spike at lag 1 and the PACF exponentially dies away. Therefore, based on these traits, the model for the variable `hatched` is a moving average process of order 1 for nonseasonal lags. However, as there is also a spike in the ACF at $k = 12$ months and the PACF shows the seasonality dying down at 24 months, we also have a moving average model at the seasonal lag. Our final model for the variable `hatched` is ARIMA(0,1,1)(0,1,1), which is short-hand for saying it is a moving average model using lagged variables at one month and 12 months.

The ACF and PACF for the differenced `eggs` data look similar to the ACF and PACF for differenced `hatched` data. That indicates that the model to use for `eggs` is the same as for `hatched`: ARIMA(0,1,1)(0,1,1). Thus the change in `eggs` and the change in `hatched` follow the same model, when looked at individually.

8.12.5   Vector Autoregression

The question of which variable leads or precedes in time in the movement of two stationary time series has been studied often in Economics. For example, in the context of predicting stock market prices, one question can be whether the price of a stock that trades in both the United States and, let's say, Germany, is such that the United States price leads the German price or the German price leads the United States price during overlapping trading periods (i.e., during the hours both markets are simultaneously open).

Questions such as this can be answered with a technique used in econometric analysis called vector autoregression (VAR). It is a method that can help us determine the time precedence between variables. If we find that one variable consistently precedes another in time, that would be evidence supporting a possible causal relationship.

In our case, we want to see whether the number of broilers hatched causes the number of broiler eggs or the number of broiler eggs causes the number of broilers hatched. For this, we need two regression equations: one to regress broilers hatched on broiler eggs and the other to regress broiler eggs on broilers hatched. As this is a time series model, we want to estimate these two regression equations taking into account that there could be causality in either direction. So, we estimate the two equations together with a common variance-covariance matrix for both. This is different from separately estimating them.

Vector autoregression is a technique for analyzing the dynamic properties like that relation we are trying to understand. Least squares regression is used to examine the autocorrelation due to the time-dependence of each of the variables. First, the analysis is performed with X as the dependent variable and its historical values and the Y values as the independent variables. Then, the process is repeated with Y as the dependent variable with its historical values and the X values as independent variables. Comparing the results of the two regressions can show the direction of possible causality.

A simple vector autoregression for our problem would be a model such as this:

$$X_{1t} = \Phi_{11}X_{1,t-1} + \Phi_{12}X_{2,t-1} + \varepsilon_{1t}$$
$$X_{2t} = \Phi_{21}X_{1,t-1} + \Phi_{22}X_{2,t-1} + \varepsilon_{2t},$$

where X_{1t} denotes the change in broilers hatched and X_{2t} denotes the change in broiler eggs, both variables are stationary with a mean of zero, and Φ_{ij} are constants we estimate by regression.

Looking at the preceding model, we notice that if Φ_{12} is zero, but Φ_{21} is not zero, there is no feedback from X_2 to X_1. Thus, the change in broilers hatched does not depend on the lagged value of broiler eggs, but broiler eggs depend on the lagged value of broilers hatched. This would indicate that broilers hatched are a leading indicator for broiler eggs.

8.12.6   Estimated Model

Based on the structure found in the ACF and PACF, the bivariate VAR model fitted to the time series by the authors using SAS is

$$hatched_t = -0.2391 hatched^*_{t-1} - 0.0043 eggs_{t-1} - 0.3768 hatched^*_{t-12} - 0.0956 eggs_{t-12}$$
$$eggs_t = 0.1237 hatched^*_{t-1} - 0.3614 eggs^*_{t-1} + 0.0543 hatched_{t-12} - 0.5001 eggs^*_{t-12}$$

where $hatched_t$ is the value at time t of the seasonal difference of the first difference for the variable *hatched* and $eggs_t$ is the value at time t of the seasonal difference of the first difference for the variable *eggs*. Statistically significant effects are indicated by an asterisk in a superscript. The p-values of the coefficients correspond to the test of the null hypothesis that a coefficient is equal to zero. A p-value ≥ 0.05 means the coefficient is not significantly different from zero.

Looking at the statistically significant effects, we can see that *hatched* last month (one-month lag) affects *eggs* this month, but no lag of *eggs* affects *hatched* in the present month. This means that while the number of broilers previously *hatched* affects the number of broiler *eggs* in incubators now, the number of broiler *eggs* incubating previously does not affect the number of broilers *hatched* now.

In economic-speak, that means that the number of chickens *hatched* is a leading indicator for the number of *eggs*, but the number of *eggs* is not a leading indicator for the number of chickens *hatched*. So, in our dilemma of causality, the chicken comes first!

The conclusion makes sense in the economic context. A downturn in broilers is probably an indication of a sluggish chicken meat market, perhaps due to factors such as a recession in the economy, maybe some pandemic of avian flu, or some other economic factor. If this is the case, it does not make economic sense to keep the number of eggs in incubators at the previous level. Why incubate eggs that will give chickens that will not be sold? Consequently, we would expect the number of eggs in incubators to go down. So, as the demand for chicken meat goes up or down, the number of eggs in incubators should follow.

The Chicken or the Egg?

As we have seen, time series analysis makes use of the basic concepts we learn studying introductory statistics: estimation, test of hypotheses, p-values, and regression. We just adapt the basic principles to the circumstances present in a time series modeling problem.

For chicken farmers, it is useful to know that the number of chickens hatched is a leading indicator for the number of eggs in incubators. For all of us concerned with the dilemma of causality, it is nice to see how quantitative methods can help us answer a classic dilemma: which comes first, the chicken or the egg? Conditional on the vector autoregressive model that we used, in the economic decision chain of the United States poultry industry, the data indicate that chickens come first.

8.12.7 Exercises

Exercise 8.17 Import the SAS data set `chicken-egg-data.sas7bdat` into R after reading the case study, and conduct a unit root test and a cointegration test of the two variables in it before differencing. After obtaining the results of those tests, would you say that we are justified in concluding, as we did in the case study, that the chicken came first? Explain your test results and the reasoning that determined your conclusion.

Exercise 8.18 Import the SAS data set `chicken-egg data.sas7bdat` into R after reading the case study, and do the differencing recommended. Estimate the VAR model presented in this case study and determine whether the model is stable. Obtain the IRFs.

9 Classical Regression with ARMA Residuals

9.1   Introduction

Classical linear regression with time series data is used to:

- understand the historical relationship of variables;
- use that understanding to predict the value of the dependent variable in the future based on future values of the independent variables.

As in VAR, there is more than one variable involved. However, classical linear regression differs from the relations studied in Chapter 8 in that in classical linear regression there is one equation and the independent variables are not also endogenous variables.

Classical linear regression is a method for incorporating external information into time series modeling. Regression has always been studied as a method that helps us explain a variable Y (called dependent, response, outcome, target, or output, depending on the field of application) by one covariate or a vector of covariates, $\mathbf{X}$ (also known as predictors, independent variables, explanatory variables, features, exogenous variables, or input). Y is also called an *endogenous variable* if the exogenous variables $\mathbf{X}$ are not affected by Y, that is, if $\mathbf{X}$ is not also endogenous. The reader may have heard the regression model referred to as *a mapping* of a set of variables into a target variable. Very likely that has occurred either in a mathematics or in a machine learning setting.

This chapter builds on the reader's familiarity with classical linear regression and applies it to the study of linear regression with time series data. Readers needing a refresher or an introduction to classical linear regression are directed to the appendix of this chapter, Section 9.9, or to some of the numerous textbooks or practical guides on classical regression analysis, such as [1, 8, 47, 154, 92, 9, 35, 56, 51].

Time series data routinely leads to the violation of one of the assumptions of the classical linear regression, the assumption of independence of the error terms. Autocorrelated residuals pose problems to the conclusions of a standard regression analysis done with Ordinary Least Squares (OLS). Such autocorrelation can seriously bias significance tests. Relationships that OLS says are statistically significant may not really be so. The appendix in Section 9.9 talks about the theoretical implications of not satisfying the assumption of no autocorrelation in the error term and reviews classical methods of diagnosing autocorrelations. Readers not familiar with this topic will find the discussion in the appendix useful before continuing reading the chapter.

A classical procedure for overcoming autocorrelation difficulties called *generalized least squares* consists of modeling the error process using the Box–Jenkins models studied in Chapters 5 and 6, thus bridging the gap between regression and classical time series analysis with Box–Jenkins methodology. Residuals are modeled with ARMA models, and autocorrelation is diagnosed with the ACF and PACF of the residuals. The reader will find an in-depth discussion of this subject in the monograph by Ostrom [142]. Section 9.2 is dedicated to generalized least squares, using an example.

There are several approaches to do classical regression analysis of time series:

(a) *Regression analysis per se, or causal analysis*, which uses other related variables to explain the movement of a time series. Assumes that the nature of the relation does not change through time. Section 9.2 is concerned with this approach.
(b) *Time series regression (or classical regression as smoother)*, which uses a polynomial trend function and dummy variables to capture the trend and seasonal patterns in the time series. This too assumes that the nature of the signal does not change through time. We saw this approach in Chapter 2 as a special example of "smoothers." Section 9.3 addresses time series regression again.
(c) *A mixture of causal and time series regression*, which combines both of the approaches just mentioned. Many of the modern functional models used in supervised machine learning are examples of this case.

The Box–Jenkins methodology also encompasses regression-like models. These, however, differ significantly in their model construction approach from the classical regression models. Because they have played an important role in Engineering, we introduce briefly *transfer function models* in Section 9.4 and *intervention models* in Section 9.5. These models have rather complicated identification procedures and are rarely used outside Engineering.

The IoT and the increasing volume, velocity and variety of data produced pose challenges for regression analysis. Supervised machine learning methods are basically derived from regression, but they address those challenges. Gradient boosting, for example, is an algorithm that uses additive modeling to nudge an approximate model toward a good one by adding simple submodels to a composite model. The simple models are called "weak learners" and are added in sequence until we are happy with the final model. Random forest methods try to achieve the same goal but in reverse. Our discussion of regression would not be complete without addressing those methods, but we postpone that until Chapter 10.

9.2 Causal Regression Analysis of Time Series

Causal regression does not just use features of the time series such as trend or seasonal variables as explanatory variables but also uses other domain variables believed to be related to the dependent variable. Sometimes these models are trying to "explain" and sometimes they are just used for forecasting. In either case, it is believed that the more information available, the better the outcome of both of those.

Example 9.1   The search for an explanation for the fluctuations in the economy over time has led many scholars to try to specify models of the consumer behavior process. Chatterjee and Price [21] presented a simplified version of the quantity theory of money that explained consumer behavior as a reaction to the money stock over time, namely

$$Y_t = \beta_0 + \beta_1 X_t + \varepsilon_t, \tag{9.1}$$

where Y_t denotes US quarterly aggregate consumer expenditures at time t from 1952 to 1956 and X_t denotes US quarterly money stock at time t, both measured in billions of current US dollars. ε_t denotes a random term, β_1 is the money multiplier, and β_0 is the average expenditures when there is no stock. β_0 and β_1 are unknown regression parameters estimated with data, and the subscript t indicates that X_t and Y_t are time series.

In a classical regression analysis aimed at understanding a process, the estimation and meaning of the regression parameters are the goal.

To investigate the degree to which Equation 9.1 accurately describes the relation between consumer expenditures and the money stock, Chatterjee and Price [21] applied classical simple linear regression to expenditure and stock time series data using OLS. The results of the regression analysis using 1952–6 data are as follows:

$$\widehat{y}_t = -154.7192 + 2.3004 x_t, \tag{9.2}$$

where the carat denotes a parameter *estimate* when corresponding to Greek letter parameters, or an estimate of the expected value of Y_t for each value of X_t, when referring to Y_t. We see that $\widehat{\beta}_1 = 2.3004$ and $\widehat{\beta}_0 = -154.7192$.

As can be seen from the estimated version of Equation 9.1, seen in Equation 9.2, the money stock had a positive impact on consumer expenditures during 1952–6. This effect is statistically significant. A \$1 billion increase in the money stock results in a \$2.3004 billion increase in consumer expenditures ($se = 0.1146$, $t = 20.080$, p-value ≈ 0). The model as a whole explains 95.73 percent of the total variance ($R^2 = 95.73\%$).

Those results make sense if all the assumptions of the linear regression model are satisfied. If the individual error terms are not independent but instead are related to each other in a systematic fashion (i.e., they are autocorrelated), any substantive conclusion is doubtful. To overcome the problem, and hence to estimate the model in an appropriate manner, it is necessary to characterize the nature of the systematic relationship, if any exists, that has not been captured by the model, and then incorporate this information into the estimation procedure. That involves looking at the residuals,

$$\widehat{\varepsilon}_t = \widehat{y}_t - (-154.7192 + 2.3004 x_t),$$

carefully to diagnose the problem.

The residual patterns obtained from our estimated model 9.2, can be seen in Figure 9.1.

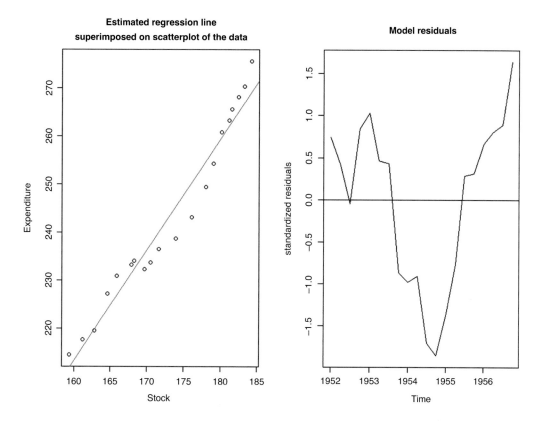

Figure 9.1 The consumer expenditures and money stock model residuals (shown in the "Money residuals" plot) are correlated.

"For time series data, the most meaningful plot for model diagnostic is the plot of the standardized residuals against time" (Chatterjee and Price [21]) if the practitioner does not know time series analysis. For the practitioner who knows how to interpret the ACF, and for those that get a little lost deducing violation of assumptions by looking at the residual plot, the ACF is more informative than the time plot.

The plot of standardized residuals when doing regression with time series data must have time in the horizontal axis, like all time plots do. As we said in Chapter 1, the most informative way of seeing time series of any kind, and residuals from a model fitted to time series data are themselves a time series, is with a time plot. We accompany the time plot with the ACF of the residuals, and we can start successfully the diagnosis of autocorrelation.

Figure 9.1, showing the residuals from model 9.2, reveals that residuals of the same sign occur in clusters or bunches. The prominent pattern is that the first seven residuals are positive, the next seven are negative, the last six are positive. This is a sign of

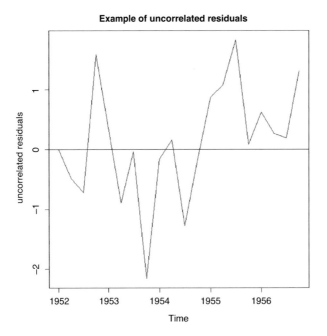

Example of uncorrelated residuals

Figure 9.2 Residuals that follow the uncorrelated assumption.

positive autocorrelation.[1] Descriptive time series statistics, such as the correlogram, can be used to study the nature of that autocorrelation. Knowing classical time series and the tools presented earlier in this book provides a great toolbox when studying regression with time series. We can see in the ACF obtained by running program *Ch9expstock.R* with data set *expenditures.csv* that the residuals are autocorrelated.   □

9.2.1   Autocorrelated Regression Residuals

Of utmost importance to the classical linear regression model is the error term ε_t in Equation 9.1. An error term is included in a model for one of the following reasons: (a) we do not know all the factors that are related to a given dependent variable, and we recognize that some factors have been omitted in the equation; (b) error is related to the collection and measurement of data; (c) there is randomness in every outcome, which can be adequately characterized by the inclusion of a random variable in the model. These three factors are summarized by the error term, and we often assume that their effects are small, and independent (i.e., not autocorrelated).

---

[1] If large positive errors are followed by other positive errors, or negative errors are followed by negative errors, there is positive autocorrelation. If positive errors are followed by negative errors, then there is negative autocorrelation. If there is no pattern as to what the next error will be, then the errors are uncorrelated. When the observations have a natural sequential order, the correlation is referred to as autocorrelation, as we have been seeing throughout this book.

Given the presence of a random term, it is necessary to characterize the relationship between consumer expenditures and the money stock in stochastic terms. That is, for every X_t there exists a probability distribution of ε_t and therefore a probability distribution of Y_t. Because of the probabilistic nature of the equation, the initial specification of the model must include some assumptions about the probability distribution of the error terms. These are related to its mean, variance, and covariance. The following three assumptions are usually made:

- the error term has a mean of zero;
- the error term has a constant variance over all observations; and
- the error terms corresponding to different points in time are independent and consequently uncorrelated.

In other words, the a priori assumption of linear regression is that the error terms follow a white noise process. As we have seen in earlier chapters, this means a correlogram like that of white noise. When the observations from different points in time are not independent, one of the assumptions is violated. When this occurs, we say that the error process is *serially correlated* or *autocorrelated*.

The error term ε_t cannot be observed, and hence it must be distinguished from the residual, $\widehat{\varepsilon}_t$. Errors are associated with the true regression model, while residuals arise from the estimation procedure. The residual measures the deviation of each observed y_t from its fitted value, $\widehat{y}_t$.

> The residuals of the estimated regression model are looked at to determine whether the model assumption that the error terms are independent holds. The ACF of the residuals helps detect the violation of that assumption.

If the random factors of one period have no effect on those in subsequent periods, we would expect to find the pattern of residuals, $y_t - \widehat{\beta}_0 - \widehat{\beta}_1 x_t$, similar to Figure 9.2; that is, the observed residuals are randomly scattered around the regression line. That interpretation is sometimes hard for the beginner. A more straightforward way to detect the "no autocorrelation" assumption is the ACF of the residuals, because by now we know what the ACF of white noise looks like.

9.2.2   Consequences of Autocorrelation

> Computer software such as R routinely outputs estimates of the parameters of a regression model and their standard error assuming that the errors are independent. Before we interpret the output, we must make sure that the assumption of independence holds. If it does not, then the results should not be interpreted. A correction of some sort must be made, depending on the nature of the autocorrelation.

We see in the appendix in Section 9.9 the theoretical consequences of autocorrelation on the standard errors of the estimates of the linear regression model. In order to assess the specific consequences of autocorrelation, we must first specify the nature of the relationship between successive error terms. One possible process followed by the random term is a first-order autoregression, AR(1), where the error at time t is autocorrelated with the error of time $t - 1$. Mathematically,

$$\varepsilon_t = \rho\varepsilon_{t-1} + W_t, \quad |\rho| < 1, \tag{9.3}$$

where ε_t is the error term at time t, ε_{t-1} the error term at time $t - 1$, ρ is a regression coefficient that happens to also be the correlation coefficient between ε_t and ε_{t-1}, and W_t is normally distributed with zero mean, constant variance and zero autocorrelation. We studied the theoretical properties of the AR(1) stochastic process in Chapter 5.

Regarding classical linear regression with error terms that are AR(1), given the presence of first-order autocorrelation, it can be shown mathematically that, while there is no effect of autocorrelation on the parameter estimates, there are problems with the estimated standard errors of the parameter estimates. This is shown in several regression books [1].

The standard errors are used to generate the t-ratio,

$$t_{n-2} = \frac{\widehat{\beta_1} - 0}{se(\widehat{\beta_1})},$$

which is then used to assess the precision and significance of the individual estimates. Now, if the formulas used for the variances are no longer appropriate, that t-ratio is no longer valid. This creates problems for the investigator who wishes to test hypotheses relating to the strength and direction of the relationship between consumer expenditures and money stock over time.

In the expenditures-stock example, there is evidence of positive autocorrelation ($\widehat{\rho} = 0.8227$) in Equation 9.3, and therefore the conclusions reached in Section 9.2 are not valid. As we can see in the appendix, Section 9.9, this means that $se(\widehat{\beta_1})$ will underestimate the true standard error and the t-ratio will be inflated. The simple regression results overestimate the "fit" of the model. Thus, it seems clear that one could be led to erroneously conclude that the money stock has a significant influence on consumer expenditures when that might not be the case. Obviously, some correction needs to be introduced to reach the right conclusion or we might be guilty of accepting the quantity theory of money when, in fact, some other model might be true. In the statistical literature, this is known as making a Type II error.

Within the formal approach of classical linear regression modeling, the procedure followed is to avoid this Type II error by modeling the autocorrelation in the error terms and use that information to reestimate the assumed model 9.1, under the new assumption that the model for the residuals is AR(1) or some other ARMA model.

> Formalizing the relationship between autocorrelated error terms is formalizing the relationship between the observations in a time series, and that, if we have been reading the earlier chapters, is something we are experts on by now. We want a time series model for the residuals.

The purpose of this section has been to highlight the nature of the problem, and to provide some justification for studying autocorrelation of residuals in a linear regression model. The material learned in earlier chapters should prepare the reader to diagnose the autocorrelation pattern and to estimate an appropriate time series model for the residuals. In the remainder of this section, we will propose what to do with the residual model in order to obtain the right estimates of the linear regression model.

9.2.3 Generalized Least Squares as a Solution

When the residual plot is as seen in Figure 9.1, the Runs test and Durbin–Watson test seen in Section 9.9 will also indicate the presence of AR(1) correlated errors, and we will not be able to do tests of hypotheses or confidence intervals with the simple linear regression model. One approach used by classical linear regression is to incorporate the autocorrelation information in the estimation. The estimated regression equation should be refitted with a correction for the autocorrelation.

One method is the Cochrane and Orcutt method. This method assumes a specific form for the process generating the autocorrelated residuals (i.e., AR(1) or some other ARMA model), transforms the regression model with autocorrelated errors into one that satisfies all the assumptions of the simple model, and then applies OLS to the transformed data. This method is usually referred to as Generalized Least Squares (GLS). The estimates are known as the GLS estimates. We explain in what follows how this is done when the autocorrelation of the residuals has been previously identified as following an AR(1) model, as in the stock and expenditure regression example that we are discussing.

With model

$$Y_t = \beta_0 + \beta_1 X_t + \varepsilon_t,$$

it follows that

$$\varepsilon_t = Y_t - \beta_0 - \beta_1 X_t$$

and

$$\varepsilon_{t-1} = Y_{t-1} - \beta_0 - \beta_1 X_{t-1},$$

which, when substituted into the model that we identified for the residuals,

$$\varepsilon_t = \rho \varepsilon_{t-1} + w_t, \qquad |\rho| < 1,$$

gives

$$Y_t - \beta_0 - \beta_1 X_t = \rho(Y_{t-1} - \beta_0 - \beta_1 X_{t-1}) + W_t, \qquad |\rho| < 1.$$

Rearranging,

$$Y_t - \rho Y_{t-1} = \beta_0(1 - \rho) + \beta_1(X_t - \rho X_{t-1}) + W_t$$

or

$$Y_t^* = \beta_0^* + \beta_1^* X_t^* + W_t,$$

where

$$Y_t^* = Y_t - \rho Y_{t-1};$$
$$X_t^* = X_t - \rho X_{t-1};$$
$$\beta_0^* = \beta_0(1 - \rho);$$
$$\beta_1^* = \beta_1.$$

If the W_t is white noise, the last model represents a linear model with uncorrelated errors. In practice, we must double-check that. That assumption implies that the AR(1) model for the residuals, when fitted to the residuals, gives white noise residuals. If indeed W_t are white noise, we run OLS using the transformed variables Y^* and X^*. The estimates of the parameters in the original equations can then be obtained indirectly by undoing the transformation done, and they are

$$\widehat{\beta_0} = \frac{\widehat{\beta_0^*}}{1 - \widehat{\rho}}; \qquad \widehat{\beta_1} = \widehat{\beta_1^*}.$$

Therefore, when the residuals in the original model have an autoregressive structure, we can transform both sides of the equation and obtain transformed variables that satisfy the assumptions of uncorrelated errors.[2]

Also worth pointing out is how the least squares derivation is made. The sum of squared residuals, which is the objective function that we minimize to obtain the parameter estimates, is

$$S(\beta_0, \beta_1) = \sum_{t=2}^{n} (y_t - \rho y_{t-1} - \beta_0(1 - \rho) - \beta_1(x_t - \rho x_{t-1}))^2,$$

and is not linear. This implies that the solution will require using a search routine such as, for example, Newton's algorithm. The new standard error formulas will contain the information on the autocorrelation:

$$se(\widehat{\beta_1}) = \frac{s}{\sqrt{\sum (x_t - \widehat{\rho} x_{t-1} - \bar{x}(1 - \widehat{\rho}))^2}},$$

where

$$s = \sqrt{\frac{S(\widehat{\rho}, \widehat{w}_{t-1}\beta_0, \widehat{w}_{t-1}\beta_1)}{n - 2}}.$$

The value of ρ is unknown and has to be estimated from the data, which is done when fitting the AR(1) to the residuals. Users of linear regression not familiar with time series analysis use the sample correlation coefficient between $\widehat{\varepsilon}_t$ and $\widehat{\varepsilon}_{t-1}$, namely[3]

$$\widehat{\rho} = \frac{\sum_{t=2}^{n} \widehat{\varepsilon}_t \widehat{\varepsilon}_{t-1}}{\sum_{t=2}^{n} \widehat{\varepsilon}_t^2}.$$

[2]Notice that the derivation could start differently,

$$Y_t = \beta_0 + \beta_1 X_t + \rho \varepsilon_{t-1} + W_t = \beta_0 + \beta_1 X_t + \rho(Y_{t-1} - \beta_0 - \beta_1 X_{t-1}) + \varepsilon_t,$$

and arrive at the same conclusions. It is an alternative derivation.

[3]As indicated in Chapter 3, the correlation obtained with R's cor() differs slightly from the one obtained with acf().

An approximate relation between the Durbin–Watson statistic d (see the appendix, Section 9.9) and ρ is

$$d \approx 2(1 - \widehat{\rho}).$$

The closer the sample value of d is to 2, the firmer the evidence that there is no auto-correlation present in the error. Evidence of first-order autocorrelation is indicated by the departure of d from 2.

From the R operations done earlier, we know that $\widehat{\rho} = 0.8227$. Substituting in the transformed model and using least squares again, we obtain

$$\widehat{y}_t^* = -136.31900 + 2.22013x_t^*,$$

and from this, we obtain

$$\widehat{\beta}_0 = \frac{-136.319}{1 - 0.8227} = -768.8607; \qquad \widehat{\beta}_1 = 2.22013.$$

The standard error of the slope parameter is now 0.366, much bigger than with the original model. Thus we are now taking into account the fact that autocorrelation makes us more uncertain about the slope parameter.

We check the residuals of the fitted transformed model and they are not autocorre-lated now. Those residuals are shown in Figure 9.2.

We used the gls () function of the nlme package [155] in R to do the work we just did in one single programming statement (after attaching the data to read the variables directly from the imported data):

```
install.packages("nlme")
library(nlme)
modelgls=gls(expenditure~stock,
                correlation=corARMA(c(0.8227),
                p=1))
summary(modelgls)
plot(y=residuals(modelgls,
                type="normalized"),
                x=as.vector(time(stock)), type="l")
abline(h=0)
# Note that type is the lowercase of L, not the number 1.
# Check that final residuals are white noise
acf(ts(residuals(modelgls,type="normalized")))
```

The gls () function is more useful when we have more independent variables and bigger ARMA autocorrelation models are used for the residuals. We need to first fit the ARMA model to the residuals and extract the coefficients to put in the gls () function, but that is easier than having to do the math to see what the transformed model looks like.[4]

---

[4]gls () assumes that $\varepsilon \sim MVN(0, \sigma^2\Omega)$, and $\sigma^2\Omega$ represents the correlation structure of the error term. When we just plot the residuals of the fitted model, we do not get any information on the correlation terms included in Ω. The information is not very useful to determine whether the model we fitted to the residuals is appropriate. It is for this reason that we need to use the normalized residuals. When using gls () models and the library nlme () in R to fit models with autocorrelated residuals, we use

```
residuals(glsmodel, type="normalized")
```

9.2.4   Autocorrelation as Symptom, Not a Malaise

All of the conclusions and work we have done so far regarding stocks and expenditures lead us to believe that there is autocorrelation in these data. But it could be that this conclusion is incorrect. The observed autocorrelation could jut be a symptom that would be better interpreted initially as a general indication of some form of model misspecification.

> In fact, it is always better to explore fully the possibility of some additional predictor variables before yielding to an autoregressive model for the error structure. It is more satisfying and probably more useful to be able to understand the source of apparent autocorrelation in terms of an additional variable, as the learning might be much higher that way. The transformation that corrects for pure autocorrelation by fitting an ARMA model may be viewed as an action of last resort [21].

Samprit Chatterjee and his coauthor Bertram Price, in their book *Regression Analysis by Example* [21], distinguish between two types of autocorrelation in linear regression modeling and present methods for dealing with each. The first type of autocorrelation may be referred to as pure autocorrelation, which we have considered in this chapter. It is due to the temporal proximity of the observations. The methods of correcting for pure autocorrelation if we want to adhere to linear regression modeling involved a transformation of the data and application of the method of GLS. The other type is only autocorrelation in appearance, which we are considering in this section. It is due to the omission of a variable that should be in the model. If successive values of the omitted variable are correlated, the errors from the estimated model will appear to be correlated. Once this variable is added to the model, the apparent autocorrelation problem disappears.

Chatterjee and Price [21] present an example where autocorrelation appears artificially because of the omission of another explanatory variable. A real estate association wants to have a better understanding of the relationship between housing starts (the dependent variable) and population growth (the independent variable). They are interested in being able to predict future construction activity. Their approach is to develop annual data on regional housing starts and try to relate this to potential home buyers in the region. The latter are proxied by the size of the 22- through 44-year-old population group in the region (in thousands). Their goal was to get a simple regression of housing starts and population. Then, using some method that they developed for projecting population changes, they would be able to estimate corresponding changes in the requirements of new houses.

The regression results gave an $R^2 = 0.9252$ and an increase in housing starts of about 70,000 per million increase in population. However, the residual plot showed

---

and the residuals for the gls model will incorporate the structure of the residual correlation. The default is "type=response" residuals. The type="pearson" gives standardized residuals (residuals divided by standard errors). The type="normalized" gives the standardized residuals premultiplied by the inverse square-root factor of the estimated correlation matrix.

that there was correlation among the residuals and the Durbin–Watson statistic was 0.621.

The authors then considered that the availability of mortgage money for the region might also explain housing starts. Adding that variable to the linear regression model, the autocorrelation in the residuals disappeared. The DW went up to 2.29 and the graph of the residuals against time showed no autocorrelation. The population effect is statistically significant, but not as strong as in the first model. In a certain sense, the effect of changes in the availability of mortgage money for a fixed level of population is more important than a similar change in population. That could be seen by regressing the variables in standard deviation units. That is, subtract the mean and divide by the standard deviation all variables and run the regression again with the transformed variables. This showed that a unit increase in the standardized value of the population (an increase of one standard deviation) led to an increase in housing starts of 0.4439 standard deviation, but an increase of one standard deviation in the index increased housing starts by 0.5693 standard deviation [21].

The example illustrates that a large value of R^2 does not necessarily confirm that the data has been fitted and explained well. Any pairs of variables that show trends over time are usually highly correlated. Second, the Durbin–Watson statistic as well as the residual plots may indicate the presence of autocorrelation among the residuals when in fact the residuals are independent, but the omission of a variable or variables has given rise to the observed situation. The Durbin–Watson statistic may have a significant value when some other model assumption is violated.

The data set for this topic is *housingchaterjee.csv*. You are asked to replicate these results in an exercise.

9.2.5 Exercises

Exercise 9.1 Download the data set *housingchaterjee.csv* and replicate the analysis described in Section 9.2.4. Compare the regression with and without the availability of mortgage money. Compare the residual plots of the two estimated models. Which results are reliable? Which results are not? Why?

Exercise 9.2 The data set *faithful*, which comes with R, is about the Old Faithful geyser in Yellowstone National Park, Wyoming, USA. This is a very famous geyser that tourists often wait to see erupt. The eruptions occur several times during the day.

Each line of the data set *faithful* contains how long an eruption is and how long people have to wait for the next eruption.

The variable eruptions measures how long the eruption is, and the variable waiting measures the waiting time. If an eruption has been long, then people usually will have to wait longer for the next one. The data set is not like the time series that we have studied because the time intervals between measurements are not the same, but the measurements are in ordered sequence and there are no gaps, meaning that no eruptions were missed.

(a) In a classical simple linear regression model with dependent variable `waiting` and independent variable `eruptions`, what time series model would you use for the residuals of that regression, if any? Explain how you reach your conclusion.

(b) If the residuals need modeling, does the model you suggest, when fitted to the residuals, provide white noise residuals of the residuals model?

(c) If you found that the residuals of the regression model were autocorrelated, do you think that we could avoid that by adding another variable? What variable?

Exercise 9.3 One way to model supply response for an agricultural sugar crop is to specify a model in which area planted (acres) depends on price. When the price of the crop's output is high, farmers plant more of that crop than when its price is low. Letting A denote area planted at time t, and P denote output price at time t, and assuming a log-log (constant elasticity) functional form, an area response model of this type can be written as

$$Y_t = \beta_0 + \beta_1 X_t + \varepsilon_t,$$

where $Y_t = \log(A)_t$ and $X_t = \log(P)_t$.

Information on the area elasticity β_1 is useful for government planning. It is important to know whether existing sugar processing mills are likely to be able to handle predicted output, whether there is likely to be excess milling capacity, and whether a pricing policy linking production, processing and consumption is desirable.

We can test this theory with cross-section data or with time series data. The data used here will be 34 annual observations on area and price. Our task is to use these data to estimate the parameters of the preceding model.

A time plot of the logged variables shows that log of area increases over time, indicating that the tendency for farmers has been to plant more of the crop over time.

When you have a variable that could be explaining the dependent variable, statisticians like to use it. They call this an structural model, as opposed to the simple model of pure trend and or seasonal. Sometimes, the variable takes care of all the autocorrelations. After fitting the model to the data by least squares regression, we obtain the following fitted model:

$$\widehat{y}_t = 6.1113 + 0.9706 x_t.$$

The estimated model confirms the theory that price increases lead to higher area being used, with constant elasticity of 0.9706. Both the intercept and the slope are significantly different from 0 (p-value = 0) and the standard error of the slope estimate is 0.1106. A 95 percent confidence interval for the slope parameter is (0.715, 1.227). Only 70.63 percent of the variability in log area is explained by price, which makes sense; the price of substitutes, the income of the farmers and other variables are supposedly affecting them.

We would take the preceding interpretation of the model results at face value if everything was technically good with the model, but the Durbin–Watson statistic indicates that the residuals have first-order positive autocorrelation (DW = 1.2912; p-value = 0.009801), and that, if true, gives us standard errors for the estimates of the model parameters that are inaccurate (too small). So, we must confirm the autocorrelation first.

A plot of the relation between residuals of the model fitted at time t and residuals at time $t + 1$ shows that they are positively related ($r = 0.7811033$) and have mean 0. If we look at the residual plot from this model, we see that there are some long runs of positive and long runs of negative residuals, indicating autocorrelation. To see autocorrelation, we need to plot the residuals against time.

A linear regression model of this relation is

$$\widehat{\varepsilon}_t = 0.7760\widehat{\varepsilon}_{t-1},$$

with the estimate of this slope parameter having p-value 0. Because all the exploration leads us to conclude that we have first-order autocorrelation in the residuals, we can conclude that the model we fitted is inappropriate. Because the model did not take into account this problem, the standard error of the slope parameter is understating the true error. The true standard error should have been computed using the fact that the model is instead

$$Y_t = \beta_0 + \beta_1 X_t + \rho\varepsilon_{t-1} + W_t.$$

So, we will transform our model to reflect this fact. Simplifying and using $\widehat{\rho} = 0.7760$, the model reduces to

$$(y_t - 0.7760y_{t-1}) = \beta_0(1 - 0.7760) + \beta_1(x_t - 0.7760x_{t-1}) + w_t.$$

We can fit this model by simple least squares. The fitted model is

$$(y_t - 0.7760y_{t-1}) = 1.36719 + 1.03081(x_t - 0.7760x_{t-1}).$$

But $1.3679 = \widehat{\beta}_0(1 - 0.7760) \rightarrow \widehat{\beta}_0 = \frac{1.3679}{1 - 0.7760} = 6.106696$.

The slope estimate is slightly different from the one of the original model, but it should have been the same. (The difference is just due to computation rounding by the software.)

The standard error of the estimate of the slope parameter is now 0.23999, larger than without the transformation. This is more accurate now, as it should be, and higher. The 95 percent confidence interval for the slope parameter is $(0.634, 1.308)$.

(a) Confirm the statements made in the preceding discussion by using the code provided in program *Bangladesh.R* and data set *Bangladesh*.
(b) Show manually how to compute the DW statistic using the residuals of `area.model`.

9.3   Time Series Regression

We saw in Chapter 2 that the components of a decomposition of a time series, additive or multiplicative, could be used as covariates in a regression model. Because classical decomposition uses moving averages, the trend cannot be forecasted, and that posed a limitation of the classical decomposition model. But there was an alternative, modeling both trend and seasonality in an alternative way. In this section, we revisit that method: regression, where the time series terms are polynomials for trends and dummy variables represent seasonality. This model could be appropriate when there is only a trend and a seasonal effect in the data and those are very similar over time. We will revisit the model with polynomials plus dummies with an example.

The simplest model is the linear trend one:

$$\mu_t = \beta_0 + \beta_1 t,$$

where t is time and μ_t is the expected value of the series at time t. This model is saying that the average of the time series increases or decreases with time. This results in the regression model for the dependent variable Y_t:

$$Y_t = \beta_0 + \beta_1 t + W_t,$$

where Y_t is the value of the time series at time t, W_t is, as usual, assumed independent and identically distributed with mean 0, variance constant uncorrelated Gaussian noise. And t is the time index $(1, 2, 3, \ldots)$. We can fit the trend regression model to the data using ordinary least squares. We fit this model to a small data set with program *ch9regdummyexample1.R*. The fitted model is

$$\widehat{y}_t = 0.0481 + 0.5146t.$$

We could use this model to predict the average value of the series in the future. Or we could use the model to detrend the data. That is, we could remove the trend so that we can see what is going on beyond that understandably increasing trend. The detrended data is obtained by subtracting the fitted trend from the original data. These residuals show some periodicity, time dependence, and the mean increases on the first quarter of each year, therefore more modeling needs to be done. It indicates that the mean of the series will periodically increase in the summer, for example.

So it seems that the data has a seasonal property that needs to be taken into account. We need to include all this in the regression model. We have two ways of fitting this model: by defining a dummy variable for each season, or by putting a constant in the model and adding a dummy variable for each of the seasons except one.

A better model that incorporates the seasonal for quarterly data would look like this:

$$Y_t = \beta_0 t + \beta_1 D_1 + \beta_2 D_2 + \beta_3 D_3 + \beta_4 D_4 + W_t,$$

or like this:

$$Y_t = Constant + \beta_0 t + \beta_2 D_2 + \beta_3 D_3 + \beta_4 D_4 + W_t,$$

where $D_i = 1$ if the variable corresponds to quarter i, $i = 1, 2, 3, 4$. With this model we could forecast the future.

We could compare these forecasts with those obtained with other competing models. But we would need a measure of the goodness of the forecast. The root mean square error is a good way to do that. As we have discussed elsewhere in the book, it is customary in time series to report the root mean square error of the forecast when comparing forecasting performance of several models.

When we fit a model with dummy variables, if we have a constant term, we will have to leave out one of the dummy variables (to avoid a serious multicollinearity problem, noninvertible matrices, etc.). First, we are going to fit the model without putting a constant in the model (if we put -1 in the `lm()` command, we are doing just that). By including `factor(quarters)`, R will create a dummy (0, 1 variable) for each quarter. For example, for quarter 1, the value will be 1 if the observations is for quarter 1, 0 otherwise. When we are forecasting with a regression model, the only way we can forecast out of sample is by using future values of the independent variables. In the model that we fitted, the only independent variables are the dummies for the quarters and the trend polynomial. We create out-of-sample data for the dummies and for the t. The out-of-sample independent variables must be in the form of a data frame.

Example 9.2 As we repeated throughout this book, when data are collected monthly it is possible that there is a seasonal effect. Consider the *rooms* data set. The program for what follows in this example is *Roomsreg.R*.

We view the data over time using nice time series plots.

To check the validity of the model for forecasting out of sample, we need to do all the analysis from now on removing the last year of data. That way, we will be able to compare our forecast with the true values of the data when doing out-of-sample forecasting. We see that the amplitude of the peaks is becoming larger over time. There is an increasing width in the scatter. That is an indication of changing variance over time. We consider ways to correct that.

We plot all three transformations to see which one stabilizes the variance better.

We assume a model for the data and fit this model as a preliminary modeling step. We assume the usual about the error terms (normal, constant variance, no autocorrelation of any order, mean 0). This is a model with an additive (constant) seasonal variation. This implies that the magnitude of the seasonal swing is independent of the trend.

To fit this model to the natural log of the rooms data, we create the time variable for the trend and the dummy variables for the months.

We fit the following model:

$$lnrooms1_t = \beta_0 t + \beta_1 D1 + \beta_2 D2 + \beta_3 D3 + \beta_4 D4 + \beta_5 D5 + \beta_6 D6 + \beta_7 D7 + \beta_8 D8 + \beta_9 D9 + \beta_{10} D10 + \beta_{11} D11 + \varepsilon_t.$$

With the results obtained, we can produce the fitted values for the in-sample and then predict the values for the out-of-sample months that we did not include in the data. We also do point forecasts of the original raw data (without logs). We also obtain prediction intervals and confidence intervals for in-sample and out-of-sample.

The root mean square error of the regression is the measure of prediction error in-sample that we use.

To do out-of-sample predictions, we create first the values of the independent variables for the out-of-sample predictions that we will make. Once the predictions have been obtained, we compute the root mean square error of the out-of-sample forecast (RMSE).

We learned in earlier chapters to do an analysis of residuals using the ACF and the Ljung–Box white noise test. So, there is no excuse not to use these tests when do regression analysis of any kind.

Moreover, because now we know how to model an ARMA process based on the ACF and PACF, we have no excuse not to model residuals using one of these models when we are doing regression analysis. We are not restricted anymore to using just the Durbin–Watson test to check for autocorrelation in the residuals or to do scatterplots.

It will still be good practice, however, to do a plot of the regression residuals against time.

The commands in R that allow the modeling with residuals that are correlated can be found in the library nlme, as in the section on gls. These are the steps to follow:

- First, fit your model using regression as usual.
- Second, analyze the residuals using the correlogram, the Ljung–Box test and the time plot.
- Third, if the residuals are not white noise, model them, making use of the information in the sample ACF and the sample PACF. Fit the model until you get white noise residuals for the residuals.
- Use the tentative model of the residuals to fit a new model to the data with the gls().

Note: if the sample ACF of the residuals has the appearance of that of a series that has a trend, that means that the model has not captured the trend well, so we need to restart the model specification with a new trend. Or maybe if the trend is stochastic, the data must be differenced. That is totally separate from what we are doing here now.

At the nonseasonal level, the residual ACF dies away quickly, and the residual PACF has spikes at lags 1 and 3 and possibly 5. For the seasonal model, we identify an AR model at lag 12.

We now estimate that model for the residuals. After we fit this model to the residuals, the residuals of the residual model become white noise, based on the sample ACF. The Ljung–Box test confirms that, too. Now, we incorporate that information into the general model.   □

9.3.1   Exercises

Exercise 9.4   The goal of this exercise is to obtain a regression model with dummies for the seasons and a polynomial trend with residuals that have an ARMA structure that you have identified. The following steps should guide you in the process. You may use program *ch9electricitydummy.R*:

(a) This question is based on the electricity production series. Download the data set to your folder. Open Rstudio and then change the working directory to that folder.
(b) Do a time plot of the electricity series and give two reasons why a log-transformation may be appropriate for the electricity series. Include the plot.

(c) Fit a seasonal indicator model (not harmonic) with a quadratic trend to the (natural) logarithm of the series. Do the linear regression model using the raw data without making it a time series object this time. Is there some problem? Why? What problem? Select the appropriate graph(s) to illustrate the problem and include it. Explain what you are doing. Write the equation of the fitted model with the coefficients.

(d) Plot the correlogram and (only if needed, i.e., if there is autocorrelation) the partial correlogram of the residuals from your model. Comment on it. Include it.

(e) If there is correlation structure left in your residuals, fit an AR model to the residuals. Write the model fitted to the residuals.

(f) Plot the correlogram of the residuals of the ARMA model, and comment.

(g) If ready to do so, use the nlme library and gls() to fit a complete model that takes into account your residuals' structure. Write the regression model fitted. Comment on the value of the regression coefficients and the standard errors. Compare with the model fitted earlier.

(h) Use the fitted model to forecast electricity production for the years 1991:2000. Explain here what the code I am giving you is doing to achieve that. Explain in your own words, not by repeating the code.

9.4 Regression Using the Box–Jenkins Methodology

Autocorrelation in the residual terms is an indication that past values of the variables affect the dependent variable. Hence it is not surprising that in the history of the classical linear regression model, covariates that consist of lagged values of the predictors and of the dependent variable have played a major role and still do. They are known as distributed lag models or dynamic models. For theoretical discussion on these, see [142]. The models that we present in this section, *transfer function models*, were proposed by Box and Jenkins [7]. They solve the problems of those other models by adopting methodologies that we have learned in this book.

In regression using the Box–Jenkins methodology, we use lagged values of variables to predict another variable with the ARIMA or Box–Jenkins approach. At first glance, the method appears like a regular multiple regression model with lagged values of the variables explaining the dependent variable. However, the process of identifying, estimating and diagnosing is remotely related to the distributed lag model or dynamic models approach. The steps in the identification, fitting and diagnostics of transfer function models are:

- Identify the model for the input series, the X variable.
- Apply the input model to the output series, the Y variable.
- Prewhiten, that is, subtract the model from the series, and identify a preliminary model.
- Estimate of the preliminary model.
- Identify a model for the residuals.

We suppose that pairs of observations of an input X and an output Y from some dynamic system are available at equally spaced intervals of time. The inertia of the system can be represented by a linear filter (i.e., a model) of the following form:

$$Y_t = v_0 X_t + v_1 X_{t-1} + v_2 X_{t-2} + \cdots ,$$

in which the output deviation from some equilibrium at some time t is represented as a linear aggregate of input deviations from equilibrium at times $t, t-1, \ldots$.

The weights $v_1, v_2, \ldots$ are called the *impulse response function* of the system. This is because the $v_j's$ may be regarded as the output (value of Y_t) or response at times $j \geq 0$ to a unit pulse input at time $t = 0$, that is, to an input X_t such that $X_t = 1$ if $t = 0$, $X_t = 0$ otherwise. Impulse response functions can take many shapes. When there is no immediate response, one or more of the initial vs, say v_0, v_1, v_{t-1}, will be equal to zero.

If we differenced the input and output series once, the same model prevails, that is, measuring the change has the same model as the absolute values.

9.4.1   Stability Condition

- If the sum of all the vs is less than infinity, the system (the model) is stable. All the models we study are assumed to be stable, and therefore, we impose that condition and we must check that it is satisfied. The stability condition implies that a finite incremental change in the input results in a noninfinitesimal change in the output.
- The requirement of stability is satisfied if the condition of stationarity for the AR part is satisfied.

9.4.2   General Transfer Function Model

- It makes sense to imagine that the response variable (the output) depends also on values of the response variable in past times and the AR random term:

$$Y_t = \delta_1 Y_{t-1} + \delta_2 Y_{t-2} + \cdots + v_0 X_t + v_1 X_{t-1} + v_2 X_{t-2} + \cdots + Z_t.$$

9.4.3   The Practice of Transfer Function Modeling

Before identification, modeling and diagnosing in transfer modeling, there are some steps that need to be done carefully. The time series must be stationary and the cross-correlogram must be inspected, to start with.

- The two observed series x_t and y_t must be made stationary before starting interpreting the correlogram. So, the sample ACF of observed x_t and y_t must be looked at first of all and appropriate pre-transformations and differencing transformations must be made until the sample ACF dies down quickly or has a cutoff kind of behavior for each. We will call these stationary x and y series x^* and y^*.

- After that, the cross-correlogram between y^* and x^* must be analyzed. We need at least 50 pairs of observations to obtain a useful estimate of the cross-correlation function. If the cross-correlogram dies quickly or has cutoff behavior, then there is stationarity. Significant sample cross-correlations are those that have spikes significantly different from 0 (they lie outside the two standard error bands). If there are patterns in the cross-correlations indicating nonstationarity, the series must be transformed further.

We use the following rules about the cross-correlogram to arrive at the identification:

(a) Make sure there are no spikes at negative lags. This means past values of y_t^* affect future x_t^*. Transfer function models do not apply in this case.
(b) Identify the positive lag where the first spike appears. Our first spike appears at lag 3, so we know that y_t^* depends on x_{t-3}^* to start with and perhaps other past lags.
(c) To determine the other past lags of x^*, look at how many lags there are between the first spike and the lag at which decay starts. It takes two lags in our case. So, we know now that y_t^* depends on x_{t-3}^*, x_{t-4}^* and x_{t-5}^*.
(d) To determine the dependence of y^* on past values of itself, we look at what happens in the cross-correlation function after lags $3 + 2 = 5$. If the sample cross correlation dies down in a damped exponential function, we set the lag dependence at 1, that is, y_t^* depends on y_{t-1}^*. If it dies in a damped sine-wave fashion, we set the lag at 2, that is, Y_t^* depends on Y_{t-1}^* and Y_{t-2}^*.

Identification
- Identify an ARMA model for the input series x_t^* that you made stationary and apply this model to y_t^*.
- Get the residuals from the model for x_t^* and the residuals of the model for y_t^*. This is called prewhitening the series.
- Find the cross-correlogram between the residuals of the x_t^* and the y_t^* models. This cross-correlogram allows us to find the impulse response function.

Estimation
- Use the estimates of the impulse response function (or the cross-correlation) to make initial guesses of the orders of the actual y_t^* and x_t^* in the models.
- With the latter, get initial estimates of the parameters. The software does that.
- Estimate the parameters for the y_t^* and x_t^* parts of the model.
- With this last estimated model, look at the residuals to determine what is the ARMA model for the residuals.
- Fit the final model that incorporates the ARMA structure for the residuals.

Example 9.3 In an investigation on adaptive optimization, a gas furnace was employed in which air and methane combined to form a mixture of gases containing CO_2 (carbon dioxide) [7]. The air feed was kept constant, but the methane feed rate could be varied in any desired manner and the resulting CO_2 concentration in the off-gases measured.

The continuous data were collected to provide information about the dynamics of the system over a region of interest. The input was coded, and that is why it appears with negative and positive signs. There are 296 successive pairs of observations (x_t, y_t) read off from the continuous records at 9-second intervals. Program *transfer.R* and data set *gas.csv* are used in what follows.

The estimated auto and crosscorrelation functions of the x_t and the y_t series damp out fairly quickly, confirming that no differencing was necessary. The identification and fitting of the input x_t indicated that it is well described by a third-order autoregressive process:

$$x_t = 1.97x_{t-1} - 1.37x_{t-2} + 0.34x_{t-3} + \widehat{w}_{t(a)}.$$

All the slope parameters of this model have p-values less than 0.001. The white noise test gives mixed results. The estimate of the residual variance is $\widehat{\sigma}^2_{W(a)} = 0.0353$.

Now we apply this model to the Y_t, that is,

$$y_t = 1.97y_{t-1} - 1.37y_{t-2} + 0.34y_{t-3} + \widehat{w}_{t(b)}.$$

The estimate of the residual variance for this model is $\widehat{\sigma}^2_{w_t(b)} = 0.131438$.

We are now interested in the residuals from these two models. That is what prewhitening is. We estimate the cross correlation between the $\widehat{w}_{t(a)}$ and the $\widehat{w}_{t(b)}$ residuals, that is, $r_{ab}(k)$, for lags $k = 1, 2, 3 \ldots$.

We estimate the impulse response function using the following formula:

$$\widehat{v}_k = \frac{r_{ab}(k)\widehat{\sigma}_{w_t(b)}}{\widehat{\sigma}_{w_t(a)}}.$$

The first three vs in the impulse response function have very small values compared to their standard errors, suggesting that there are two periods of delay, that is, the effect of x_t on y_t takes three periods to start. Then, the effect ends after period 8.

The hardest part now is to identify a lag for the x_t and y_t part of the model. For sure, x_t will start having an effect at lag $t - 3$, so that is a start. But the computer does this preliminary estimation. We don't see this.

Preliminary estimation (i.e., using the theoretical relations between the vs and the parameters of the model) suggests an initial model:

$$\widehat{y}_t = 0.57y_{t-1} - 0.02y_{t-2} + 0.53y_{t-3} + 0.33y_{t-4} + 0.51y_{t-5}.$$

But the computer does this preliminary estimation. We don't see this.

Next, we must estimate the actual transfer function model using those preliminary estimates, obtaining

$$\widehat{y}_t = 0.15411y_{t-1} - 0.2774y_{t-2} + 0.628y_{t-3} + 0.472y_{t-4} + 0.7366y_{t-5}.$$

The point now is to look at the sample ACF of the residual to determine an ARMA model for the residuals. It looks like, after fitting the model, the residuals follow an

AR(2) model. So now we fit a model that incorporates that assumption about the residuals and everything else we have found out. Since the preceding model has such horrible p-values, possibly because of the high correlation between the parameters at lag 1 and 2 for the y_t part, we try to remove the second lag for the y_t part. The final model is

$$\widehat{y}_t = 5.3293 + 0.54842Y_{t-1} - 0.53552X_{t-3} + 0.376X_{t-4} + 0.51894X_{t-5}$$
$$+ 1.53292Z_{t-1} - 0.63297Z_{t-2}.$$

The residuals for this final model should be white noise. If they are not, we would start all over again. Luckily, the residuals for this model are white noise according to the Ljung–Box test and the sample ACF. □

9.5   Intervention Models

Abrupt interruptions are not a natural feature of time series data. They are usually attributed to some exogenous intervention, such as an epidemic, a war, a drastic change in policy, natural disasters, strikes, an earthquake and other unexpected events. The magnitude and abruptness of the effect renders all other interpretations impossible [116]. Intervention models, also known as interrupted time series regression models and as quasi-experimental time series models [117], are used when exceptional external events, called interventions, affect the variable to be forecasted. The statistical model is, basically,

$$Y_t = model_{before} + model_{after} + W_T.$$

where $model_{before}$ is a model before the intervention, and $model_{after}$ is the model after the intervention. The null hypothesis for this model is that the model before equals the model after.

We use special types of functions called step functions and impulse functions to build intervention models. In general, we can express an intervention model as $Y_t = V_t + N_t$, where V_t denotes the response to one or more interventions (the change in the mean function due to the intervention), and N_t denotes the model for the random part of the series ($N_t = Y_t - V_t$), the model, when there is no intervention, for the unperturbed process modeled as some ARIMA process. Before the intervention, V_t is identically zero.

Sometimes trends, seasonality and the random component of a time series hide the intervention and its effects. That is the reason why a full-fledged time series approach with ARIMA models is necessary.

The following example, featured in Bowerman and O'Connell [8], McCleary and colleagues [116], McDowall and colleagues [117] and other time series books, illustrates clearly what an intervention is and how to model it.

Example 9.4   In March 1974, the Cincinnati Bell Telephone Company initiated a policy intended to reduce the frequency of local directory assistance calls. According to this policy, each subscriber is allowed three such calls each month and then is charged 20 cents for each additional such call. Prior to March 1974, there had been no such charge. A data set that consists of the monthly average number of directory assistance calls per day (Sundays excluded) made by Cincinnati Bell subscribers from January 1962 to December 1976 is available for analysis in file *cincinnati.csv*. There are 180 observations in this data set, and the number of calls in March 1974 is observation 147. The source of this data set is an article by A. J. McSweeny in 1978 [119]. The effect of the new charge was to reduce substantially the number of calls. Program `telephone.R` contains the code for this example.

To estimate the size of this reduction and to develop a model for forecasting future directory assistance calls, special types of dummy variables called step functions and impulse functions are used to build intervention models. Bowerman and O'Connor describe the general steps to follow:

Step 1

- Find a Box–Jenkins model describing the time series values observed before the intervention. We will do this for the directory data.
- Consider the 146 observations before the intervention. Taking the seasonal difference of the regular difference of the y_t values produces stationary y_t^* values. The stationary values can be described by a seasonal moving average model. A model describing calls before the intervention is $Y_t^* = W_t + \beta_1 W_{t-12}$, where $Y_t^* = (1 - B^{12})(1 - B)Y_t$.

Step 2

- Using appropriately defined dummy variables, find a regression model describing the intervention:
$Y_t = CS_t + W_t + \beta_1 W_{t-12}.$
The S_t is called a step function and is defined as follows:
$S_t = 0$ if $t < 147$ (before March 1974)
$S_t = 1$ if $t > 147$ (March 1974 and after).
The constant C should be negative and represents the decrease in the expected number of monthly average calls per day associated with the change.

Step 2 indicates that there are two models. The one before the intervention is $Y_t^* = \beta_1 W_{t-12} + W_t$, and the one after the intervention is $Y_t^* = C + \beta_1 W_{t-12} + W_t$. The C is the decrease in the expected number of monthly average calls per day associated with the change.

Step 3

- Modify the regression model of step 2 by (i) differencing the dummy variable in the way that the y_t values were differenced in step 1; (ii) describing the error term W_t in the regression model by using the Box–Jenkins model describing the stationary Y_t^* of step 1. The modified model describing the monthly average calls per day is
$Y_t^* = CS_t^* + \beta_1 W_{t-12} + W_t,$

where

$Y_t^* =$ seasonal difference of first difference of Y_t or $Y_t - Y_{t-1} - Y_{t-12} + Y_{t-13}$,

$S_t^* =$ seasonal difference of first difference of S_t or $S_t - S_{t-1} - S_{t-12} + S_{t-13}$.

Bowerman and O'Connell [8] estimated, using the statistical software SAS, that $C = -399.82474$. So, they estimated that the effect of the charge was to reduce the expected number of monthly average calls per day by about 40,000. The final model is

$$\hat{y}_t^* = -399.824 s_t^* - 0.8565 w_{t-12},$$

where

$y_t^* =$ seasonal difference of first difference of y_t or $y_t - y_{t-1} - y_{t-12} + y_{t-13}$,

$s_t^* =$ seasonal difference of first difference of s_t or $s_t - s_{t-1} - s_{t-12} + s_{t-13}$. □

9.5.1 Different Intervention Assumptions

The intervention in the calls data was of the following kind: there was an abrupt change, and its effect was permanent.

The change could be an abrupt start and abrupt decay, or it could be that the intervention affects the mean function gradually, with its effect noticed only in the long run. There are many other possible scenarios, and each of them has a different step function.

Example 9.5 Cryer and Chan [32] studied the log of monthly airline passenger-miles in the United States from January 1996 through May 2005. The time series is seasonal. Air traffic suddenly dropped in September 2001, after the September 11, 2001 terrorist attack. Air traffic gradually regained the losses as time went on. This, the authors say, is an example of an intervention that results in a change in the trend of a time series, like the Cincinnati example seen earlier. The terrorist attack in September 2001 had lingering depressing effects on air traffic. The data set is *airmiles.data*, and the R program to follow this example while reading it is *airmiles.R*.

The authors considered that the intervention may be specified as an AR(1) process with pulse input at September 2001. But, they claim, the unexpected turn of events in September 2001 had a strong instantaneous chilling effect on air traffic. Thus, they modeled the intervention effect (the 9/11 effect) as

$$V_t = C_1 S_t^T + \frac{C_2}{1 - C_2 B} S_t^T,$$

where T denotes September 2001. In this specification, $C_1 + C_2$ represents the instantaneous 9/11 effect, and for $k \geq 11$, $C_1(C_2)^k$ gives the 9/11 effect k months afterward.

The model for the log of the data before the intervention was identified, after fitting, diagnostics, new identification and new fitting was an $ARIMA(0, 1, 1)(0, 1, 1)_{12}$. Cryer and Chan also noticed some additive outliers in December 1996, January 1997, and December 2002. Outliers are considered interventions of an unknown nature that have a pulse response function (affect instantaneously, but the effect disappears immediately afterward), and they were added to the model.

The model fit was good. The impulse response function indicated that air traffic regained its losses toward the end of 2003.   □

9.5.2   Spline Regression

A spline model might be more appropriate than an intervention model if the intervention results in a damping off rather than a drop-off, because a spline model will capture the slope change smoothly.

Spline regression is a general technique for fitting and smoothing the twists and turns of a time line (the changing slope of a regression line). It is simplest when the spline knots are few and known in advance. They are also known as piecewise linear regression.

The basic idea is to model the intervention, not as a jump, but as a smooth transition:

$$T_t = \beta_0 + \beta_1 t + \beta_2 D(t - (t_i)) + W_t.$$

where $D(t - t_i) = 1, 2, 3, \ldots$ is a count of the number of years since the intervention. A model like this gives two linear regressions without a jump in the trend.

Marsh and Cormier [114] give an in-depth introduction to spline models.

9.6   Problems

Problem 9.1   In January 1983, the UK government introduced legislation that made the wearing of seat belts compulsory for those in the front seats of cars. The aim of the law was to reduce the numbers killed and seriously injured in accidents. As there are now data available for before and after the event, we may reasonably attempt to see what, if any, effect the law had on injuries. The obvious and simple question is, "Did the law have any effect?" The data is *ksi.txt* and contains the numbers of accidents by months for periods two years before and two years after the introduction of the law. These data were analyzed using rather sophisticated methods; however, there is a lot we can do using basic statistical tools already familiar to you. The start date is January 1981 and the end date is December 1984. Being critical of what we are doing, complete the following parts.

(a) After making the data a time series object, plot the data, seasonally adjust it and obtain a trend estimate after seasonally adjusting and plotting it, all using commands in Chapter 1 of the textbook. According to this basic simple visual revelation, was the law effective? Comment critically on what you see.

(b) The data set *ksi-2* contains accidents and a new dummy variable that reflects the two time periods: 1 for after the law was enacted, 0 for before. It also has a time variable. Using standard regression techniques, estimate two trend lines and see if there is a statistically significant difference between the two trend lines (regression of the data against time and the dummy variable). We will not seasonally adjust the data now. Is there a statistically significant difference in the trends. Why? What in the regression tells you that? If they are significantly different, write the two trend lines fitted. Check the correlogram of the residuals of the regression. Interpret what you see. What should we do to make use of this regression model properly (i.e., still keep using the regression model we are suggesting)? Do what is needed to improve and what you already know how to do based on what you learned in this chapter.

(c) Difference the data and repeat the analysis. The differenced data measures the change in the variable. Do your results change much? Comment.

Problem 9.2 One of the time series in the UCI Machine Learning Repository is the Hourly Minneapolis-St Paul, MN traffic volume for westbound I-94, which has not only traffic but also weather and holiday features from 2012–18 (Hourly Interstate 94 Westbound traffic volume for MN DoT ATR station 301, roughly midway between Minneapolis and St Paul, MN) [66]. The hourly weather features and holidays are included in the data set for impacts on traffic volume. The traffic data is from the Minnesota Department of Transportation and the weather data is from OpenWeatherMap. Next to the URL of the data in [66], there is a video on anomaly detection that you could watch before approaching the data set.

The variables in this data set are as follows:

- holiday: Categorical US National holidays plus regional holiday, Minnesota State Fair
- temp: Numeric Average temp in kelvin
- rain_1h: Numeric Amount in mm of rain that occurred in the hour
- snow_1h: Numeric Amount in mm of snow that occurred in the hour
- clouds_all: Numeric Percentage of cloud cover
- weather_main: Categorical Short textual description of the current weather
- weather_description: Categorical Longer textual description of the current weather
- date_time: DateTime Hour of the data collected in local CST time
- traffic_volume: Numeric Hourly I-94 ATR 301 reported westbound traffic volume

(a) Explore the data and see what things need to be cleaned and fixed before proceeding to an exploratory analysis of the data.

(b) Use some of the approaches employed in the case studies of Chapters 4 and 3 to explore the seasonal features and trend of the data and reach a conclusion as to which features are important to include in a multiple regression model.

(c) Decide which model of the ones discussed in this chapter would be appropriate to fit to the data set to explain the traffic volume.

Problem 9.3 Bowerman and O'Connell [8], in their chapter 6, exercise 1, page 342, consider annual US lumber production from 1947 to 1976 (in millions of board feet). The data were obtained from the US Department of Commerce Survey of Current Business and can be found in data set *lumber.csv*.

(a) Plot the data against time. Discuss why the time plot of the data indicates that the following model is reasonable:

$$y_t = \beta_0 + \varepsilon_t.$$

(b) Use R to obtain the output of a regression analysis of this data. That is, estimate the model.

(c) Test for positive autocorrelation by using the Durbin–Watson statistic and setting $\alpha = 0.05$. Does the residual plot seem consistent with the outcome of the test?

Problem 9.4   The past 20 monthly sales figures for a new type of watch sold at Lambert's Discount Stores are in file *watch.csv*.

(a) Do a time plot and discuss why the time plot indicates that the following model is reasonable:

$$y_y = \beta_0 + \beta_1 t + w_t,$$

where $t = 1, 2, 3, \ldots 21, 22$ and 23. Show how the point forecasts are calculated.

(b) Test for positive autocorrelation by using the Durbin–Watson statistic with $\alpha = 0.05$. Does the residual plot seem consistent with the outcome of the test?

Problem 9.5   A rocket motor is manufactured by bonding a chemical igniter with propellant inside the metal housing. It is speculated that the longer the motors sit before usage, the more the bond deteriorates, with deterioration measured by sheer strength of the propellant. If we are interested in determining the effect of age on sheer strength, the dependent variable of the regression will be sheer strength and the independent variable will be age.

Check the residuals with a plot of the standardized residuals against age, and do a Durbin–Watson test of the residuals to determine whether autocorrelation of order 1, AR(1) exists in the residuals when you fit a simple linear regression model of shear strength against age. The data set is *propellant.txt*.

Problem 9.6   Chatterjee and Price [21] offer as an example of this situation the efforts of a company that produces and markets ski equipment in the United States to obtain a simple aggregate relationship of quarterly sales (S) in millions of dollars to some leading economic indicator. The indicator chosen is personal disposable income, PDI, in billions of current dollars. The initial model is

$$S_t = \beta_0 + \beta_1(PDI) + w_t.$$

Data for 10 years (40 quarters) is available. The first observation was obtained for the first quarter of 1964, the last for the fourth quarter of 1973. After fitting this model by least squares, we find that the residuals show a pattern. Residuals from the first and fourth quarters are positive, and residuals from the second and third quarters are negative for all years. Since skiing activities are affected by weather conditions, we suspect that a seasonal effect has been overlooked. The pattern of residuals suggests that there are two seasons that have some bearing on ski sales: the second and third quarters, which correspond to the warm weather season, and the fourth and first quarters, which correspond to the winter season when skiing is in full progress.

One way to characterize the seasonal effect is by defining an indicator variable (dummy variable) that takes the value 1 for each winter quarter and is set to zero for each summer quarter. The expanded data set can be found at skisales. Using the additional seasonal variable, the model is expanded to

$$S_t = \beta_0 + \beta_1 PDI_t + \beta_2 A_t + w_t,$$

where A_t is the dummy variable and β_2 is the parameter that will represent the seasonal effect. Specifically the model represents the assumptions that sales can be approximated by a linear function of PDI, one line for the winter season and one for the summer season. The lines are parallel: that is, the marginal effect of changes in PDI is the same in both seasons. The level of sales as reflected by the intercept is different in each season. Since Z_t is always 1 in the cold weather quarters and zero for warm quarters, we have

$$S_t = (\beta_0 + \beta_2) + \beta_1 (PDI)_t + w_t$$

for the winter season, and

$$S_t = \beta_0 + \beta_1 (PDI)_t + w_t$$

for the summer season.

Looking at the graph of the residuals against time we can see that the seasonal patterns have been removed. The precision of the estimated marginal effect of PDI increased. Also, the seasonal effect has been quantified, and we can say that for a fixed level of PDI the winter season brings between $4,759,484 and $6,169,116 over the summer season (with confidence coefficient equal to 95 percent).

(a) Analyze the data and replicate the results obtained.
(b) Test for first-order autocorrelation in the residuals of the model without the dummy variable, using the DW test. What did you learn regarding the usefulness of the DW to detect higher-order autocorrelations?

Problem 9.7   The data set is the number of hours of calculator use over time.

(a) First, enter the data and look at the data distribution. Does the assumption of normality look reasonable?

```
calculator=c(197,211,203,247,239,269,308,262,258,
256,261,288,296,276,305,308,356,393,363,386,443,308,
358,384)
hist(calculator,main="histogram of calculator data")
```

For the rest of the problem, use the program in *calculator.R*.
(b) Now view the data over time using nice time series plots. Describe the time plot.
(c) Calculate the regression line. Write here the fitted model and summarize the findings, justifying them with the output.

(d) Execute the Durbin–Watson test of AR(1) in the residuals, but this time we will look at all possible DW alternative hypotheses. Write down the conclusions of the DW tests.

Problem 9.8   Read the mink and muskrat data *mink.dat* that appeared in [85]. This data set contains annual mink and muskrat trappings in the Hudson Bay Area for the years 1848 to 1911. We are interested in designing a model to predict mink trappings. We wonder whether using the information provided in the muskrat trappings will give us better predictors of mink trappings than using the mink time series alone (better in the mean square prediction error sense) and using an ARIMA$(p, d, q)(P, D, Q)$ model. What do you think? Using the period 1948–99 as the training data, and the period 1900–11 as the forecast period, provide an answer supported by statistical analysis. Write your answer in the form of a report and support it along the way with graphs, results, and tables.

Problem 9.9   Cowpertwait and Metcalfe [30] fitted a state-space model to forecast the salinity of the Murray river. According to these authors, "the Murray River supplies about half of South Australia's urban water needs and, in dry years, this can increase to as much as ninety percent. Other sources of water in South Australia are bore water and recycled water, although both tend to have high salinity. The World Health Organisation (WHO) recommendation for the upper limit of salinity of potable water is 500 mg/l (approximately 800 EC), but the domestic grey water system and some industrial and irrigation users can tolerate higher levels of salinity. The low rainfall and increasing population makes the efficient use of water resources a priority, and there are water-blending schemes that aim to maximise the use of recycled water. The average monthly salinity, measured by electrical conductivity (EC; microSiemens per centimetre at 25C) and flow (Gigalitres per month) at Chowilla on the Murray River, have been calculated from data provided by the Government of South Australia (…) Predictions of salinity are needed for the recycled water schemes to be operated efficiently."

Cowpertwait and Metcalfe used a data set that we will call *Murray.txt*, which contains 81 months of flows and salt concentrations at Chowilla to help in that prediction.

The data is monthly and starts in January 1996. The following variables are used:

Variable name in data set	Description
Rain (mm)	Precipitation at that location in inches
Flow (ggl/meter) (L)	Water flow
Temp (centigrades)	Temperature
Chowilla (S)	Salinity at the Chowilla station (ms/cm)
Renmark	Salinity of the Renmark wetlands adjacent to the river

Of interest in this problem is to forecast with the best possible models the variable S. You will use exponential smoothing, univariate a-theoretical models that only use the past history of this variable (ARIMA), multivariate models that use the past history of all the variables only (VAR), or regression models that use the other variables as covariates, if you can justify that there is a causal reason to do so with cross-correlation functions. Then you will use one of the machine learning methods, `prophet`, seen in the case study of Chapter 6.

For fitting purposes, you will not use the last six observations. So use for model fitting all except the last six observations. Thus, you will use only 75 observations as training set and the last six as test set. Keep the last six to compare your forecasts with the actual values and compute the RMSE of those six forecasts. Fit those models, and compare their forecasting performance using the RMSE. You report should contain the following:

(a) INTRO. Intro to the problem, talk about the data, plot the data, display the first five lines of the data set and say that you will be analyzing the forecasting performance of several competing forecasting models with the intention of choosing the best one among them and obtaining an average forecast from the best of them are the following: Describe the data set, explore relations, check correlations, do needed tests, explore the seasonality, trends, all the things we usually do.

(b) Fit the following models, explaining what you are doing.

(1) ARIMA or SARIMA modeling

- Identification
- Estimation. Write final estimated model with standard errors, p-values.
- Diagnostics, tests
- Forecasts

(2) Exponential Smoothing. Fitting, diagnostics, forecasting.

(3) Time Series regression with autocorrelated errors using seasonal dummies, time trends and other covariates that could be causally related to Murray. Fitting, diagnostics, corrections for autocorrelations if needed, forecast. Write final estimated model, standard errors, p-values, tests.

(4) Vector autoregression of all the variables and impulse response functions. Fitting, diagnostics, forecasting. Justify model. Write final estimated model, interpret.

(5) Prophet. Fitting, diagnostic and forecasting.

(6) Conclusion. Do theoretical models that use other variables forecast better than atheoretical ones? Discuss your forecasts and how they compare in terms of RMSE. Include a table like this one containing all the forecasts for each method and the average of the forecast for each year (see section on consensus forecasts in Chapter 6):

Time	ARIMA	Exp sm	VAR	Reg	Average	true S
April 2002		forecast	forecast	forecast	average of row	
May 2002	...	...				
....	..					
RMSE						

where

$$RMSE = \sqrt{\frac{\sum(\text{forecast} - \text{true S})^2}{6}}.$$

(7) References.
(8) Appendixes of program with all the output and separated by model clearly.

Problem 9.10 Consider the data set *returns.txt*. This data set contains the returns per month (in cents per pound) obtained from young chicken packaged and ready to go. The return represents the monthly profit in this part of the chicken industry. The data extends from January 1967 to December 2002. During all this period there were some major events that could have affected returns in this industry. One of them is that the industry changed drastically, with all the operations in the chicken process becoming vertically integrated in a few large firms that took care of everything from hatching eggs, raising the chicken to packing the chicken meat and eggs. There was also the oil crisis in the early seventies.

(a) Do the data in "*returns.txt*" reveal whether those events affected the industry at all? Explain. Illustrate your answer with the appropriate statistical analysis.
(b) Find the sample autocorrelations (up to lag 20) and plot them.

9.7   Quiz

Question 9.1 The following quarterly univariate data, Y_t, is the subject of this problem. You may see the time plot of the data in Figure 9.3.

```
y=c(3.3602,-3.1769,0.3484,7.469,4.4963,-0.4621,
0.7218,6.9484,5.2374,2.9242,4.7006,11.2793,5.1637,
1.5441,12.121,9.6588,8.0922,3.9653,11.4177,13.2088)
```

We fitted the following linear regression model to this data.

$$\hat{y} = 1.09 + 0.46\,time - 4.77Q2 - 0.335Q3 + 3.052Q4,$$

where t is time $(1, 2, 3, \ldots)$ and Q is a dummy variable for the quarter. Q2, for example, is 1 if the observation is for quarter 2. The first quarter in the data set is quarter 1.
The values of the first two residuals are as follows.

The time series

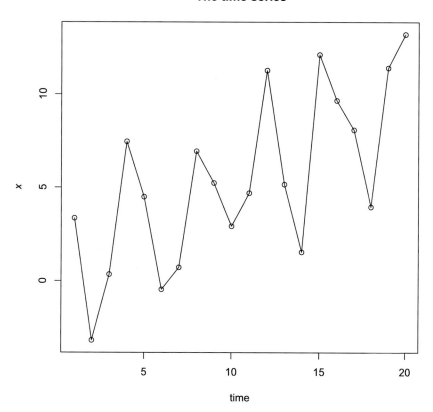

Figure 9.3 The time plot of a time serie used in Question 9.1.

(a) 3.3602, −3.1769
(b) 0, 3
(c) 1.5, −0.356
(d) 2.1431, −3.1454
(e) −5.324, 2.179

Question 9.2 Which of the following could be used to detect autocorrelation present in the residuals of a linear regression model? Check all that apply.

(a) Runs test
(b) Durbin–Watson test
(c) sample ACF
(d) lag plots

Question 9.3 Solutions to correct for the autocorrelation of the residuals of a linear regression model have consisted of which of the following? (Check all that apply.)

(a) Use generalized least squares instead of OLS to estimate the parameters of the regression model.

(b) Add more independent variables that might be correlated with the dependent variable.

(c) Ignore the autocorrelation and adjust the standard error to account for it.

(d) Approach the problem by using Box–Jenkins transfer function models or intervention models.

Question 9.4   If using GLS to estimate the parameters of a regression model when there are autocorrelated errors and using the `gls` function of R, which types of residuals should be checked after the correction?

(a) The default residuals created after fitting the model with the `gls` function of the `nlme` package.

(b) The residuals of the AR model fitted to the residuals.

(c) The normalized residuals.

(d) The residuals before correcting for the autocorrelation.

Question 9.5   The model fitted to the residuals of a linear regression model that presented a problem of autocorrelation should itself have residuals that are

(a) autocorrelated

(b) a function of the independent variables in the model

(c) a moving average process of order 1, MA(1)

(d) white noise

Question 9.6   Time series proper regression models are models that (check all that apply)

(a) incorporate a model for the trend

(b) incorporate indicator variables for the seasonal effects

(c) incorporate independent variables that are not trends or seasonals

(d) only include past values of the dependent variable as independent variables

Question 9.7   The correction for autocorrelation of residuals in a regression model is important in order to avoid which of the following?

(a) underestimated standard errors

(b) overestimated standard errors

(c) finding too many parameters that are not significant significant

(d) finding too many nonsignificant parameters significant

9.8   Case Study: Forecasting at Scale

The authors of the R package `forecast` [78] have created a new package based on the `tidyverse` package in R to address the needs of modern forecasting in business. As in the case of the `forecast` package, the authors have created an open source book to accompany the execution of the functions in the package. The text is introductory and accessible. The package requires familiarity with the `tidyverse` package, so the learning curve for this package is steeper than for the authors' `forecast` package, which required familiarity with Base R. The package and library `fpp3` must be downloaded in order to run the examples in the book. Other libraries that need to be installed are `fable`, `tsibble`, `tsibbledata` and `feasts`. In fact, most of the programs written in Base R for this book have been translated to this new packages language and can be found at the book's website. After doing this case study, they will be easier to understand.

This case study is different from the ones we have done in other chapters. The reader will focus on Section 11 of [76] and study what the authors are doing to forecast aggregate and disaggregated time series. If needed, the reader can review the book from the beginning to become familiarized with the R packages used, which are all relatively new, some of them not yet in CRAN. The reader does not need to become an expert, but rather see how this new package does things that we have been doing in this book in past chapters.

Once familiarized with the package a little bit, focus on the examples in Section 11.1 and see what is common to all the examples. The authors talk about forecasting large numbers of disaggregated time series and the aggregate series when the goal is that the aggregate forecast be the total of the individual forecasts. You are not being asked to become an expert on such endeavor, but rather to look at the examples in 11.1 and the details of 11.4 and think about what variables you could use as dependent and independent variables were you to do a multiple regression model. Express your thoughts by doing the exercises.

Think that this problem of aggregate versus disaggregate forecasting is faced daily by major retailers, high-tech companies and governments. Walmart and Amazon must predict thousands of product lines' demands from different regions of the world and at the same time obtain aggregate forecasts. Thus, the topic of Section 11.1 in [76] is something that is commonly found in the workplace nowadays.

9.8.1 Exercises

Exercise 9.5 If you had to write a regression model to estimate the most aggregate level of Australian tourism over time, which would be the most appropriate variables to include as independent variables in your regression (based on what you read in section 11 of [76])? Which would be the dependent variable?

Exercise 9.6 If you had to write a regression model to estimate the most disaggregated level of Australian tourism over time, which would be the most appropriate variables to include in your regression? Which would be the dependent variable? Which would be the independent variables?

Exercise 9.7 In section 7 of [76] there is a discussion on regression modeling at a very introductory level, but some nice R tools are introduced to enhance the visuals. There are also some terms that we have not talked about in this chapter. What do the authors mean by ex ante forecast? In what context do they apply it? What is, according to the authors of [76], scenario-based forecasting? Where in our book have we talked about something similar?

9.9   Appendix: Review of Introductory Classical Regression

This appendix serves as a review of simple linear regression, and a classical approach to the correction for autocorrelation that has been used in the past. This material is presented in many books on regression modeling ([1], 177–182; 307–310; 313–315; [21]).

For this review, we assume that we will be applying the classical regression to time series data, although the theory also applies to other types of data.

The main tenet of a linear regression model is that y_t, the observed value of a random variable Y_t in a random sample ($t = 1, 2, \ldots$), can be decomposed into two parts: a systematic part, $\mu_t \mid x_t$, the expected value when another variable X_t takes a particular value x_t, and a part that cannot be explained by those circumstances, but whose effect is small, ε_t.

A scatterplot of values of the observed y_t against x_t, for all values of t, where the trend looks linear, would suggest that a linear model can be used to explain the following relation:

$$y_t = \mu_t \mid x_t + \varepsilon_t,$$

where $\mu_t \mid x_t = \beta_0 + \beta_1 x_t$ is the systematic part, the part that can be explained by the other random variable, and ε_t is the part that cannot be explained by those circumstances. The regression model attempts to estimate the μ_t by estimating the values of β_0 and β_1. Thus, the model is usually presented in this form:

$$y_t = \beta_0 + \beta_1 x_t + \varepsilon_t.$$

The full specification of the simple regression model consists of the following assumptions that, if satisfied, make the Ordinary Least Squares estimators of the β_0 and β_1 parameters BLUE (Best Linear Unbiased Estimators):[5]

- Linearity: the relationship between X_t and Y_t is linear.
- Nonstochastic X: $E[\varepsilon_t X_t] = 0$, meaning ε_t and X_t are uncorrelated.
- The error term has mean 0: $E[\varepsilon_t] = 0$.
- The error term has constant variance: $E[\varepsilon_t^2] = \sigma^2$.
- The error term at time t is uncorrelated with the error at any $t - k$: $Cov[\varepsilon_t, \varepsilon_{t-k}] = E[\varepsilon_t \varepsilon_{t-k}] = 0, \quad k \neq 0$.

---

[5]OLS is the method usually learned in an introduction to regression book; but, of course, there are other methods. Depending on the discipline, introductory regression books bear a diverse array of titles. For example, in Economics, the titles would usually be "Introduction to Econometrics," (introduction to the use of regression in Economics). In Psychology, the title would be "Introduction to Psychometrics," (introduction to the use of regression in Psychology).

Provided that these assumptions hold, it is possible to estimate optimally (i.e., minimizing some loss function) the regression parameters and their variances. OLS consists of conducting the following optimization problem:

$$S(\beta_0, \beta_1) = \sum_{i=1}^{n}(y_i - \beta_0 - \beta_1 x_i)^2, \tag{9.1}$$

$$\partial S/\partial \beta_0 = 2n\beta_0 - 2\sum y_i + 2\beta_1 \sum x_i = 0,$$

$$\partial S/\partial \beta_1 = 2\beta_1 \sum x_i^2 - 2\beta_0 \sum x_i - 2\sum x_i y_i = 0, \tag{9.2}$$

and solving for the two unknowns, β_0 and β_1. The solutions are

$$\widehat{\beta_1} = \frac{n\sum x_i y_i - \sum x_i \sum y_i}{n\sum x_i^2 - (\sum x_i)^2}$$

and

$$\widehat{\beta_0} = \bar{y} - \widehat{\beta_1}\bar{x}.$$

The $\widehat{\beta_0}$ and $\widehat{\beta_1}$ are only approximations to the true parameters β_0 and β_1. How good are they? Mathematical statistics helps us evaluate their expected value, standard error and distribution if we make an additional assumption. By adding the following assumption, we have the classical normal linear regression model.

• The error term, ε_t, is normally distributed.

This assumption is necessary only for the statistical testing of the model (e.g., significance tests such as the t-test and F test), confidence intervals and prediction intervals. The t-test statistic is conducted using

$$t = \frac{\widehat{\beta_1} - \beta_1}{se(\widehat{\beta_1})},$$

which is t-distributed with $n - 2$ degrees of freedom under the normality assumption for ε_t.

Under the assumption of normality, $\widehat{\beta_0}$ is normally distributed with $E(\widehat{\beta_0}) = \beta_0$ and standard error:

$$se(\widehat{\beta_0}) = \frac{\sigma^2 \sum x_i^2}{n\sum (x_i - \bar{x})^2}. \tag{9.3}$$

$\widehat{\beta_1}$ is normally distributed with $E(\widehat{\beta_1}) = \beta_1$ and standard error

$$se(\widehat{\beta_1}) = \frac{\sigma^2}{\sum (x_i - \bar{x})^2}. \tag{9.4}$$

The estimator for σ^2 is

$$\widehat{\sigma}^2 = s^2 = \frac{\sum_t (y_t - \widehat{\beta_0} - \widehat{\beta_1})}{n - 2}, \tag{9.5}$$

which we plug into all formulas for estimates in which σ^2 appears.

The estimators are optimal in the sense that they are unbiased, efficient and consistent. They are BLUE under the assumptions made. Unbiased estimator means that the expected value of the estimator is equal to the true value of the parameter. An estimator is relatively efficient if it has a smaller variance than any other estimator. An estimator is consistent if both the variance and the bias approach zero as the sample size approaches infinity. Together, these properties mean that an estimator is centered around the true value, not too far from that value on average, and converges to the true value as the sample size increases. These are all very desirable properties if we want to trust the estimator as our guide to the true value of the parameter.

9.9.1   Confidence Intervals for $E(Y_t \mid X_t)$

Recall that when we fit a simple regression line to data we are obtaining an estimate of $\mu_t \mid x_t$. For any x_t, say x_0, the estimate obtained is

$$\widehat{\mu}_0 \mid x_0 = \widehat{E(Y \mid x_0)} = \widehat{\beta}_0 + \widehat{\beta}_1 x_0.$$

The $\widehat{\mu}_0 \mid x_0$ is a random variable, because it depends on random variables $\widehat{\beta}_0, \widehat{\beta}_1$:

$$\mathrm{Var}(\widehat{\mu}_0) = \mathrm{Var}(\widehat{\beta}_0) + x_0^2 \mathrm{Var}(\widehat{\beta}_1) + 2x_0 \mathrm{Cov}(\widehat{\beta}_0, \widehat{\beta}_1) \qquad (9.6)$$

$$= \frac{\sigma^2 \sum x_i^2}{n \sum (x_i - \bar{x})^2} + x_0^2 \frac{\sigma^2}{\sum (x_i - \bar{x})^2} - 2x_0 \frac{\bar{x}\sigma^2}{\sum (x_i - \bar{x})^2}$$

$$= \sigma^2 \left[\frac{1}{n} + \frac{(x_0 - \bar{x})^2}{\sum (x_i - \bar{x})^2} \right].$$

The standard error is just the square root of the variance. Notice that $\widehat{\mu}_o \mid x_0$ is what we usually call $\widehat{y}_o$.

The sampling distribution of the $\widehat{y}_o$ is normal if the assumptions of the linear model hold. A $100(1 - \alpha)\%$ confidence interval for $E(Y \mid X)$ when the independent variable is X_0 is

$$\widehat{\mu}_0 \mid x_0 \pm t_{n-2}^{\alpha/2} \widehat{\sigma} \sqrt{\frac{1}{n} + \frac{(x_0 - \bar{x})^2}{\sum (x_t - \bar{x})^2}}.$$

9.9.2   Prediction Intervals for Y

Given x_0, we might want to estimate not the expected value of Y for that value of X, but rather the actual value of Y for that value of X. That involves an extra dose of uncertainty. The point estimate will be the same, but the prediction intervals will be wider than the confidence intervals:

$$se(\widehat{y}_0 \mid x_0) = \widehat{\sigma} \sqrt{1 + \frac{1}{n} + \frac{(x_0 - \bar{x})^2}{\sum (x_t - \bar{x})^2}}.$$

In the model used for starting the derivation, we must include the ε_t, which has variance σ^2.

The prediction interval then will be

$$\widehat{\mu}_0 \mid X_0 \pm t_{n-2}^{\alpha/2} s \sqrt{1 + \frac{1}{n} + \frac{(X_0 - \bar{X})^2}{\sum (X_i - \bar{X})^2}}.$$

Usually, we would obtain the prediction and confidence intervals with the data we have, for each value of the independent variable. This will give an idea of the goodness of the model.

9.9.3 Forecasting

One type of prediction is the prediction we make of the data that we have used to fit the model. If we are planning to predict what will happen outside that time range, then the prediction is out-of-sample.

We first can validate the model by splitting the sample into the training set and the test set. The training set is to "learn" the model, that is, to train the model for this data. The test set then will be used to see how well the model helps predict data that was not used to fit the model. An indication that the model will work for data not used to learn the model is that observations in the test set are actually inside the prediction and confidence intervals. This will help us validate the model for prediction.

A different issue is forecasting. In forecasting, we must forecast the independent variable first, and then apply the model to that estimated value. This fact has made regression appear at a disadvantage when competing with other methods. Uncertainty increases when forecasting out-of-sample. If we must add to that uncertainty the uncertainty from having to forecast the predictors, then we have much larger uncertainty.

9.9.4 When the Errors Are Autocorrelated

When the assumption of non-autocorrelation of the errors is violated, then, from a theoretical point of view, equations 9.3, 9.4 and 9.5 are no longer appropriate.

The presence of autocorrelation has several effects on the analysis using ordinary least squares. These are summarized as follows:

- Least squares estimates of the regression coefficients are unbiased but are not efficient in the sense that they no longer have minimum variance (they are not BLUE).
- The estimate of σ^2 and the standard errors of the regression coefficients may be seriously understated; that is, from the data the estimated standard errors would be much smaller than they actually are, giving a spurious impression of accuracy.
- The confidence intervals and the various tests of significance commonly employed would no longer be strictly valid.

The presence of autocorrelation can be a problem of serious concern for the preceding reasons and should not be ignored. For that reason, it is important to know what to do when there is autocorrelation. There are several options, but we will distinguish two very general ones: (i) correct for the autocorrelation somehow but continue with the linear regression model; (ii) use some of the time series models learned in this book instead of using classical linear regression. This chapter discusses the former.

Before we take any action regarding corrections, we need to make sure that there is autocorrelation.

In the next section, we review classical approaches to detecting autocorrelation in the context of the linear regression literature.

9.9.5　Formal Classical Statistical Tests for Autocorrelated Residuals

Since autocorrelation poses such serious problems for the use of simple OLS, it is extremely important to test for its presence in a given time series. If one does not know or is not willing to assume any form of autocorrelation, it is necessary to turn to the time series for information. Residual plots and tests are needed. We talk in this appendix about classical approaches to test for autocorrelation of residuals.

9.9.6　Runs Test and Durbin–Watson Statistic

Some people do a formal runs test to see if there is autocorrelation in the residuals. Geologists and hydrologists like to use runs tests. They are not used much by time series statisticians, but they are intuitive and useful. Others who are willing to assume that the error term might be autoregressive of order one, might use the Durbin–Watson (DW) test. In this section, we study the runs test and the DW test.

Runs Test

To do a Runs test, we do a sequence plot, that is, plots of the signs of the residuals and counting the number of runs in a plot. In our present example, the sequence plot of the residuals is $+++++++-------+++++$ and it indicates three runs. With n_1 residuals being positive and n_2 residuals negative, under the null hypothesis of randomness, the expected number of runs and its variance would be

$$H_0: \mu = \frac{2n_1 n_2}{n_1 + n_2} + 1; \qquad \sigma^2 = \frac{2n_1 n_2 (2n_1 n_2 - n_1 - n_2)}{(n_1 + n_2)^2 (n_1 + n_2 - 1)}.$$

In our case studied in Section 9.2, $n_1 = 13$ and $n_2 = 7$, giving the expected number of runs under randomness to be 10.1 and a standard deviation of 1.97. The observed number of runs is three. The deviation of 5.1 is more than twice the standard deviation, indicating a significant departure from randomness. This formal runs test procedure merely confirms the conclusion arrived at visually from looking at Figure 9.1 that there is a pattern in the residuals.

The runs test we have used is approximate and assumes large n_1 and n_2. It should not be used for small values of n_1 and n_2 (less than 10 each). Books on nonparametric statistics explain exact runs tests. So, we should not rely much on it in our case here since we have only 20 observations in the data set.

Besides the graphical analysis, which can be confirmed by the runs test, autocorrelated errors (if they are of the AR(1) order) can also be detected by the Durbin–Watson (DW) test.

Durbin–Watson Statistic to Test for AR(1) Autocorrelation in the Residuals

The Durbin–Watson statistic, d, is the basis of a popular test of autocorrelation in regression analysis. The test is based on the assumption that successive residuals are correlated following an autoregressive time series stochastic process of order 1:

$$\varepsilon_t = \rho \varepsilon_{t-1} + W_t, \quad |\rho| < 1,$$

where ρ is the correlation coefficient between ε_t and ε_{t-1}, and W_t is normally white noise.

The Durbin–Watson statistic is defined as follows:

$$d = \frac{\sum_{t=2}^{n}(\widehat{\varepsilon}_t - \widehat{\varepsilon}_{t-1})^2}{\sum_{t=2}^{n}\widehat{\varepsilon}_t^2},$$

where $\widehat{\varepsilon}_t, t = 1, 2, \ldots$ are the residuals after you fit the model. The distribution of this statistic if the null hypothesis is true follows a linear combination of χ^2 distributions.

This test is implemented in the `lmtest` package in R. The statistic d is used for doing one of the following tests:

- $H_0: \rho = 0;$ $H_a: \rho > 0$

  - If $d < d_{L,\alpha}$, reject H_0
  - If $d > d_{U,\alpha}$, do not reject H_0
  - $d_{L,\alpha} < d < d_{U,\alpha}$, test is inconclusive

- $H_0: \rho = 0;$ $H_a: \rho < 0$

  - If $4 - d < d_{L,\alpha}$, reject H_0
  - If $4 - d > d_{U,\alpha}$, do not reject H_0
  - $d_{L,\alpha} < 4 - d < d_{U,\alpha}$, test is inconclusive

- $H_0: \rho = 0;$ $H_a: \rho \neq 0$

  - If $d < d_{L,\alpha/2}$ and $4 - d < d_{L,\alpha/2}$, reject H_0
  - If $d > d_{U,\alpha/2}$ and $4 - d > d_{U,\alpha/2}$, do not reject H_0
  - $d_{L,\alpha/2} < d < d_{U,\alpha/2}$ and $d_{L,\alpha/2} < 4 - d < d_{U,\alpha/2}$, test is inconclusive

The value of d is close to 2 when the autocorrelation is 0, and close to 0 when the autocorrelation is 1. The values of (d_L, d_U) were tabulated by Durbin and Watson. The L and the U refer to the table for the DW test. If using software, the software will give the p-value of the test, and those conditions just listed will not be needed.

When R code or other statistical software code is not available, tables are needed to determine statistical significance. For each level of significance, α, those tables will have lower (L) critical values and upper (U) critical values. We have listed those values below each type of test.

A problem with the DW is that it gives evidence only of first-order autocorrelation. If the pattern of time dependence is higher than first order, the plot of residuals will still be informative, but the DW statistic will not. It is possible that the DW statistic does not suggest autocorrelation, but there is autocorrelation of a higher order. For example, if you omit a dummy variable for the seasonal component, the DW will not complain about autocorrelation, but the errors will tend to have a pattern of negative residuals for some quarters and positive for others, suggesting autocorrelation. In this book, we studied other time series models that could be considered for the residuals.

10 Machine Learning Methods for Time Series

10.1 Introduction

> Machine Learning (ML) is the study of computer algorithms that teach computers how to use data to solve a problem [61]. An algorithm to do regression, for example, uses data to estimate the parameters of the model and to forecast. It is one of the several predictive models in ML. When the model is "fitted" to training data, ML says that the model is "learned." The whole process is called "supervised learning." In some sense, all the model fitting and forecasting done in this book is a ML task.

There is much more to ML than regression, of course. But advanced ML methods, such as for example Neural Networks (NN), were not originally developed for time series data, although even in their infancy they have shown some promise for the analysis of time series [132]. The volume, velocity and variety of time series data that we are confronted with nowadays is contributing to an increasing incidence of application of those methods in time series data analysis. They assist us in making sense of massive amounts of time series data that sensors and the IoT have allowed us to record.

We distinguish between supervised and unsupervised ML methods. We have introduced unsupervised ML concepts such as clustering in exercises, examples and case studies throughout the book. Unsupervised methods concern grouping of time series by similarity without knowing a label for the groups. We have seen clustering, with k-means, to divide the time series into clusters or groups of similar members. In some cases, like the rivers Example 1.14 we can label the rivers. In others, as when we clustered stochastic models in Example 3.11, labeling would be harder. Clustering is a method used for knowledge discovery only. Once we know the groups, it is up to us how we want to label them. Clustering provides an insight into the natural groupings found in data [102]. Supervised ML methods require that we know the label of the groups in order to classify new objects into a particular group. Classifier algorithms allow us to do that. An advantage of supervised ML is that it can be used for prediction. Supervised ML methods concern the prediction of one or more variables using other variables, a topic that we explored with statistical methods in Chapters 7, 8, 9.

> Usually, handling ML models for regression require some basic background in training and tuning machine learning models in general [99] [132].

In order to do ML for time series, the reader would benefit from some acquaintance with optimization algorithms that are simple, such as, for example, Newton algorithm, and from knowledge of the language of ML. When doing time series with ML methods, the reader will need to control the stopping rule, the number of times that the algorithm will be allowed to run and the criterion used for stopping, which are things that a basic Newton algorithm requires. For example, when we did cluster analysis using summary features of the time series in Chapters 1, 3 and 5, we needed to specify the number of clusters to consider. That is an argument of the algorithm (a parameter, in ML language). We used a function in R that makes other arguments by default, but if we wanted to do better, we would have had to enter more arguments in the k-means function. And in Chapter 2 we left `prophet` use some default arguments, but we could perhaps have improved it by changing the arguments. Numerical optimization with R at the level of Chapter 7 in Braun and Murdoch's book [11] or a textbook on numerical optimization at that level would suffice to get the general idea.

> ML algorithms act like black boxes of complex mathematics that makes the algorithm function, and that makes the careful fine-tuning of orders fed to them crucial in order to obtain good performance.

Familiarity with the meaning of words used in ML also is needed, particularly if the reader's training in data science was done in a Statistics learning environment. ML practitioners come from different backgrounds. A language has been forged in which an unknown word to us in ML denotes something that we learned in Statistics using a totally different name in a totally different context. For example, the orders that we give to the algorithms are called "parameters" by ML practitioners, a term that, in Statistics, is reserved for the statistical model parameters. What an ML practitioner calls parameters we call arguments of functions in the R software. Most algorithms come with default arguments (aka parameters in ML), but quite often changing the arguments will result in improved performance. To give another example, variables in raw data are interchangeably called "features" by practitioners of ML, when the same ML uses the term features for summaries of time series. To add one more example, fitting a model to a data set is called training in ML. Knowing the different meanings of the same words used in both ML and Statistics will make the task of becoming familiar with ML for time series easier for someone that has been trained in Statistics or that plans to use both statistical time series and ML in time series analysis. Accessible introductions to ML that use R are offered by Lantz [102].

This chapter is a small survey of supervised ML methodology for time series, accompanied by a few examples found in the available literature that uses R for the analysis. We will start with ML-based regression in Section 10.2. In that section we will talk about Random Forests (RF) and Gradient Boosting (GB). Then in Section

10.3 we will introduce Deep Learning (DL). Unsupervised ML is then taken in Section 10.5. We end the chapter with an application to learning about blood glucose and insulin pumps in Section 10.6. The prediction of blood glucose level for individual patients is a very active area of ML research for time series data in health applications. The reader will find additional examples in the book's website.

10.2   Machine Learning-Based Regression

Machine Learning-based regression (ML-based regression) is the process of creating time series forecasting models with ML models. On the plus side, ML-based regression follows the same process that we use to forecast with the linear regression model. That is, we must do exploratory analysis of the time series data to determine what features (variables) might be important to predict the dependent variable, we fit (learn) a model to a training set and test the forecasting performance of the model with a test set. We input independent variables and predict the dependent variable. The practitioner with subject matter knowledge will decide which variables to feed to the algorithm and thus improve its performance (although in ML-based regression that is not considered necessary). But the reader should once more be aware that ML-based regression, like other ML methods, was not originally developed for time series–specific data, unlike the other models we have studied in the other chapters of this book. We have been alerting the reader to this fact throughout the ML examples we have introduced in earlier chapters.

There are some drawbacks in using ML-based regression models. They are a little more complex than the methods analyzed so far in this book in that they require more technical knowledge. Quantifying the uncertainty in the models fitted or the forecast produced by an ML model is not as easy as with the classical methods. In fact, it is not an expected outcome of ML-based regression. Despite their higher complexity for a beginner, ML-based regressions present many advantages for forecasting when the data is of high quality and there are intrinsic relationships between the independent variables and the dependent variable. Readers interested in implementing ML-based regression are recommended to first undergo an introduction to ML and perhaps also some numerical optimization. Software offers default settings, but some knowledge of what optimizations entail will result in better models.

In this section, we will be talking about Random Forest (RF) and Gradient Boosting (GB). First, we will summarize the methods; then we will summarize the procedures typically followed when applying those types of algorithms to predict time series, and finally we will give an example.

10.2.1   Random Forest

Random Forest (RF) algorithms are decision tree methods. To understand RF algorithms it helps if the reader has some familiarity with decision trees. Decision trees mirror the way humans make decisions: one step at a time, and in a highly nonlinear fashion, thinking about how one variable should affect our decision, then another, much like a flow chart [132, 105]. A good introduction to how decision trees are used in ML is [102]. Growing a tree for regression involves deciding which features to choose and what conditions to use for splitting, how many times to split, and when to stop the algorithm. The splitting is of the time series variables' values.

Random Forest is one of the most popular ML algorithms for ML-based regression. The name comes from the fact that the algorithm builds many decision trees, that is, it is based on an ensemble of multiple tree models. The forest is the collection of all tree models. None of them is particularly good, but when combined the average of the ensemble performs well. Each model randomly generates a training set using bootstrap sampling of the raw data, a methodology called *bagging* or *bootstrap aggregating*. After the forest is built, the algorithm ensembles the prediction of all the trees in the forest into one output. This combination of randomizing the input for each tree model and then averaging their results reduces the likelihood of overfitting the model [99]. As we saw in Chapter 6, basing our predictions on an ensemble of several model predictions is a very common practice in time series, one that gives very good results.

Using this method for time series involves thinking carefully about how to represent the time series so that the temporal relationship is not lost. Like any algorithm, RF will require, in addition to the data set, some inputs from the user, inputs that will control what the RF does, how long it runs, how many trees to consider, how many branches to have in each tree, what performance criterion to use and other model building arguments. In turn, the algorithms are tailored to offer us the importance of each input in predicting the dependent variable, the model coefficients, some diagnostics about the model performance, the root mean square error of the fit as a function of the number of trees considered, and the predictive performance of the final model.

According to Nielsen [132], RF is not a good tool for working with time series in its random form.

10.2.2 Gradient Boosting

Gradient Boosting (GB) is an ML-regression approach that combines many simple models (also known as weak models or weak learners) into a single composite model [23]. In contrast with RF, GB creates the models sequentially, with the idea that later models correct the mistakes of earlier models by trying to predict the residuals of the earlier model. Data misfit by earlier models is more heavily weighted by later models. The first tree will try to match the data directly. The second tree will attempt to fit the residuals of the first model. The third tree will try to fit the residuals left after the second tree is fit, and so on. The algorithm minimizes a loss function that includes a penalty for model complexity, pretty much like the AIC penalizes for too many coefficients in ARMA models. Gradient boosted trees have been very successful at forecasting in Kaggle competitions and have often outperformed traditional statistical models.

Parr and Howard [146] contain a very gentle and intuitive introduction to gradient boosting regression for beginners. They illustrate how the regression model is trained by adding weak models gradually until we are happy with the fit of the trained model to the data. The final model is the addition of many, say M weak models $f(x)$ of the dependent variable as a function of the independent variable:

$$\widehat{y} = \sum_{m=1}^{M} f_m(x).$$

Boosting constructs and adds weak models in a stagewise fashion, one after the other, each one chosen to improve the overall model performance. Algorithmically, boosting models are expressed as a recursion:

$$F_m(x) = F_{m-1} + f_m(x),$$

where F represents the composite model.

The boosting strategy is greedy in the sense that choosing $f_m(x)$ never alters previous weak models. Some optimizing criterion function must be used to choose M [146]. The training is done gradually on the residual vectors, and this optimizes the overall model for the squared error loss function. Training on the residual vector optimizes the absolute error loss function.

Krispin [99], chapter 12, is a very accessible explanation of these models for beginners.

10.2.3 Familiar Procedures in ML-Based Regression Methods

When reviewing examples of ML prediction of time series with R, the reader will notice some common elements in the methodology followed.

- First of all, the raw time series data sets are the same as the data sets we start with in statistical time series analysis. They are subject to exploratory analysis like that done throughout this book to determine what seems and what does not seem important in the series. Cleaning, deciding what window of the data to use, correlation analysis, decomposition, all that is done. Users of ML methods applied to time series who have not received training in the material of the chapters in this book will not do very well in their ML-based predictions on average.
- ML practitioners will use the term "features" to refer to variables in the data set or variables that are created after the exploratory analysis. They will say that they extract features of the raw data set to convert the data set to an ML data set. For example, if a time stamp contains month, day, hour, year of the observation, and the exploratory analysis of the data reveals daily, weekly and monthly seasonality, new variables (features) of the date variable will be created that consist of one variable representing the month, one variable representing the day, one variable representing the hour, and, of course, one variable representing the year. It is also common to create other features such as, for example, assigning a code to days that are weekends or holidays, as the assumptions are that those days are different. Familiarity with R's handling of dates, and the library `lubridate` to extract those features of the time stamps helps in this task. We have practiced earlier in the book to handle dates and extracting features as well.

- The time series will be partitioned into a training set that will be used to build the model and a test set to determine the forecasting performance or predictive capabilities of the model when faced with new inputs, as we have done throughout this book when we modeled the time series.
- A performance evaluation criterion or criteria will be decided. This could be the RMSE, MAD, or MAPE, among others. The MAPE is used often by ML practitioners.
- The model(s) will be built (learned, fitted) using the training set, and the performance of the model(s) will be evaluated using the performance evaluation criterion with the training set.
- The model will be used to predict out-of-sample, that is, the values corresponding to the period of the test set and then the test set data will be compared to the forecast. The performance evaluation criterion will be used to determine the performance.
- The model performance criterion, for example, the MAPE of the prediction in the training set, and the MAPE of the prediction in the test set will be compared. If the latter is larger than the former, the conclusion will be that there was "overfitting" and that therefore the analysis should be repeated. At this point, a reevaluation of the variables entered to build the model is conducted. Analyzing the importance (contribution to variance) of the variables is done, and only those that contribute the most to the variability of the variable forecasted are kept for the next model fitting.
- Repeat the process until the MAPE in the training set and the MAPE in the test set are similar. Hopefully, the MAPE has also been reduced with this process in both.
- Benchmarking with a multiple classical time series regression is standard practice in ML-based regression.

Basically, the only new thing in ML regression is the number of times the regression is performed on the variables that have been selected to be in the model.

Running a regression many times, sometimes a thousand times, and for a large data set, is very demanding of a computing system. For this reason, the algorithms are optimized to perform such tasks at scale for the data, and the number of routines to do.

Of the two, RF and GB, GB models are preferred for prediction in the sense that they have the ability to weed out irrelevant or noisy variables and focus on the most important ones [132]. Gradient boosting methods for ML-based regression have outperformed traditional statistical methods with big data.

But make no mistake: GB and RF methods require a lot of work. Garbage in, garbage out. Not only do we need to feed the algorithm good data, but we also must think, after fitting a model and testing it with the test data, whether other arguments would have improved the performance metric, whether we should have included other types of variables.

Another thing to keep in mind is that since ML-based regression was not designed for time series data, it is important to adopt a critical approach to the evaluation of the models obtained. Having residuals that are not autocorrelated is as important with ML-based regression as in classical regression. And aiming at having confidence intervals for the forecasts should be a goal.

10.2.4 Software for ML-Based Regression

As is often the case in R, there are several packages to conduct ML-based regression. One of the packages for RF regression is `randomForest`.

The package `h2o` allows us to perform both RF and GB regression.

Example 10.1 In the example that follows we summarize the example provided by Krispin [99] for the analysis of a simple time series. Krispin's own package `TSstudio` is combined with the `h2o` package.

Krispin [99] studied the US monthly total vehicle sales (in thousands of units) using the R package `h2o` [104]. The package requires the installation of the Java language on the computer to be used for the analysis. The data set can be found in the package `TSstudio` [100], and it is a `ts` object with start date of January 1976 and end date of December 2019. The code is available in Krispin's github location and is accessible from our own *MLreg.R* program.

It is illustrative to follow the procedure used by Krispin to realize that even though we use ML methods, we still need to visualize and inspect and preprocess the data sets. All the insights gained from studying the basics of time series analysis in previous chapters of this book are needed for ML tools as well.

The insights gained from preliminary exploratory analysis help us identify the main characteristics of the series. For any ML analysis we need to:

- Identify the frequency, start and ending points of the series.
- Decompose the time series to become familiar with the cycles, trends, seasonality.
- Conduct a thorough seasonality analysis, particularly when the time series presents several seasonalities.
- Study the autocorrelation analysis of the series.
- Make the decision as to whether the whole historical time series will be used for the analysis or not.

After inspection, the data is found to have cycles and monthly seasonality. Krispin detects a change in trend and decides to use the data from January 2010 to December 2019 for subsequent analysis, since only 12 months ahead will be forecasted. For longer-term forecasting the author recommends using the whole history of the time series.

Features Engineering

Features engineering for vehicle sales is kept to a minimum. The trend component, a quadratic polynomial in time, the month as a factor and the twelve-month-lagged values of vehicle sales are made separate features. These are new variables created from the other variables and are columns of approximately the same length as the original data files in the flat data file. This results in a data frame that starts in January 2011 (because we lose 12 observations due to lagging the data 12 months).

Training, Testing Data and Inputs for Forecast

The forecast horizon will be 12 months. So the author partitions the data into a training set that goes from January 2011 to December 2018. The test set is the remaining months of 2019. The author plans to evaluate the model's performance using the MAPE score on the testing partition.

> Krispin warns, like other authors, that the main characteristic of ML models is the tendency to overfit on the training set. This will manifest itself in the ratio between the error score on the testing set and the training set, which will be very different. The features need to be created for the future months.

As in other regression approaches seen in this chapter, ML models need to create the inputs for the forecast themselves, for the predictions outside the training set. Krispin forecasts the sales during the months of 2019, so a data frame corresponding to that period, with the predictors or independent variables, must be fed to the model for the forecasting.

Starting an h2o Cluster

The training, test and forecast input data sets need to load to the h2o cluster.

Training the Model

Krispin uses cross-validation to train the model. Cross-validation is used widely in ML. It reduces the chance of overfitting. The training data set is split, randomly, into K folders. The model is then trained K times, each time leaving out one folder and training with $K-1$ folders. Throughout the process the model tunes the model parameters. The final model is tested on the testing partition.

RF Model

The h2o.randomForest function in R is used to train and tune the RF model. Some arguments must be given to the function in order to do a reasonable RF model. Krispin uses 500 trees and five-folder CV training. The author tells the model to stop when the RMSE is 0.0001.

A variance importance plot tells the importance of each of the variables in the fitting process. Not surprisingly the author finds that lagged-12-month sales are the most important contributor.

The output of the model contains the model and information about the model's parameters and performance. Only 44 of the 500 trees allowed were used. A plot indicating how the RMSE decreases as the number of trees increases can also be obtained.

Performance in the Testing Set

The model was created in order to forecast. Thus looking at the performance in the test set is done after model fitting in the training set.

At this point, the author lets us know that we can do further model searching and optimization of the model search in order to improve the accuracy of the training set model and the prediction. Indeed, after the model search is performed, the MAPE for the forecast accuracy decreases considerably when using the best model.

Gradient Boosting Model and DL Model

Krispin [99] also fitted a gradient boosting model to the time series. As in the random forest model, several arguments of the function had to be adjusted. A neural network model was also fitted, and this one performed better than the other two models. For that reason, we proceed to discuss Deep Learning (DL) in the next section, which is based on Neural Networks. □

10.2.5 Exercises

Exercise 10.1 Use program *MLreg.R* and the hourly energy consumption time series of the case study in Chapter 3 and repeat the analysis described in this section but for this data using the AEP time series in that data set. Compare the performance of RF and GB in the training and test set. Which method leads to a better forecast? Address all the aspects of the process mentioned in this section and write a small report that contains sections with those aspects.

10.3 Deep Learning

In their 1991 book, Lawrence and Luedeking [103] characterize Neural Networks as "the hottest technological development since the transistor." And that is why one should know about neural networks. The authors predicted that they would be a common household item by the year 2000. We are in 2021 and we don't hear households talk about them yet. However, many lives are being decided by this hot technology[107]. Neural networks allow entities to be creative about problem-solving because they do not require the use of rules or math; they only need examples to learn from. Their origin dates to the 1940s. Neural networks simulate biological behavior rather than sequentially process rules or math as Artificial Intelligence (AI) methods do. The area of Machine Learning that uses neural networks is called *Deep Learning*. Deep learning is often applied to problems where the input data and output data are well defined, yet the process that relates the input to the output is not easy to define.

Deep learning for time series is a relatively new endeavor but a promising one [132]. The name given to the most prevalent neural networks that are time oriented is *Recurrent Neural Networks* or RNN. Not all problems lend themselves to neural

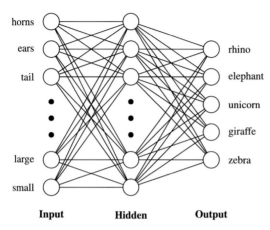

horns

ears   rhino

tail   elephant

                                                        unicorn

                                                        giraffe

large   zebra

small

Input   Hidden   Output

Figure 10.1   A sketch of the layers of a neural network.

network solutions. But it is believed that forecasting with neural networks has a lot of advantages compared to the more classical methods. Of the several types of RNN available, the *Long Short-term Memory Networks* (LSTM) are the most appropriate for sequential data like time series.

An artificial neural network (unlike the neural network of the brain) is a model that simulates a biological neural network. A neural network simulator, a program that creates a model of neurons and the connections between them and then trains this network, is the basis of applications. There are commercial neural network simulators that make neural network design quite simple.

All neural networks learn by association. A very simple example of this is learning to identify an animal by associating the inputs "large", "tail", "ears" and other terms with the output "elephant."

The neurons in a neural network are usually organized into three layers: input, hidden, and output. Figure 10.1 contains a sketch of what a simple network might look like. Sometimes more than one hidden layer is used. Information flows from the input layer to the hidden layer to the output layer. Every neuron in one layer is connected to all the neurons in the next layer. In passing from one neuron to another via an edge (the line connecting two neurons), a value is multiplied by that edge's weight and then usually passes through some kind of nonlinear activation function that can fit highly complex, nonlinear data. This function combined with the weights and the inputs provides the output. If y represents the output, x_i the inputs, w_i the weights, and f the activation function, then a simple model of what goes on is this:

$$y(x) = f\left(\sum_{i=1}^{n} w_i x_i\right).$$

A *training algorithm* decides the structure of the weights.

There are many different ways neural networks can learn. The most popular learning method is by example and repetition, also called *back propagation*. For example, in order to predict the next day's price for a particular stock, we may decide to collect daily historical information for the Dow Jones industrial average, the consumer price index, the price of certain commodities, the price of the stock, and the price of competitor's stocks [103]. The data must be captured into a computer file and translated into a data form that the network can understand. The network is then trained with the data. With a trained network, we can present to the network an unknown data set and the network will be able to provide outputs. In fact, neural networks have been found to work well in deciding when to buy or sell stocks. They are seen as a special case of pattern recognition in which the element of time is involved in the input patterns. However, neural networks don't view time the same way that we do.

> The way time enters NN is different from how it enters our statistical models. Features are extracted, and although the data remains sorted in time, it is not the temporal sequence but features from it that help train the network.

Since all neural networks learn from examples rather than from rules or mathematical formulas, a way to train a neural network to do forecasting consists of giving the neural network inputs and outputs. That is, give the network past values of a time series and other time series believed to affect the one of interest, and then future values corresponding to the same time series. Train the network by connecting each input to all the different time series that might be related to the target time series. A network that is not trained will make mistakes, but once it learns to recognize the patterns, it won't and it will not need to have the future values. We will give the network the inputs, and the network will forecast the future values by itself. Neural networks were used for financial forecasting already in the early 1990s. Data that comes from sensors or some external mechanical device needs to be made a computer file before we can use it to train the network.

Example 10.2   Lawrence and Luedekin [103] show what needs to be provided in order to construct a financial predictor for those who are not experts in finance. This NN predicts the Dow Jones Industrial Average one month in advance (the output layer, one neuron), using past financial information (the input layer, several neurons).
   Some of the inputs to the NN are:

Neuron number	Input	Values allowed
1	Consumer price index this month (CPI(t))	200 to 300
2	Price of crude oil this month (Oil(t)) ($)	15 to 30
3	Annual inflation rate this month (Infl(t)) (%)	4 to 12
4	Dow change this month Dow(t)	−100 to 100
5	Consumer price index last month	200 to 300
6	Price of crude oil last month ($)	15 to 30

Neuron number	Input	Values allowed
7	Annual inflation rate last month (%) 4 to 12	
8	Dow change last month	−100 to 100
9 to 21	Current month (1 if current, 0 otherwise)	Jan, Feb....

The output is the Dow average (Dow(t+1), change next month in values from −100 to 100).

For data that shifts over time, the change in the variable is more appropriate to use than the actual value. That is why the change in the Dow is considered.

Preparing the patterns for the NN to learn

In order for the NN to learn from our inputs, we must create examples of patterns that we will feed the network. We rearrange the inputs as follows (we do not include all the months variables):

Jan	F	.	D	CPI(t)	CPI(t-1)	Oil(t)	Oil(t-1)	Inf(t)	Inf(t-1)	Dow(t)	Dow(t-1)	.	**Dow(t+1)**
1	0	...	0	229	220	20	21.9	5.5	5.45	−5	−3	...	−12
0	1	...	0	235	229	19.8	20	5.44	5.5	−12	−5	...	−6
...	...	...	...	...	...	...	...	...	...	...	...	...	...
...	...	...	...	...	...	...	...	...	...	...	...	...	...
0	0	...	1	244	235	19.6	19.8	5.51	5.44	−6	−12	...	−16

Notice that the patterns are just a data set of features that we created from the data available to us.

Defining the NN

We already decided that the NN will predict Dow(t+1) and, therefore, that will be the output. We already decided the information that will help the network make the predictions, and we will provide that information to the network. We give the network not only past values of the Dow but values and past values of other variables. Different data types are preferred when using NN because the NN will learn to pay attention to the items that matter and ignore the ones that do not. Our network uses numeric input and output. One neuron is assigned to each input item for a total of 21 neurons. And one neuron is assigned to the output, which is the predicted change in the Dow Jones (Dow(t+1)).

Hidden Layers

We must determine the number of hidden layers to include. The authors of this example think that 15 hidden layers would be a good number to start with. Since all the values of the data, except the month, are continuous values, the authors recommend a continuous neuron transfer function.

Training the NN

Each line of the features data file, excepting the last column, after normalizing (using z scores or converting to −1 to 1 values, or some other form of normalization) is fed to the NN. Each line is an example of a pattern or fact. With these patterns, as many as the lines of data we can create with the available data set, we train the NN. For each pattern, the NN will output a Dow(t+1).

The NN must receive those patterns in random order. That is, we would not first feed line 1, and then line 2, and then line 3 and so on, but rather the lines would be selected at random. That way, the NN will learn better. Data sets such as financial data tend to follow trends that shift, so the network might learn incorrectly too soon that the Dow is going up, just to have to learn later that it is going down. By randomizing, we avoid that.

Diagnosing Accuracy

It could be that the NN does not predict the right Dow(t+1) in some cases during the training. If that happens, then there are some optional courses of action. More variables could be added. Alternatively, the cases could be looked at closely. Perhaps there are outliers, and we should take them out and retrain. There is still one more option, which consists of requiring a lower level of accuracy in the prediction, in the sense that a deviation of one point from the true value, for example, is considered good enough.   □

We will now continue the analysis started with the example of Krispin and explain how the author did the NN prediction.

10.3.1 More about the Layers: Network Architecture

The NN that we described in Example 10.2 is a single-layer NN. All the inputs in the input layer connect to the output layer. As a user of NN, the reader will have to tell the algorithm how many hidden layers to include in the NN (the depth of the NN) and how many nodes to include in each hidden layer. How the layers are connected or the type of architecture to use will also need to be chosen. The author of Example 10.1 thought that 15 hidden layers would be enough. How many layers to include is a trial-and-error decision. However, if there are nonlinear relations to think of in the data at hand, the recommendation is to include a few hidden layers to allow the NN to have better performance.

There are several approaches to choosing the number of layers and nodes. One of them is the grid approach, which consists of creating a grid of pairs of numbers of layers and number of nodes and trying each of them. It is hard to find the optimal configuration mathematically.

10.3.2 Interpretable Machine Learning

There is much debate nowadays about the need to have ML models that are interpretable. An interpretable ML model is constrained in model form so that it is either useful to someone or obeys structural knowledge of the domain [164, 84]. An interpretable ML model, which transparently conveys inside the black box, is preferred to having to explain the black box. An example of an interpretable ML model that the reader can play with is the entry to the FICO Explainable ML Challenge 2018 [165], where a video shows what is involved in making the model interpretable. According to [164], visualization can make the interpretability possible, not only for small models but for complex models.

Rudin's presentation and the video at [165] were not designed for time series. But similar procedures could be used. In order to do that, it would be helpful to first have familiarity with ML models and their applications to time series data.

10.4   Big Data

ML methods have gained popularity because of big data (in volume, velocity and variety). Many commonly used data analysis tools in R and Python are not efficient or well implemented for big data. Unix command-line commands can do a lot with a large data set that would take tons of memory and time in R. A second best alternative is R's `data.table`, an underused but highly performant data frame package that has fantastic time functionality [132]. It has substantially better performance with big data than tools available in Python.

R's `data.table` has function `fread()` that enables partial reading of files. `fread` is similar to `read.table` but faster and more convenient. Arguments `sep`, `colClasses` and `nrows` are automatically detected. Dates are read as character data. They can be converted to date format with the `fasttime` package or standard Base R function.

10.4.1   Exercises

Exercise 10.2   Using the Birmingham data set in the case study of Chapter 4, prepare a data set of features that would allow us to implement a NN to the data. The objective is to predict the occupancy in the Broad Street parking structure using the occupancy of nearby parking structures. Then use `AutoML` of the `h2o` package. Compare the result with the predictions that would be obtained with RF and GB.

Exercise 10.3   Using the same input as used in the RF and GB example discussed in Example 10.1, Krispin [99] used the `AutoML` of the `h2o` package and this procedure chose DL models with different tuning settings. `AutoML` can scale up while having multiple time series to forecast with minimal intervention from the user.

10.5   Unsupervised Learning and Feature Generation

As defined in [132], feature generation is the process of finding a quantitative way to encapsulate the most important traits of time series data into just a few numeric values and categorical labels. This was initially proposed because machine learning was not tailored for time series; and formatting the data that way, with a few features for each time series, made it possible to use machine learning methods.

Code that automatically generates a large number of features of a time series exists, but it presents some dangers. First of all, it may generate features that are not relevant for the time series at hand. We have been introducing features generation gradually in this book. But consider, for example, generating features for the data in the diabetes data graph. It would not make sense to use the features that we use for financial time series to summarize the features of data generated by a synchronized continuous glucose monitor and insulin pump. Instead, we would recur to the medical literature on diabetes management to learn what are the features that help doctors and patients determine diagnoses and quality of management. And using automatic feature generation for the data sets we considered in earlier chapters would give a lot of nonsense.

The background knowledge about the subject matter measured and the domain knowledge in the area should dictate the features selected. As Nielsen [132] suggests, stationarity, the length of the data (which affects having more or less extreme min and max), the error propensity of the measuring instrument and the purpose of feature generation are important factors to consider.

There are open-source, automatic time series feature generation software packages that are useful in a variety of domains, and there are domain-specific time series feature generation software packages. We will focus on what R has to offer regarding time series features generation that is useful in a variety of domains. The most prominent package is `tsfeatures` [79], developed by Rob Hyndman and colleagues Once time series features have been created, it is important to select the most important ones among those generated. R offers recursive feature elimination (RFE).

For further reading on time series feature generation, see the bibliography of Nielsen [132]. In addition to that, the `tsfeatures` package is explained in [208], using time series that come with R.

10.5.1   Feature-Based Machine Learning

In this approach to time series, we do not assume a stochastic model for the time series as we have been doing throughout the book. We instead focus on identifying patterns that help us classify the time series and unsupervised learning in the form of clustering analysis for time series. We have also been doing some of this throughout the book. Random forests and gradient boosting decision trees can also be used for classification using features. The reader will find examples of those in the book's website.

The method closest to using all the values of the time series to classify time series is Dynamic Time Warping (DTW), which is a metric used to measure similarity between time series. This method is apt for clustering time series whose most salient feature is their overall shape. The units or scale or length of the time series can be different. Every point in one time series must be matched with at least one point of the other time series.

10.6   Case Study: Smart Medical Devices

In [174] we studied the blood glucose measurements obtained with a CGM for a diabetic patient. In this case study, we focus not on one patient, but on the larger task of using the same readings from thousands of patients to learn about how the closed loop technology sensor-pump works. According to the overview of the history of time series by Aileen Nielsen [132], time series made their way into medicine when the first practical electrocardiograms (ECGs) were invented in 1901. Soon after, in 1924, the electroencephalogram (EEG) was introduced into medicine. Both ECGs and EEGs produce electrical signals or time series, and for a long time they were the predominant time series analyzed in medicine. Nowadays, ECG and EEG time series classification has been and still is an active area of research in medicine. AI is now embedded in EEG and ECGs to provide diagnosis recommendations to healthcare providers. The accuracy of the AI-enabled device's recommendation needs to be statistically validated [177].

The proliferation of wearable sensors and "smart" electronic medical devices routinely provide measurements from both sick and healthy people. To make advances in medical devices and healthcare in general, appropriate analysis of the massive amounts of data produced by those devices is needed. Time series analysis methods are needed for most of these devices, as they collect time-stamped data. Specialized departments at academic institutions are starting to focus on computational medicine, and lucrative businesses are pursuing profitable discoveries with these data.

In the area of diabetes management, the development of continuous subcutaneous insulin infusion systems has allowed for refinement in the delivery of insulin, while continuous glucose monitors provide patients and clinicians with a better understanding of the minute-to-minute glucose variability, leading to the titration of insulin delivery based on this variability when applicable. Merging of these devices has resulted in sensor-augmented insulin pump therapy, which continuously produces data. Programmable insulin administration in basal and bolus fashion is integrated and augmented with glucose biosensors to provide real-time, data-driven glycemic control and early detection of hypoglycemia, in some cases involving over 200 measurements of several variables per day. Patients upload their insulin pump and glucose sensor data to a website provided by the medical instrument provider. The data helps the medical instrument provider improve its products, helps doctors manage the diabetes of their patients more efficiently, and helps patients monitor and calibrate their pumps. The combination of technologies is particularly helpful to manage the diabetes of Type I diabetic patients (T1D patients). More information about insulin pump/sensor diabetes management and the technology involved can be found at [18, 176].

Scheiner [176] tells us the features of glucose sensor and insulin pump data that are important to learn from the data. According to this author, because CGM devices generate glucose values around the clock, the data are not skewed by checking more often during periods of high or low glucose or by only checking prior to meals. The statistics generated by CGM devices are thus more valid than those garnered from blood glucose meters. The following are some of the variables to look for, according to [176]:

- The mean (average) sensor glucose represents a fairly true average, albeit slightly lower than reality due to a natural tendency for the systems to err on the side of lower rather than higher values, as well as the prolonged lag time that occurs when recovering from hypoglycemia. Adding 2.3 percent to a CGM average is a good way to correct for these system deficiencies. A1c can then be estimated by using the eAG equation ([CGM average +46.7]/28.7).
- Another useful statistic is the standard deviation (SD). A high SD means that there are many glucose values that are significantly above or below the average. A low SD indicates a relatively low number of outliers. An SD that is less than 33 percent of the average is generally desirable. An SD that is more than 50 percent of the average indicates excessive variability. Because glucose variability may be associated with both risk of long-term complications and hindering short-term quality of life, efforts to minimize, measure, and manage glycemic excursions are worthwhile in addition to management of hemoglobin A1c.
- A highly practical statistic is the percentage of time spent above, below, and within one's target glucose range. The target range can be customized on each software program and should be individualized for each patient. A major goal of diabetes treatment is to spend as much time as possible within the target range and as little time above and below target. Measuring time in range can be useful for evaluating therapeutic changes and motivating patients to continue with behaviors that are producing desired results.
- The magnitude (and timing) of postprandial glucose peaks.
- Effectiveness of (or need for) mealtime insulin doses.
- Correction factor/insulin sensitivity.
- Whether basal insulin doses are set properly.
- Duration of the insulin action curve.
- Patterns of hypoglycemia.
- Result of treatment of hypoglycemia.
- Effect of glucose-lowering medications.
- Impact of lifestyle events and activities.
- System abuses or behaviors that may be sabotaging one's control.

The features just mentioned would not be appropriate for other contexts, but in the case at hand, were we to generate features classification or prediction models, those would be the ones we would use.

A very active area of machine learning research for time series data in health applications is the prediction of blood glucose levels for individual patients. Nielsen [132], p. 384, contains a very nice complete case study using blood glucose sensor data from a "Dexcom" device using R. The website of this book contains an illustration of this example. We refer the reader to that example.

References

[1] Abraham, B. and Ledolter, J. *Introduction to Regression Modeling*. Thomson, Brooks Cole (2006).

[2] Bagnall, A., Lines, J., Vickers, W. and Keogh, E. The UEA & UCR Time Series Classification Repository. www.timeseriesclassification.com.

[3] Bates, J. M. and Granger, C. W. J. The Combination of Forecasts. *Organizational Research Quarterly* 20(4): 451–468 (1969). https://perma.cc/9AEE-QZ2J.

[4] Bojer, C. S. and Meldgaard, J. P. Kaggle Forecasting Competitions: An Overlooked Learning Opportunity. *International Journal of Forecasting* 37(2): 587–603 (2020). DOI: https://doi.org/10.1016/j.ijforecast.2020.07.007.

[5] Bojer, C. and Meldgaard, J. P. The M5: A Preview from Prior Competitions. Foresight. *The International Journal of Applied Forecasting* 58(Summer): 17–23 (2020).

[6] Bollerslev, T. Generalized Autoregressive Conditional Heteroskedasticity. *Journal of Econometrics* 31: 307–327 (1986).

[7] Box, G. E. P., Jenkins, G. M. and Reinsel, G. C. *Time Series Analysis. Forecasting and Control*. Third edition. Prentice Hall International, Inc. (1994).

[8] Bowerman, B. L. and O'Connell, R. T. *Forecasting and Time Series. An Applied Approach*. Third edition. Duxbury (1993).

[9] Bowerman, B. L., Murphy, E. S. and O'Connell, R. T. *Regression Analysis: Unified Concepts, Practical Applications, and Computer Implementation*. Business Expert Press (2015).

[10] Brand, P. T. and Williams, J. T. *Multiple Time Series Models*. Quantitative Applications in the Social Sciences, volume 148. SAGE (2006).

[11] Braun, W. J. and Murdoch, D. J. *A First Course in Statistical Programming with R*. Second edition. Cambridge University Press (2016).

[12] Brillinger, D. R. *Time Series. Data Analysis and Theory*. SIAM (2001).

[13] Brockwell, P. J. and Davis, R. A. *Time Series: Theory and Methods*. Springer-Verlag (1987).

[14] Bureau of Labor Statistics. Unemployment Statistics. www.bls.gov/opub/ted/2018/unemployment-rate-2-1-percent-for-college-grads-3-9-percent-for-high-school-grads-in-august-2018.htm?view_full.

[15] Bureau of Labor Statistics. Frequently Asked Questions about Seasonality. www.bls.gov/cps/seasfaq.htm.

[16] Carcione, J. M., Santos, J. E., Bagaini, C. and Ba, J. A. Simulation of a COVID-19 Epidemic Based on a Deterministic SEIR Model. *Frontiers in Public Health* 8:230(May 2020). DOI: https://doi.org/10.3389/fpubh.2020.00230.

[17] Carlton, M. A. and Devore, J. L. *Probability with Applications in Engineering, Science, and Technology*. Second edition. Springer-Verlag (2017).

[18] Cengiz, E., Sherr, J. L., Weinzimer, S. A. and Tamborlane, W. V. New-Generation Diabetes Management: Glucose Sensor-Augmented Insulin Pump Therapy. *Expert Review of Medical Devices*. 8(4):449–458 (July 2011). DOI: https://doi.org/10.1586/erd.11.22.

[19] Chan, K. S. and Ripley, B. TSA: Time Series Analysis. R package version 1.3.(2020). https://CRAN.R-project.org/package=TSA.

[20] Chatfield, C. *Time Series Forecasting*. Chapman & Hall/CRC (2000).

[21] Chatterjee, S. and Price, B. *Regression Analysis by Example*. Wiley & Sons (1977).

[22] Chen, G., Glen, D. R., Saad, Z. S. et al. Vector Autoregression, Structural Equation Modeling, and Their Synthesis in Neuroimaging Data Analysis. *Computers in Biology and Medicine* 41(12): 1142–1155 (December 2011).

[23] Chen, T., He, T., Benesty, M. et al. xgboost: Extreme Gradient Boosting. R package version 1.2.0.1(2020). https://CRAN.R-project.org/package=xgboost.

[24] Cimadomo, J., Giannone, D., Lenza, M., Monti, F. and Sokol, A. Nowcasting with Large Bayesian Vector Autoregressions. Paper presented at the NBER-NSF Seminar on Bayesian Inference in Econometrics and Statistics (SBIES). Washington University, St. Louis, Olin School of Business (virtual seminar). Also as working paper at European Central Bank, Working Paper Series No. 2453(2020). www.ecb.europa.eu/pub/pdf/scpwps/ecb.wp2453 465cb8b18a.en.pdf.

[25] Citizen Science. www.citizenscience.gov.

[26] Cleveland, W. S. and Devlin, S. J. Locally Weighted Regression: An Approach to Regression Analysis by Local Fitting. *Journal of the American Statistical Association* 83: 596–610 (1988) DOI: http://doi.org/10.2307/2289282.

[27] Cleveland, R. B., Cleveland, W. S., McRae, J. E. and Terpenning, I. STL: A Seasonal-Trend Decomposition Procedure Based on Loess. *Journal of Official Statistics* 6: 3–33, (1990).

[28] Cobb, G. W. The Problem of the Nile: Conditional Solution to a Changepoint Problem. *Biometrika* 65(2): 243–251 (August 1978).

[29] COVID-19 ForecastHub. https://covid19forecasthub.org/doc/.

[30] Cowpertwait, P. S. P. and Metcalfe, A. V. *Introductory Time Series with R*. Springer-Verlag (2009).

[31] Cressie, N. A. C. *Statistics for Spatial Data*. Revised edition. Wiley & Sons (1993).

[32] Cryer, J. D. and Chan, K. S. *Time Series Analysis. With Applications in R*. Second edition. Springer-Verlag (2008).

[33] 2019 Data Expo. American Statistical Association. http://stat-computing.org/dataexpo/2009/.

[34] De Veaux, R. D., Velleman, P. F. and Bock, D. E. *Stats: Data and Models*. Fifth edition. Pearson (2020).

[35] Derryberry, D. *Basic Data Analysis for Time Series with R*. Wiley (2014).

[36] Dickey, D. A. and Fuller, W. A. Distribution of the Estimators for Autoregressive Time Series with a Unit Root. *Journal of the American Statistical Association* 74(366): 427–431 (1979).

[37] Diez, D., Cetinkaya, M. and Barr, D. *OpenIntro Statistics*. Fourth edition. OpenIntro (2019).

[38] Diggle, P. J. *Time Series. A Biostatistical Introduction*. Clarendon Press (2004).

[39] Dua D. and Graff, C. UCI Machine Learning Repository http://archive.ics.uci.edu/ml. University of California, School of Information and Computer Science, (2019). https://archive.ics.uci.edu/ml/datasets.php and http://archive.ics.uci.edu/ml/datasets.php.

[40] Dunning, T. and Friedman, E. *Time Series Databases. New Ways to Store and Access Data*. O'Reilly Media (2014).

[41] Dunnin, T. and Friedman, E. *Practical Machine Learning: A New Look at Anomaly Detection*. O'Reilly Media (2018).

[42] Dutta-Roy, T. Basics of Audio File Processing in R. https://medium.com/@taposhdr/basics-of-audio-file-processing-in-r-81c31a387e8e.

[43] Engle, R. F.: Autoregressive Conditional Heteroscedasticity with Estimates of the Variance of United Kingdom inflation. *Econometrica* 50(4): 987–1008 (July 1982).

[44] European Commission. Eurostat. Your Key to European Statistics. https://ec.europa.eu/eurostat/web/covid-19/visualisation http://ec.europa.eu/eurostat/web/main/home.

[45] Ewing, B. T., Riggs, K. and Ewing, K. L. Time Series Analysis of a Predator-Prey System: Application of VAR and Generalized Impulse Response Function. *Ecological Economics* 60(3): 605–612 (2007).

[46] Facebook. Prophet. Forecasting at Scale. https://facebook.github.io/prophet/.

[47] Faraway, J. J. *Linear Models with R*. Chapman & Hall/CRC (2005).

[48] Fildes, R., Schaer, O. and Svetunkov, I. Software Survey: Forecasting 2018. www.informs.org/ORMS-Today/Public-Articles/June-Volume-45-Number-3/Software-Survey-Forecasting-2018.

[49] FRED. Federal Reserve Bank of St. Louis. https://fred.stlouisfed.org/.

[50] Frye, C. and Mehrota, V. Forecasting 2016 (Software Survey). *OR/MS Today*, Vol. 43, No. 3 (June 2016). http://viewer.zmags.com/publication/085442e2#/085442e2/46.

[51] Gelman, A. Statistical Modeling, Causal Inference and Social Science. https://statmodeling.stat.columbia.edu/2020/05/18/hey-i-think-somethings-wrong-with-this-graph/.

[52] Gould, R., Wong, R. and Ryan, C. *Introductory Statistics. Exploring the World through Data*. Third edition. Pearson (2020).

[53] trends.google.com. Google Trends. http://trends.google.com/trends.

[54] Granger, C. W. J. and Hatanaka, M. *Spectral Analysis of Economic Time Series*. Princeton University Press (1964).

[55] Granger, C. W. J. and Newbold, P. *Forecasting Economic Time Series*. Second edition. Academic Press (1983).

[56] Graybill, F. A. *Matrices with Applications in Statistics*. Second edition. Duxbury Press (2001).

[57] Gilbert, P. D. Time Series Database Interface (TSdbi). Guide and Illustrations. https://cran.microsoft.com/snapshot/2022-09-01/web/packages/TSdbi/TSdbi.pdf.

[58] Guorong, D., Li, X., Fan, J. and Shen, Y. Brief Analysis of the ARIMA Model on the COVID-19 in Italy. DOI: https://doi.org/10.1101/2020.04.08.20058636.

[59] Han, J. and Kamber, M. *Data Mining: Concepts and Techniques*. Morgan Kaufman Publishers (2001).

[60] Harvey, A. C. *Time Series Models*. Second edition. Massachusetts Institute of Technology Press (1993).

[61] Hastie, T., Tibshirani, R. and Friedman, J. *The Elements of Statistical Learning. Data Mining, Inference, and Prediction*. Second edition. Springer-Verlag (2019).

[62] Hays, J. M. Forecasting Computer Usage. *Journal of Statistics Education* 11:1(2003). www.jse.amstat.org/v11n1/datasets.hays.html.

[63] Hogarth, R. M. and Soyer, E. Using Simulated Experience to Make Sense of Big Data. *MIT Sloan Management Review* 56: 49–54 (2015). https://sloanreview.mit.edu/article/using-simulated-experience-to-make-sense-of-big-data/.

[64] Hogarth, R. M. and Soyer, E. Improving Judgments and Decisions by Experiencing Simulated Outcomes. In E. A. Wilhelms & V. F. Reyna (Eds.), *Neuroeconomics, Judgment, and Decision Making* (pp. 254–273). Psychology Press (2015). https://psycnet.apa.org/record/2014-31822-013.

[65] Hogarth, R. M. and Soyer, E. Providing Information for Decision Making: Contrasting Description and Simulation. *Journal of Applied Research in Memory and Cognition* 4(3): 221–228 (2015). DOI: https://doi.org/10.1016/j.jarmac.2014.01.005 www.sciencedirect.com/science/article/pii/S2211368114000060.

[66] Hogue, J. Anomaly Detection, A to Z. https://github.com/dreyco676/Anomaly_Detection_A_to_Z/blob/master/Anomaly%20Detection%20A%20to%20Z.pdf. https://archive.ics.uci.edu/ml/datasets/Metro+Interstate+Traffic+Volume#.

[67] Horton, N. J., Baumer, B. S. and Wickham, H. Setting the Stage for Data Science: Integration of Data Management Skills in Introductory and Second Courses in Statistics. *Chance*, Vol. 28, 2015, 40–49. https://chance.amstat.org/2015/04/databases/.

[68] Hylleberg, S. *Seasonality in Regression*. Academic Press Inc. (1986).

[69] Hyndman, R. CRAN Task View: Time Series Analysis. https://cran.r-project.org/web/views/TimeSeries.html.

[70] Hyndman, R. Github site for `forecast`. https://github.com/robjhyndman/forecast.

[71] Hyndman, R. *Seasonal Periods*. Hyndsight (2014). https://robjhyndman.com/hyndsight/seasonal-periods/.

[72] Hyndman, R. Tidy Time Series Analysis in R. Tsibble, Feasts and Fable. Sept 2019. Seminar. https://robjhyndman.com/seminars/tidyverts/.

[73] Hyndman, R. Time Series Forecasting. Paper presented at R-Studio-conference 2020. https://github.com/rstudio-conf-2020/time-series-forecasting.

[74] Hyndman, R. `tscompdata` R package. https://github.com/robjhyndman/tscompdata.

[75] Hyndman, R. and Athanasopoulos, G. *Forecasting: Principles and Practice*. Second edition. OTexts (2018). OTexts.com/fpp2. https://otexts.com/fpp2/case-studies.html.

[76] Hyndman, R. and Athanasopoulos, G. *Forecasting: Principles and Practice* (with `fable` package) (2021). OTexts.com/fpp3.

[77] Hyndman, R., Athanasopoulos, G., Bergmeir, C. et al. `forecast`: Forecasting Functions for Time Series and Linear Models. R package version 8.13. (2020) https://pkg.robjhyndman.com/forecast/.

[78] Hyndman, R. and Khandakar, Y. Automatic Time Series Forecasting: The forecast Package for R. *Journal of Statistical Software* 26(3): 1–22 (2008). www.jstatsoft.org/article/view/v027i03.

[79] Hyndman, R., Kang, Y., Manso, P. M. et al. (2020). `tsfeatures`: Time Series Feature Extraction. R package version 1.0.2. https://CRAN.R-project.org/package=tsfeatures https://cran.r-project.org/web/packages/tsfeatures/vignettes/tsfeatures.html.

[80] Hyndman, R. J., Koehler, A. B., Ord, J. K. and Snyder, R. D. *Forecasting with Exponential Smoothing. The State Space Approach*. Springer-Verlag (2008).

[81] International Institute of Forecasters. M5 Forecasting Competition (2020). https://forecasters.org/resources/time-series-data/.

[82] Kaggle. M5-Forecasting-Accuracy. Estimate the Unit Sales of Walmart Retail Goods. www.kaggle.com/c/m5-forecasting-accuracy/data.

[83] International Institute of Forecasters. https://forecasters.org/.

[84] James, G., Witten, D., Hastie, T. and Tibshirani, R. *Introduction to Statistical Learning with R*. Springer-Verlag (2017).

[85] Janacek, G. *Practical Time Series*. Arnold (2001).

[86] Kaggle. Competitions Documentation. www.kaggle.com/docs/competitions.

[87] Kaggle Datasets. Hourly Energy Consumption. www.kaggle.com/robikscube/hourly-energy-consumption.

[88] Kaggle Search. Wikipedia. www.kaggle.com/search?q=wikipedia.

[89] Kaggle. Novel Corona Virus 2019 Dataset. www.kaggle.com/sudalairajkumar/novel corona-virus-2019-dataset (accessed on May 4, 2020).

[90] Kaushik, S. Beginner's Guide on Web Scraping with R (Using rvest) with Hands-on Example. www.analyticsvidhya.com/blog/2017/03/beginners-guide-on-web-scraping-in-r-using-rvest-with-hands-on-knowledge.

[91] Kay, S. M. *Modern Spectral Estimation*. Prentice Hall (1987).

[92] Kennedy, P. *A Guide to Econometrics*. Massachusetts Institute of Technology Press (1998).

[93] Keyes, O. and Lewis, J. `pageviews`: An API Client for Wikimedia Traffic Data. R package version 0.5.0. https://CRAN.R-project.org/package=pageviews.

[94] Kilian, L. and Lutkepohl, H. *Structural Vector Autoregressive Analysis*. Cambridge University Press (2018).

[95] Klein, W. K. Timing Is All: Elections and the Duration of United States Business Cycles. *Journal of Money, Credit and Banking* 28(1): 84–101 (1996).

[96] Kluwer, W. Blue Chip Economic Indicators. Top Analysts' Forecasts of the U.S. Economic Outlook for the Year Ahead. Vol. 43, No. 1, January 10, 2018. https://lrus.wolterskluwer.com/media/2444/bcei0118email.pdf.

[97] Kluwer, W. Blue Chip Financial Forecasts. Top Analysts' Forecasts of U.S. and Foreign Interest Rates, Currency Values and the Factors that Influence Them. Vol. 37, No. 1, January 1, 2018. https://lrus.wolterskluwer.com/media/2446/bcff0118email.pdf.

[98] Koopmans, L. H. *The Spectral Analysis of Time Series*. Academic Press (1974).

[99] Krispin, R. *Hands-on Time Series Analysis with R*. Packt (2019).

[100] Krispin, R. TSstudio: Functions for Time Series Analysis and Forecasting. R package version 0.1.6. https://CRAN.R-project.org/package=TSstudio.

[101] Kwak-Hefferan, E. About Old Faithful, Yellowstone's Famous Geyser. Yellowstone National Park. April 27, 2020. www.yellowstonepark.com/things-to-do/about-old-faithful.

[102] Lantz, B. *Machine Learning with R. Expert Techniques for Predictive Modeling*. Third edition. Packt (2019).

[103] Lawrence, J. and Luedeking, S. *Introduction to Neural Networks. Computer Simulations of Biological Intelligence*. Third edition. California Scientific Software (1991).

[104] LeDell, E., Gill, N., Aiello S. et al. h2o: R Interface for the "H2O" Scalable Machine Learning Platform. R package version 3.32.0.1 (2020). https://CRAN.R-project.org/package=h2o.

[105] Liaw, A. and Wiener, M. Classification and Regression by randomForest, *R News* 2(3): 18–22 (2002). https://CRAN.R-project.org/doc/Rnews/.

[106] Li, C. and Maheu, J. M. A Multivariate GARCH-Jump Mixture Model. Paper presented at the NBER-NSF NBEIS Conference. Washington University Olin School of Business (2020). https://apps.olin.wustl.edu/conf/SBIES/files/pdf/2020/80.pdf.

[107] Lim, B. and Zohren, S. *Time Series Forecasting with Deep Learning: A Survey*. The Royal Society Publishing (2020). https://arxiv.org/pdf/2004.13408.pdf.

[108] Lock, H. L., Lock, P. F., Lock, J., Lock, E. F. and Lock, D. *Statistics. Unlocking the Power of Data*. Second edition. Wiley (2016).

[109] Lütkepohl, H. *New Introduction to Multiple Time Series Analysis*. Springer-Verlag (2005).

[110] Lytras, T., Gkolfinopoulou, K. and Bonovas, S. FluhHMM: A Simple and Flexible Bayesian Algorithm for Sentinel Influenza Surveillance and Outbreak Detection. *Statistical Methods in Medical Research* 28(6): 1826–1840 (2018). DOI: https://doi.org/10.1177/0962280218776685.

[111] Machine Learning Repository. UCI. https://archive.ics.uci.edu/ml/datasets/Air+Quality.

[112] Maharaj, E. A., D'Urso, P. and Caiado, J. *Time Series Clustering and Classification*. CRC Press (2019).

[113] Marple, S. L. *Digital Spectral Analysis with Applications*. Prentice Hall (1987).

[114] Marsh, L. C. and Cormier, D. R. *Spline Regression Models*. SAGE (2002).

[115] McAdams, B. H. and Rizvi, A. A. An Overview of Insulin Pumps and Glucose Sensors for the Generalist. *Journal of Clinical Medicine* 5(1): 5 (2016). DOI: https://doi.org/10.3390/jcm5010005.

[116] McCleary, R., McDowall, D. and Bartos, B. J. *Time Series Experiments*. Oxford University Press (2017).

[117] McDowall, D., McCleary, R., Meidinger, E. E. and Hay, Jr., R. A. *Interrupted Time Series*. SAGE (1980).

[118] McLachlan, G. and Peel, D. *Finite Mixture Models*. Wiley (2000).

[119] McSweeny, A. J. Effects of Response Cost on the Behavior of a Million Persons: Charging for Directory Assistance in Cincinnati. *Journal of Applied Behavior Analysis* 11(1): 47–51 (1978). DOI: https://doi.org/10.1901/jaba.1978.11-47.

[120] McTaggart, R., Daroczi, G. and Leung, C. Quandl: API Wrapper for Quandl.com. R package version 2.10.0. (2019). https://CRAN.R-project.org/package=Quandl.

[121] Meinhold, R. J. and Singpurwalla, N. D. Understanding the Kalman Filter. *The American Statistician* 37(2): 123–127 (1983).

[122] Meissner, P. wikipediatrend: Public Subject Attention via Wikipedia Page View Statistics. R package version 2.1.6. (2020). https://CRAN.R-project.org/package=wikipediatrend.

[123] Mindhive. www.web.mindhive.org/.

[124] Molinero, C. M. 1990. The Autocorrelation Function of a Time Series with a Deterministic Component. *IMA Journal of Mathematics Applied* 3: 25–30 (1991).

[125] Montgomery, D. C., Jennings, C. L. and Kulahci, M. *Introduction to Time Series Analysis and Forecasting*. Wiley (2008).

[126] Moritz S. and Bartz-Beielstein, T. imputeTS: Time Series Missing Value Imputation in R. *The R Journal* 9(1): 207–218 (2017). DOI: https://doi.org/10.32614/RJ-2017-009.

[127] Mulla, R. Hourly Energy Consumption over 10 Years of Hourly Energy Consumption Data from PJM in Megawatts. www.kaggle.com/robikscube/hourly-energy-consumption.

[128] Nasdaq Data Link. https://data.nasdaq.com.

[129] National Geographic. Geyser. www.nationalgeographic.org/encyclopedia/geyser/.

[130] NBER-NSF SBIES Conference Program. https://apps.olin.wustl.edu/conf/SBIES/Home/Default.aspx?pid=4.

[131] Newton, H. J. *Timeslab: A Time Series Analysis Laboratory*. Wadsworth & Brooks/Cole (1988).

[132] Nielsen, A. *Time Series Analysis: Prediction with Statistics and Machine Learning*. O'Reilly (2019).

[133] National Centers for Environmental Information. www.ncdc.noaa.gov/cag/national/time-series/110/tavg/.

[134] API for National Centers for Environmental Information. www.ncdc.noaa.gov/cdo-web/webservices/v2.

[135] NOAA Tides and Currents. National Centers for Environmental Information. tidesandcurrents.noaa.gov/ofs/leofs/leofs.html.

[136] National Travel and Tourism Office. ITA. https://travel.trade.gov/view/m-2017-I-001/index.asp.

[137] O'Hara-Wild, M., Hyndman, R. and Wang, E. feasts: Feature Extraction and Statistics for Time Series. R package version 0.1.6 (2020). https://CRAN.R-project.org/package=feasts.

[138] O'Hara-Wild, M., Hyndman, R. and Wang, E. fable: Forecasting Models for Tidy Time Series. R package version 0.2.1. https://CRAN.R-project.org/package=fable.

[139] Ooms, J., James, D., DebRoy, S., Wickham, H. and Horner, J. RMySQL: Database Interface and "MySQL' Driver for R. R package version 0.10.20 (2020). https://CRAN.R-project.org/package=RMySQL.

[140] OR/MS Forecasting Software Surveys. www.informs.org/ORMS-Today/OR-MS-Today-Software-Surveys/Forecasting-Software-Survey.

[141] Oracle. MySQL. https://dev.mysql.com/doc/mysql-getting-started/en/.

[142] Ostrom, Jr., C. W. *Time Series Analysis. Regression Techniques*. Second edition. SAGE (1990).

[143] Ozonoff, A., Sukpraprut, S. and Sebastiani, P. Modeling Seasonality of Influenza with Hidden Markov Models. *JSM Proceedings of the American Statistical Association, SDND Section on Defense and National Security*. August 6–10, 2006. American Statistical Assocation (2006).

[144] The Comprehensive R Archive Network. https://cran.r-project.org/web/packages/pageviews/pageviews.pdf.

[145] Papastefanopoulos, V., Linardatos, P. and Kotsiantis, S. COVID-19: A Comparison of Time Series Methods to Forecast Percentage of Active Cases per Population. *Applied Sciences*, 10 (June 3, 2020), 3880. DOI: https://doi.org/doi:10.3390/app10113880.

[146] Parr, T. and Howard, J. Gradient Boosting: Distance to Target. https://explained.ai/gradient-boosting/L2-loss.html#sec:2.1.

[147] Peck, R., Short, T. and Olsen, C. *Statistics and Data Analysis*. Cengage (2020).

[148] Peng, R., Kross, S. and Anderson, B. *Mastering Software Development in R*. bookdown (2017). https://bookdown.org/rdpeng/RProgDA/.

[149] Petris, G. (2010). An R Package for Dynamic Linear Models. *Journal of Statistical Software* 36(12): 1–16. www.jstatsoft.org/v36/i12/.

[150] Petris, G., Petrone, S. and Campagnoli, P. *Dynamic Linear Models with R*. Springer-Verlag (2009).

[151] Pfaff, B. *Analysis of Integrated and Cointegrated Time Series with R*. Second edition. Springer-Verlag (2008). www.pfaffikus.de/.

[152] Pfaff, B. VAR, SVAR and SVEC Models: Implementation within R Package vars. *Journal of Statistical Software* 27(2008). www.jstatsoft.org/v27/i04/.

[153] Phoa, F. and Sanchez, J. Modeling the Browsing Behavior of World Wide Web Users. *Open Journal of Statistics* 3(2): 145–154 (2013). www.scirp.org/journal/PaperInformation.aspx?PaperID=30712.

[154] Pickup, M. *Introduction to Time Series Analysis*. SAGE (2014).

[155] Pinheiro, J., Bates, D., DebRoy, S., Sarker, D. and RCore Team. Linear and Nonlinear Mixed Effects Models. Rpackage version 3.1-149 (2020). https://CRAN.R-project.org/package=nlme.

[156] Pole, A., West, M. and Harrison, J. *Applied Bayesian Forecasting and Time Series Analysis*. Chapman and Hall (1994).

[157] Population by Country Dataset (2020). www.kaggle.com/tanuprabhu/populationby-country-2020.

[158] Priestley, M. B. *Spectral Analysis and Time Series*. Academic Press (1981).

[159] R Core Team. R: A Language and Environment for Statistical Computing. R Foundation for Statistical Computing, Vienna, Austria (2020). www.R-project.org/.

[160] Raghavan, V. Modeling and Inferencing for Activity Profile of Terrorist Groups. *JSM Proceedings of the American Statistical Association, Section CASD* (2015).

[161] RATS (Regression Analysis of Time Series). Estima (2020). https://estima.com/ratswhy.shtml.

[162] Rice, G., Wirjanto, T. and Zhao, Y. Functional Garch-X model with an Application to Forecasting Oil Return Curves. Paper presented at the NBER-NSF SBEIS Conference, Washington University Olin School of Business, 2020. https://apps.olin.wustl.edu/conf/SBIES/files/pdf/2020/109.pdf.

[163] RStudio Team. *RStudio: Integrated Development for R*. RStudio, PBC (2020).

[164] Rudin, C. Seeing into Data and Models. Plenary talk at the Data Science, Statistics and Visualization Conference, July 23, 2020.

[165] Rudin, C., Shaposhnik, Y., Wang, T. et al. Two Layer Additive Risk Model. Duke Data Science Entry to FICO Data Challenge (video), 2018. http://dukedatasciencefico.cs.duke.edu.

[166] Russell, E. An Investigation into Statistical Approaches to Audio Restoration. Master's thesis submitted for the degree of MSci (Hons) Mathematics with Statistics at Lancaster University, 2007.

[167] Ryan, J. A. and Ulrich, J. M. xts: eXtensible Time Series, R package version 0.12.1 (2020). https://CRAN.R-project.org/package=xts.

[168] Said, S. E. and Dickey, D. A. Testing for Unit Roots in Autoregressive-Moving Average Models of Unknown Order. *Biometrika* 71(3): 599–607. DOI: https://doi.org/10.2307/2336570.

[169] Sanchez, J. Application of Classical, Bayesian and Maximum Entropy Spectrum Analysis to Nonstationary Time Series Data. In J. Skilling (Ed.), *Maximum Entropy and Bayesian Methods* (pp. 309–319). Kluwer Academic Publishers (1989).

[170] Sanchez, J. Bayesian, Autoregressive and Periodogram Spectrum Analysis as Applied to the U.S. Chicken Broiler Industry. Ph.D. thesis submitted to the Graduate School of Arts and Sciences of Washington University, 1989.

[171] Sanchez, J. and Wang, J. Which Came First: The Chicken or the Egg? *STATS* 47: 12–16 (2007). www.amstat.org/publications/stats/index.cfm?fuseaction=welcome.

[172] Sanchez, J. and Liu, C. Bayesian Hierarchical Models of the Browsing Behavior of World Wide Web Users. Computing Science and Statistics. Vol 36, p. 884. Proceedings of the 36th Symposium on the Interface. Baltimore, Maryland, May 26–29, 2004. (2008). www.interfacesymposia.org/I04/master.pdf.

[173] Sanchez, J. *Probability for Data Scientists*. First edition. Cognella (2020).

[174] Sanchez, J. (2021). CGM and Insulin Pump Data to Introduce Classical and Machine Learning Time Series Analysis Concepts to Students. Department of Statistics, UCLA. https://escholarship.org/uc/item/4qp1p4j9.

[175] Scheaffer, R. L. and Young, L. J. *Introduction to Probability and Its Applications*. Third edition. Cengage Learning (2009).

[176] Scheiner, G. CGM Retrospective Data Analysis. *Diabetes Technology & Therapeutics*. Vol. 18, No. S2 (January 19, 2016). DOI: http://doi.org/10.1089/dia.2015.0281.

[177] Sheth, M. and Erchul, D. Statistical Considerations in AI-enabled Diagnostic Devices. Presentation at the Joint Statistical Meetings, August 3, 2020. ww2.amstat.org/meetings/jsm/2020/onlineprogram/AbstractDetails.cfm?abstractid=310985.

[178] Shiavi, R. *Applied Statistical Signal Analysis*. Second edition. Academic Press (1999).

[179] Shumway, R. H. and Stoffer, D. S. *Time Series Analysis and Its Applications with R Examples*. Fourth edition. Springer-Verlag (2017).

[180] Siegrist, K. Random Services. Resources. Random. www.randomservices.org/random/.

[181] Sims, C. A. Macroeconomics and Reality. *Econometrica* 48: 1–48 (1980).

[182] Solfi, D. H., Alba, E. and Yao, X. Predicting Car Park Occupancy Rates in Smart Cities. Second International Conference, Smart-CT 2017, Malaga, Spain, June 14–16, 2017, pp. 107–117. DOI: https://doi.org/10.1007/978-3-319-59513-9_11 + Birmingham City Council. https://archive.ics.uci.edu/ml/datasets/Parking+Birmingham.

[183] Spiegelhalter, D. The Problem with Failing to Admit We Don't Know. *Scientific American*, September 19, 2019.

[184] Spiegelhalter, D. *How to Learn from Data*. Basic Books (2019).

[185] Strat, Y. L. and Carrat, F. Monitoring Epidemiologic Surveillance Data Using Hidden Markov Models. *Statistics in Medicine* 18: 3463–3478 (1999).

[186] Taylor, S. J. and Letham, B. *Forecasting at Scale*. PeerJ Preprints. September 27, 2017. DOI: https://doi.org/10.7287/peerj.preprints.3190v2. Also available at lethalletham.com/ForecastingAtScale.pdf.

[187] Taylor, S. J. and Letham, B. Automatic Forecasting Procedure. Rpackage version 0.6 (2020). https://CRAN.R-project.org/package=prophet.

[188] Teetor, P. Recipes for State Space Models in R. Scribd (2015). www.scribd.com/document/335675356/Recipes-for-State-Space-Models-in-R-Paul-Teetor.

[189] Timmerman, A. Forecast Combinations. In G. Elliot, C. W. J. Granger & A. Timmerman (Eds.), *Handbook of Economic Forecasting*, volume. 1(ch. 4). Elsevier (2006).

[190] Tintle, N., Chance, B. L., Cobb, G. W. et al. *Introduction to Statistical Investigations*. First edition. Wiley (2016).

[191] Trapletti, A. and Hornik, K. tseries: Time Series Analysis and Computational Finance. R package version 0.10-47 (2019). https://CRAN.R-project.org/package=tseries.

[192] Kung-Sik Chan and Brian Ripley (2020). TSA: Time Series Analysis. R package version 1.3. https://CRAN.R-project.org/package=TSA.

[193] Tsay, R. S. *Analysis of Financial Time Series*. Third edition. Wiley (2010).

[194] Twitter API in R. https://cfss.uchicago.edu/notes/twitter-api-practice/.

[195] Uber Movement. https://movement.uber.com/?lang=en-US.

[196] UCLA Department of Computational Medicine. https://compmed.ucla.edu/overview.

[197] United States Department of Transportation. Bureau of Transportation Statistics. 2007 Traffic Data for U.S. Airlines and Foreign Airlines U.S. Flights. BTS 16-18. www.bts .gov/newsroom/2017-traffic-data-us-airlines-and-foreign-airlines-us-flights.

[198] Census Bureau. The X-13ARIMA-SEATS Seasonal Adjustment Program. www.census .gov/srd/www/x13as/.

[199] van der Bles, A. M., van der Linden, S., Freeman, A. L. J. et al. Communicating Uncertainty about Facts, Numbers and Science. *Royal Society Open Science* 6(5) (2019). DOI: https://doi.org/10.1098/rsos.181870.

[200] Venables, W. N. and Ripley, B. D. *Modern Applied Statistics with S*. Fourth edition. Springer-Verlag (2002). www.stats.ox.ac.uk/pub/MASS4/.

[201] von Wachter, T. and Sullivan, J. Accessing and Safeguarding Administrative Data at CCPR: Census Research Data Center (RDC). http://ccrdc.ucla.edu/wp-content/uploads/ sites/63/2018/12/RDC_Presentation_Nov2018.pdf.

[202] West, M. and Harrison, J. *Bayesian Forecasting and Dynamic Models*. Springer-Verlag (1989).

[203] Wickham, H. and Grolemund, G. *R for Data Science. Import, Tidy, Transform, Visualize and Model Data*. O'Reilly (2017).

[204] Wickham et al. Welcome to the tidyverse. *Journal of Open Source Software* 4(43): 1686 (2019). DOI: https://doi.org/10.21105/joss.01686.

[205] Wild, C. Introducing Time Series Data (2016). https://new.censusatschool.org.nz/ resource/introducing-time-series-data/.

[206] Wild, C. Comparing Series (2016). https://new.censusatschool.org.nz/resource/compar-ing-series/.

[207] Wildman, Jr., R. A. Long-Term and Seasonal Trends of Wastewater Chemicals in Lake Mead: An Introduction to Time Series Decomposition. *Journal of Statistics Education* 25: 38–49 (2017). http://tandfonline.com/loi/ujse20.

[208] Yang, Y. and Hyndman, R. Introduction to the tsfeatures package. https://cran.r-project .org/web/packages/tsfeatures/vignettes/tsfeatures.html.

[209] Zeileis, A. and Grothendieck, G. zoo: S3 Infrastructure for Regular and Irregular Time Series. *Journal of Statistical Software* 14(6): 1–27 (2005). doi:10.18637/jss.v014.i06.

[210] Zelin, A. Modelling Calls and Effects. *Significance* 12(6): 34–39 (2015). https://rss .onlinelibrary.wiley.com/doi/10.1111/j.1740-9713.2015.00867.x.

[211] Zucchini, W. and MacDonald, I. L. *Hidden Markov Models for Time Series. An Introduction Using R*. CRC Press, Taylor & Francis (2009)

Index

Printed in the United States
by Baker & Taylor Publisher Services